THE LIBRARY
CITY COLLEGE PLYMOUTH

WIND ENERGY
Renewable Energy and the Environment
SECOND EDITION

ENERGY AND THE ENVIRONMENT

SERIES EDITOR
Abbas Ghassemi
New Mexico State University

PUBLISHED TITLES

Wind Energy: Renewable Energy and the Environment, Second Edition
Vaughn Nelson

Solar Radiation: Practical Modeling for Renewable Energy Applications
Daryl R. Myers

Solar and Infrared Radiation Measurements
Frank Vignola, Joseph Michalsky, and Thomas Stoffel

Forest-Based Biomass Energy: Concepts and Applications
Frank Spellman

Introduction to Renewable Energy
Vaughn Nelson

Geothermal Energy: Renewable Energy and the Environment
William E. Glassley

Solar Energy: Renewable Energy and the Environment
Robert Foster, Majid Ghassemi, Alma Cota
Jeanette Moore, and Vaughn Nelson

WIND ENERGY
Renewable Energy and the Environment
SECOND EDITION

Vaughn Nelson

CRC Press
Taylor & Francis Group
Boca Raton London New York

CRC Press is an imprint of the
Taylor & Francis Group, an **informa** business

Cover photo credits: 1 Vestas V90, north of Gruver, Texas, USA. Three MW, rotor diameter 90 m, tower height 80 m. Notice minivan at the base of the tower. Vestas, www.vestas.com (4/14/2013).

2 Telecom, Laisamis, Kenya. Bergey, 7.5 kW, rotor diameter 6.4 m, tower height 24 m. Bergey Windpower, www.bergey.com (4/14/2013).

4 Llano Estacodo Wind Ranch, 80 MW, White Deer, Texas, USA. Wind turbines, Mitsubishi, 1 MW, rotor diameter 56 m, tower height 60 m. Cielo Wind Power, www.cielo.com (4/14/2013).

Large: Turkey Track wind farm, 169.5MW, south of Sweetwater, Texas, USA. Wind turbines, GE Wind, 1.5 MW, rotor diameter 77 m, tower height 80 m. Invenergy, www.invenergyllc.com (4/14/2013).

CRC Press
Taylor & Francis Group
6000 Broken Sound Parkway NW, Suite 300
Boca Raton, FL 33487-2742

© 2014 by Taylor & Francis Group, LLC
CRC Press is an imprint of Taylor & Francis Group, an Informa business

No claim to original U.S. Government works

Printed on acid-free paper
Version Date: 20131004

International Standard Book Number-13: 978-1-4665-8159-3 (Hardback)

This book contains information obtained from authentic and highly regarded sources. Reasonable efforts have been made to publish reliable data and information, but the author and publisher cannot assume responsibility for the validity of all materials or the consequences of their use. The authors and publishers have attempted to trace the copyright holders of all material reproduced in this publication and apologize to copyright holders if permission to publish in this form has not been obtained. If any copyright material has not been acknowledged please write and let us know so we may rectify in any future reprint.

Except as permitted under U.S. Copyright Law, no part of this book may be reprinted, reproduced, transmitted, or utilized in any form by any electronic, mechanical, or other means, now known or hereafter invented, including photocopying, microfilming, and recording, or in any information storage or retrieval system, without written permission from the publishers.

For permission to photocopy or use material electronically from this work, please access www.copyright.com (http://www.copyright.com/) or contact the Copyright Clearance Center, Inc. (CCC), 222 Rosewood Drive, Danvers, MA 01923, 978-750-8400. CCC is a not-for-profit organization that provides licenses and registration for a variety of users. For organizations that have been granted a photocopy license by the CCC, a separate system of payment has been arranged.

Trademark Notice: Product or corporate names may be trademarks or registered trademarks, and are used only for identification and explanation without intent to infringe.

Visit the Taylor & Francis Web site at
http://www.taylorandfrancis.com

and the CRC Press Web site at
http://www.crcpress.com

Contents

Series Preface xi
Preface xv
Acknowledgments xvii
Series Editor xix
Author xxi

Chapter 1 Introduction 1

    1.1   History 1
        1.1.1   Dutch Windmills 1
        1.1.2   Farm Windmills 2
        1.1.3   Wind Chargers 5
        1.1.4   Generation of Electricity for Utilities 6
    1.2   Wind Farms 11
    1.3   Small Systems 13
    1.4   Community Wind 14
    Links 14
    References 14

Chapter 2 Energy 17

    2.1   General 17
        2.1.1   Renewable Energy 20
        2.1.2   Advantages and Disadvantages of Renewable Energy 20
        2.1.3   Economics 21
    2.2   Definitions of Energy and Power 22
    2.3   Energy Fundamentals 23
    2.4   Energy Dilemma and Laws of Thermodynamics 24
        2.4.1   Conservation 24
        2.4.2   Efficiency 24
    2.5   Exponential Growth 26
    2.6   Use of Fossil Fuels 28
        2.6.1   Petroleum 29
        2.6.2   Natural Gas 32
        2.6.3   Coal 33
    2.7   Nuclear Energy 34
    2.8   Mathematics of Exponential Growth 35
        2.8.1   Doubling Time 36
        2.8.2   Resource Consumption 36
    2.9   Lifetime of Finite Resource 36
    2.10   Global Warming 38
    2.11   Summary 39
    Links 40
    References 40
    Suggested Readings 41
    Questions and Activities 41

Order of Magnitude (OM) Estimates 41
Problems 42

Chapter 3 Wind Characteristics 45
    3.1   Global Circulation 45
    3.2   Extractable Limits of Wind Power 45
    3.3   Wind Power 47
    3.4   Wind Shear 49
    3.5   Wind Direction 53
    3.6   Wind Power Potential 54
    3.7   Turbulence 55
    3.8   Wind Speed Histograms 56
    3.9   Duration Curve 57
    3.10   Variations in Wind Power Potential 58
    3.11   Wind Speed Distributions 59
    3.12   General Comments 61
    Links 61
    References 61
    Questions and Activities 62
    Problems 62

Chapter 4 Wind Resource Assessment 65
    4.1   United States 66
    4.2   European Union 67
    4.3   Other Countries 69
    4.4   Ocean Winds 71
           4.4.1   Texas Gulf Coast 71
           4.4.2   World 73
    4.5   Instrumentation 73
           4.5.1   Cup and Propeller Anemometers 77
           4.5.2   Wind Direction 77
           4.5.3   Instrument Characteristics 78
           4.5.4   Measurement 79
           4.5.5   Vegetation Indicators 80
    4.6   Data Loggers 82
    4.7   Wind Measurement for Small Wind Turbines 83
    Links 84
           Maps 84
           Ocean Wind Data 84
           Data Logger, Sensor, and Tower Information and Photos 84
    References 84
    Problems 85

Chapter 5 Wind Turbines 87
    5.1   Drag Devices 87
    5.2   Lift Devices 87
    5.3   Orientation of Rotor Axis 88
    5.4   System Description 89

Contents

5.5	Aerodynamics	90
5.6	Control	93
	5.6.1 Normal Operation	95
	5.6.2 Faults	98
5.7	Energy Production	99
	5.7.1 Generator Size	99
	5.7.2 Rotor Area and Wind Map	100
	5.7.3 Manufacturer's Curve	101
5.8	Calculated Annual Energy	101
5.9	Innovative Wind Power Systems	102
5.10	Applications	107
	5.10.1 Electrical Energy	107
	5.10.2 Mechanical Energy	109
	5.10.3 Thermal Energy	111
	5.10.4 Wind Hybrid Systems	111
5.11	Summary	111
Links		111
References		111
Problems		112

Chapter 6 Design of Wind Turbines ... 115

6.1	Introduction	115
6.2	Aerodynamics	115
6.3	Mathematical Terms	116
6.4	Drag Device	117
6.5	Lift Device	118
	6.5.1 Maximum Theoretical Power	121
	6.5.2 Rotation	121
6.6	Aerodynamic Performance Prediction	122
6.7	Measured Power and Power Coefficient	128
6.8	Construction	131
	6.8.1 Blades	131
	6.8.2 Other Components of System	135
6.9	Evolution	138
6.10	Small Wind Turbines	139
References		142
Problems		143

Chapter 7 Electrical Issues ... 147

7.1	Fundamentals	147
	7.1.1 Faraday's Law of Electromagnetic Induction	150
	7.1.2 Phase Angle and Power Factor	150
7.2	Generators	152
	7.2.1 Induction Generator, Constant RPM Operation	153
	7.2.2 Doubly Fed Induction Generator, Variable RPM Operation	156
	7.2.3 Direct-Drive Generator, Variable RPM Operation	156
	7.2.4 Permanent Magnet Alternator, Variable RPM Operation	156
	7.2.5 Generator Comparisons	156
	7.2.6 Generator Examples	156

	7.3	Power Quality	158
	7.4	Electronics	159
		7.4.1 Controllers	159
		7.4.2 Power Electronics	162
		7.4.3 Inverters	163
	7.5	Lightning	163
	7.6	Resistance Dump Load	164
	Links		164
	References		164
	Problems		165

Chapter 8 Performance ... 167

	8.1	Measures of Performance	167
	8.2	Wind Statistics	169
	8.3	Wind Farm Performance	169
		8.3.1 California Wind Farms	170
		8.3.2 Wind Farms in Other States	172
		8.3.3 Other Countries	174
	8.4	Wake Effects	176
	8.5	Enertech 44	178
	8.6	Bergey Excel	179
	8.7	Water Pumping	181
		8.7.1 Farm Windmills	182
		8.7.2 Electric-to-Electric Systems	184
	8.8	Wind–Diesel and Hybrid Systems	185
	8.9	Blade Performance	186
		8.9.1 Surface Roughness	187
		8.9.2 Boundary Layer Control	189
		8.9.3 Vortex Generators	190
		8.9.4 Flow Visualization	190
	8.10	Comments	192
	Links		192
	References		193
	Problems		194

Chapter 9 Siting ... 197

	9.1	Small Wind Turbines	197
		9.1.1 Noise	200
		9.1.2 Visual Impact	201
	9.2	Wind Farms	203
		9.2.1 Long-Term Reference Stations	203
		9.2.2 Siting for Wind Farms	203
	9.3	Digital Maps	204
	9.4	Geographic Information Systems	205
	9.5	Wind Resource Screening	206
		9.5.1 Estimated Texas Wind Power (Pacific Northwest Laboratory)	207
		9.5.2 Estimated Texas Wind Power (Alternative Energy Institute)	208
		9.5.3 Wind Power for United States	210
	9.6	Numerical Models	210

Contents

9.7	Micrositing	210
9.8	Ocean Winds	215
9.9	Summary	215
Links		216
References		216
Problems		217

Chapter 10 Applications and Wind Industry 219

10.1	Utility Scale	219
10.2	Small Wind Turbines	222
10.3	Distributed Systems	227
10.4	Community Wind	229
	10.4.1 United States	229
	10.4.1.1 Minnesota	231
	10.4.1.2 Schools, Colleges, and Universities	231
	10.4.1.3 Electric Cooperatives	232
	10.4.1.4 Municipal and City Operations	232
	10.4.2 Other Countries	233
10.5	Wind–Diesel Generation	234
10.6	Village Power	238
	10.6.1 China	240
	10.6.2 Case Study: Wind Village Power System	240
10.7	Water Pumping	242
	10.7.1 Design of Wind Pumping System	243
	10.7.2 Large Systems	245
10.8	Wind Industry	246
	10.8.1 1980 through 1990	247
	10.8.2 1990 through 2000	248
	10.8.3 2000 through 2010	250
	10.8.4 2010 Onward	251
10.9	Storage	252
	10.9.1 Compressed Air Energy Storage	256
	10.9.2 Flywheels	256
	10.9.3 Batteries	257
	10.9.3.1 Lead–Acid Batteries	257
	10.9.3.2 Lithium (Li) Ion Batteries	258
	10.9.3.3 Sodium–Sulfur Batteries	258
	10.9.3.4 Flow Batteries	259
	10.9.4 Other Types of Batteries	259
	10.9.5 Hydrogen Fuel Cells	259
10.10	Comments	260
Links		261
References		261
Problems		264

Chapter 11 Institutional Issues 267

11.1	Avoided Costs	267
11.2	Utility Concerns	268
	11.2.1 Safety	268

		11.2.2	Power Quality	269
		11.2.3	Connection to Utility	269
		11.2.4	Ancillary Costs	270
	11.3	Regulations		270
	11.4	Environment		270
	11.5	Politics		274
	11.6	Incentives		275
		11.6.1	United States	275
			11.6.1.1 State Incentives	276
			11.6.1.2 Green Power	277
			11.6.1.3 Net Metering	278
		11.6.2	Other Countries	278
	11.7	Externalities		280
	11.8	Transmission		281
	Links			283
	References			283
	Problems			285

Chapter 12 Economics 287

	12.1	Factors Affecting Economics	287
	12.2	General Comments	288
	12.3	Economic Analysis	289
		12.3.1 Simple Payback	289
		12.3.2 Cost of Energy	290
		12.3.3 Value of Energy	293
	12.4	Life Cycle Costs	293
	12.5	Present Worth and Levelized Costs	295
	12.6	Externalities	296
	12.7	Wind Project Development	296
		12.7.1 Costs	300
		12.7.2 Benefits	301
		12.7.3 Sales of Electricity	302
	12.8	Hybrid Systems	303
	12.8	Summary	305
	12.9	Future Developments	306
	Links		307
	References		308
	Problems		308

Index 311

Series Preface

By 2050, the demand for energy could double or even triple as the global population rises and developing countries expand their economies. All life on earth depends on energy and the cycling of carbon. Affordable energy resources are essential for economic and social development as well as food production, water supply availability, and sustainable, healthy living. In order to avoid long term adverse and potentially irreversible impact of harvesting energy resources, we must explore all aspects of energy production and consumption including energy efficiency, clean energy, global carbon cycle, carbon sources and sinks, and biomass as well as their relationship to climate and natural resource issues. Knowledge of energy has allowed humans to flourish in numbers unimaginable to our ancestors. The world's dependence on fossil fuels began approximately two hundred years ago. Are we running out of oil? No, but we are certainly running out of the affordable oil that has powered the world economy since the 1950s. We know how to recover fossil fuels and harvest their energy for operating power plants, planes, trains, and automobiles which results in modifying the carbon cycle and additional greenhouse gas emissions. This has resulted in the debate on availability of fossil energy resources, peak oil era, and timing for anticipated end of fossil fuel era and price and environmental impact versus various renewable resources and use, carbon footprint, emission and control including cap and trade and emergence of "green power."

Our current consumption has largely relied on oil for mobile applications and coal, natural gas, nuclear or water power for stationary applications. In order to address the energy issues in a comprehensive manner, it is vital to consider the complexity of energy. Any energy resource, including oil, gas, coal, wind, biomass, etc., is an element of a complex supply chain and must be considered in the entirety as a system from production through consumption. All of the elements of the system are interrelated and interdependent. Oil, for example, requires consideration for interlinking of all of the elements including exploration, drilling, production, transportation, water usage, and production, refining, refinery products and by-products, waste, environmental impact, distribution, consumption/application, and finally emissions. Inefficiency in any part of the system has impact on the overall system, and disruption in one of these elements causes major interruption and a significant cost impact. As we have experienced in the past, interrupted exploration will result in disruption in production, restricted refining and distribution, and consumption shortages; therefore, any proposed energy solution requires careful evaluation and as such, may be one of the key barriers to implementing the proposed use of hydrogen as a mobile fuel.

Even though an admirable level of effort has gone into improving the efficiency of fuel sources for delivery and use of energy, we are faced with severe challenges on many fronts. This includes population growth, emerging economies, new and expanded usage, and limited natural resources. All energy solutions include some level of risk including technology snafus, changes in market demand, economic drivers, and others. This is particularly true when proposing energy solutions involving implementation of untested alternative energy technologies.

There are concerns that emissions from fossil fuels lead to changing climate with possibly disastrous consequences. Over the past five decades, the world's collective greenhouse gas emissions have increased significantly even as efficiency has increased resulting in extending energy benefits to more of the population. Many propose that we improve the efficiency of energy use and conserve resources to lessen greenhouse gas emissions and avoid a climate catastrophe. Using fossil fuels more efficiently has not reduced overall greenhouse gas emissions due to various reasons and it is unlikely that such initiatives will have a perceptible effect on atmospheric greenhouse gas content. While there is a debatable correlation between energy use and greenhouse gas emissions, there are effective means to produce energy, even from fossil fuels, while controlling emissions. There are also emerging technologies and engineered alternatives that will actually manage the makeup of the atmosphere, but will require significant understanding and careful use of energy.

We need to step back and reconsider our role and knowledge of energy use. The traditional approach of micromanagement of greenhouse gas emissions is not feasible or functional over a long period of time. More assertive methods to influence the carbon cycle are needed and will be emerging in the coming years. Modifications to the carbon cycle means we must look at all options in managing atmospheric greenhouse gases including various ways to produce, consume, and deal with energy. We need to be willing to face reality and search in earnest for alternative energy solutions. There appears to be technologies that could assist; however, they may all not be viable. The proposed solutions must not be in terms of a "quick approach", but a more comprehensive, long-term (ten, twenty-five, and fifty plus years) approach that is science based and utilizes aggressive research and development. The proposed solutions must be capable of being retrofitted into our existing energy chain. In the meantime, we must continually seek to increase the efficiency of converting energy into heat and power.

One of the best ways to define sustainable development is through long-term, affordable availability of limited resources including energy. There are many potential constraints to sustainable development. Foremost of these is the competition for water use in energy production, manufacturing, farming, and others versus a shortage of fresh water for consumption and development. Sustainable development is also dependent on the earth's limited amount of productive soil. In the not too distant future, it is anticipated that we will have to restore and build soil as a part of sustainable development. Hence, possible sustainable solutions must be comprehensive and based on integrating our energy use with nature's management of carbon, water, and life on earth as represented by the carbon and hydro-geological cycles. The challenges presented by the need to control atmospheric greenhouse gases are enormous and require "out of the box" thinking, innovative approaches, imagination, and bold engineering initiatives in order to achieve sustainable development. We will need to ingeniously exploit even more energy and integrate its use with control of atmospheric greenhouse gases.

The continued development and application of energy is essential to the sustainable advancement of society. Therefore, we must consider all aspects of the energy options including performance against known criteria; basic economics and benefits; efficiency; processing and utilization requirements; infrastructure requirements; subsidies and credits; waste and ecosystem, as well as unintended consequences such as impacts to natural resources and the environment. Additionally, we must include the overall changes and the emerging energy picture based on current and future efforts in renewable alternatives, modified and enhanced fossil fuels, and evaluate the energy return for the investment of funds and other natural resources such as water. In the United States, water is a precious commodity in the west in general and in the southwest in particular and has a significant impact on energy production, including alternative sources due to the nexus between energy and water and the major correlation with the environment and sustainability related issues.

A significant driver in creating this book series that is focused on alternative energy and the environment was provoked as a consequence of lecturing around the country and in the classroom on the subject of energy, environment, and natural resources such as water. While the correlation between these elements, how they relate to each other and the impact of one on the other is understood, it is not significantly debated when it comes to integration and utilization of alternative energy resources into the energy matrix. Additionally, as renewable technology implementation grows by various states, nationally and internationally, the need for informed and trained human resources continues to be a significant driver in future employment resulting in universities, community colleges, trade schools offering minors, certificate programs, and even in some cases, majors in renewable energy and sustainability. As the field grows, the demand for trained operators, engineers, designers, and architects that would be able to incorporate these technologies into their daily activity is increasing. We receive a daily deluge of flyers, emails, and texts on various short courses available for interested parties in solar, wind, geothermal, biomass, etc., under the umbrella of re-tooling an individual's

Series Preface

career and providing trained resources needed to interact with financial, governmental, and industrial organizations.

In all my interactions throughout the years in this field, I have conducted significant searches in locating integrated textbooks that explain alternative energy resources in a suitable manner and that would complement a syllabus for a potential course to be taught at the university while providing good reference material for interested parties getting involved in this field. I have been able to locate a number of books on the subject matter related to energy, energy systems, resources such as fossil, nuclear, renewable, and energy conversion, as well as specific books in the subjects of natural resource availability, use, and impact as related to energy and environment. However, specific books that are correlated and present the various subjects in detail are few and far between. We have, therefore, started a series of texts each addressing specific technology fields in the renewable energy arena. As a part of this series there are textbooks in wind, solar, geothermal, biomass, hydro, and others yet to be developed. Our texts are intended for upper-level undergraduate and graduate students and for informed readers who have a solid fundamental understanding of science and mathematics as well as individuals/organizations that are involved with design and development of the renewable energy field entities that are interested in having reference material available to their scientists and engineers, consulting organizations, and reference libraries. Each book presents fundamentals as well as a series of numerical and conceptual problems designed to stimulate creative thinking and problem solving.

The Series Editor wishes to express his deep gratitude to his wife, Maryam, who has served as a motivator and intellectual companion and too often was a victim of this effort. Her support, encouragement, patience, and involvement have been essential to the realization of this series.

Abbas Ghassemi, PhD
Las Cruces, New Mexico

Preface

Since the first edition was published, the population of the earth increased from 6.5 billion to 7.1 billion in 2013. Essentially, we have not really done much to resolve the enormous problems of over-population and over-consumption. As before, the first priority of national and global policies is to focus on conservation and efficiency, and the second priority is the need to transition from fossil fuels to renewable energy. The underdeveloped countries are in transition and their energy use, materials consumption, and emission of greenhouse gases will soon be in line with those of the developed countries. The increase in energy consumption of fossil fuels cannot continue. If it does, the world is headed for a catastrophe.

The major change since 2007 has been the large annual increase of wind energy. By the end of 2012, about 180,000 wind turbines with an installed capacity of 282 gigawatts had been installed in wind farms. About 900,000 small wind turbines (less than 100 kilowatts) with an estimated capacity of 850 megawatts were in use. Wind energy has become part of the solution for the transition to renewable energy, especially for the generation of electricity.

The second edition contains updates on wind energy installation and capacity and fossil fuel production. The section on distributed wind has been expanded and new sections on global warming, community wind, and storage have been added.

Acknowledgments

I am deeply indebted to colleagues, present and past, at the Alternative Energy Institute (AEI) of West Texas A&M University (WTAMU), and at the Wind Energy Group (program canceled in 2012) at the Agricultural Research Service, U.S. Department of Agriculture, Bushland, Texas. The students in my classes and the students who worked at AEI provided insights and feedback. Many others including numerous international researchers and interns worked with us on energy projects at AEI and USDA. Thanks also to the Instructional Innovation and Technology Laboratory at WTAMU for preparing the computer drawings.

I want to express gratitude to my wife, Beth, who has put up with me all these years. As always, she is very supportive, especially in visiting all those wind farms to obtain information and take photos and accompanying me on many trips to different parts of the world.

Series Editor

Dr. Abbas Ghassemi is the director of Institute for Energy and Environment (IEE) and professor of Chemical Engineering at New Mexico State University. As the director of IEE, he is the chief operating officer for programs in education, research, and outreach in energy resources including renewable energy, water quality and quantity, and environmental issues. He is responsible for budget and operation of the program. Dr. Ghassemi has authored and edited several textbooks and has many publications and papers in the areas of energy, water, carbon cycle including carbon generation and management, process control, thermodynamics, transport phenomena, education management, and innovative teaching methods. His research areas of interest include risk-based decision making, renewable energy and water, carbon management and sequestration, energy efficiency and pollution prevention, multiphase flow, and process control. Dr. Ghassemi serves on a number of public and private boards, editorial boards, and peer review panels and holds MS and PhD degrees in chemical engineering, with minors in statistics and mathematics, from New Mexico State University and a BS in chemical engineering, with a minor in mathematics, from University of Oklahoma.

Author

Dr. Vaughn Nelson has been involved with renewable energy, primarily wind energy, since the early 1970s. He is the author of eight books (five on CDs) and published more than fifty articles and reports. He also served as the principal investigator on numerous grants and conducted more than sixty workshops and seminars from local to international levels.

Dr. Nelson's primary work focused on wind resource assessment, education and training, applied research and development, and rural applications of wind energy. Presently he is a professor emeritus of physics and remains in close contact with the Alternative Energy Institute (AEI) at West Texas A&M University (WTAMU). He was AEI's director from its inception in 1977 through 2003. He returned as director for another year in July 2009 and retired as the dean of the Research and Information Technology Graduate School at WTAMU in 2001.

Dr. Nelson served on a number of State of Texas Committees, most notably the Texas Energy Coordination Council for eleven years. He received three awards from the American Wind Energy Association, one of which was the Lifetime Achievement Award in 2003, and was named a Texas Wind Legend by the Texas Renewable Industries Association in 2010. He also served on the boards of directors for state and national renewable energy organizations.

Dr. Nelson developed the material for a new online course in renewable energy at WTAMU in 2010 and the resulting book, *Introduction to Renewable Energy, Renewable Energy and the Environment*, was published by CRC Press in 2011.

Dr. Nelson earned a PhD in physics from the University of Kansas; an EdM from Harvard University; and a BSE from Kansas State Teachers College in Emporia. He was a member of the Departamento de Física, Universidad de Oriente, Cumana, Venezuela for two years and then a professor at WTAMU from 1969 until his retirement.

1 Introduction

Industrialized societies run on energy, and as third world countries industrialize, especially China and India with their large populations, the demand for energy is increasing. Economists look at monetary values (dollars) to explain the manufacture and exchange of goods and services. However, in the final analysis, the physical commodity is the transfer of energy units. While industrialized nations comprise only one-fourth of the population of the world, they use four fifths of the world's energy. Most of these forms are solar energies that fall into two classifications:

- Stored solar energy: fossil fuels—coal, oil, and natural gas—all of them are finite and therefore depletable.
- Renewable energy: radiation, wind, biomass, hydro, and ocean thermal and waves. Many people discount renewable solar energy; some even call it an "exotic" source of energy. However, it is the source of all food, most fibers, and heating and cooking in many parts of the world [1].

Other forms of energy are tidal (due to gravitation), geothermal (heat from the earth), and nuclear (fission and fusion). In reality, geothermal is a form of renewable energy because as heat it is replenished from below when it is released through the earth's surface.

The main sources of energy in industrialized nations are fossil fuels. Based on their widespread use combined with growing demands and an increasing world population, the need to switch to other energy sources is imminent. Whether this change will be rational or catastrophic depends on the enlightenment of the public and their leaders.

1.1 HISTORY

The use of wind as an energy source begins in antiquity. Vertical axis windmills for grinding grain were reported in Persia in the tenth century and in China in the thirteenth [2]. At one time, wind was a major source of energy for transportation (sailboats), grinding grain, and pumping water. Windmills and water mills were the largest power sources before the invention of the steam engine. Windmills, numbering in the thousands, for grinding grain and pumping drainage water were common across Europe, and some were even used for industrial purposes such as sawing wood. As the Europeans colonized the world, windmills were built everywhere according to the International Molinological Society.

The main long-term use of wind (except for sailing) has been to pump water. Besides the Dutch windmills, another famous example was the use of sail wing blades to pump water for irrigation on the island of Crete. One of the blades had a whistle on it to notify the operator to change the sail area when the winds were too high.

1.1.1 DUTCH WINDMILLS

At one time, over 9,000 windmills operated in The Netherlands. Of course, a number of different designs were used, from the early post mills to the taller mills whose tops rotated to keep the blades perpendicular to the wind. Today, the Dutch windmills are famous attractions in The Netherlands (Figure 1.1).

FIGURE 1.1 Dutch windmills at Kinderdijk World Heritage Site, The Netherlands.

FIGURE 1.2 Thatched Dutch windmill in museum. Notice water flow at bottom of windmill into canal. The author in much younger days is next to helical pump.

The machines for pumping large volumes of water from a low head were as large as 25 m in diameter and most parts were made of wood. Even the helical pump, an Archimedean screw, was made of wood (Figure 1.2). The mills were quite sophisticated in terms of the aerodynamics of the blades. A miller would rotate (yaw) the top of the windmill from the ground with a rope attached to a wooden beam on the cap so the rotor would be perpendicular to the wind. Others used small fan rotors to yaw the big rotors. The rotational speed and power were regulated by the amount of sail on the blades.

The miller and his family lived in the bottom of the windmill, and the smoke from the fireplace was vented to the upper floors to control insects. Fire was a major hazard faced by thatched windmills.

1.1.2 Farm Windmills

Farm windmills were primary factors that aided the settlement of the Great Plains of the United States [3]. From 1850 on, water pumping windmills were manufactured in the tens of thousands. The early wood machines (Figure 1.3) have largely disappeared from the landscape except for a few in isolated farmhouses and museums.

By 1900, most windmills were made of metal. They still had multiblade vanes and the blades were 3 to 5 m in diameter (Figure 1.4). Although the use of farm windmills peaked in the 1930s and 1940s, when over 6 million were in operation, these windmills are still manufactured and continue to pump water for livestock and residence uses. The American Wind Power Center and Museum in Lubbock, Texas, has an outstanding collection of farm windmills (Figures 1.5 and 1.6) from the early wood mills to the later metal types. The museum is starting to collect small operating wind

Introduction 3

FIGURE 1.3 Historical farm windmills at J.B. Buchanan farm near Spearman, Texas. The windmills were later moved to an outdoor museum in Spearman.

FIGURE 1.4 Farm windmill on the Southern High Plains in the United States.

FIGURE 1.5 Some of the many old windmills in the pavilion at the American Wind Power Center and Museum. (Photo courtesy of Coy Harris.)

FIGURE 1.6 Old windmills on towers at the American Wind Power Center and Museum in Lubbock, Texas. The Vestas V47 is in the background.

FIGURE 1.7 Small wind turbines at the American Wind Power Center and Museum in Lubbock, Texas. Parentheses indicate turbine size in meters and rated power in kilowatts. Left to right: Urban Green Technology (1.4 × 0.9, 1.0), Air Dolphin (1.2, 1.0), Honeywell (2.2, 1.0), Raum (4.0, 4.0), Windspire (3.0 × 0.6), Skystream (Southwest Windpower, 3.7, 2.1). Tower heights are around 10 m.

turbines (Figure 1.7) and already has blades for several small wind turbines plus a large 1.5-MW General Electric wind turbine on display. Electricity is provided onsite by a Vestas V47 660-kW turbine on a 40-m tower.

Most farm windmills are in Africa, Argentina, Australia, Canada, and the United States. Because farm windmills are fairly expensive, a resurgence of design changes has focused on creating less expensive systems. Another major advance is the development and commercialization of stand-alone electric–electric systems for pumping enough water for irrigation, village use, or both [4].

The farm windmill proves that wind energy is a valuable commodity, even though the proportion of the energy market is small. For example, an estimated 30,000 farm windmills operate in the Southern High Plains of the United States. Even though their individual power output is low (0.2 to 0.5 kW), they collectively provide an estimated output of 6 MW.

If the windmills for pumping water were converted to electricity from the electric grid, the transition would require around 15 MW of thermal power from a generating station and over $1 billion for transmission lines, electric pumps, and other equipment. This does not consider the dollars saved in fossil fuel with an energy equivalent of 130 million kilowatt hours (kWh) per year (equivalent to 80,000 barrels of oil per year). Because many of these windmills are thirty years old or older and maintenance costs are $250 to $400 per year, farmers and ranchers are seeking alternatives such as solar pumping rather than purchasing new windmills.

In 1888, Brush built a windmill to generate electricity. The device was based on a rotor (large number of slats) and tail vane of a large farm windmill. The wooden rotor (17 m in diameter) was connected to a direct current generator through a 50:1 step-up gearbox to produce around 12 kW in good winds. The unit operated for twenty years but the low rotational speed was too inefficient to produce electricity. For example, a wind turbine with the same-diameter rotor would produce around 100 kW.

1.1.3 Wind Chargers

As electricity became commercially practical, some isolated locations were too far from generating plants and transmission lines were too costly. Therefore, a number of manufacturers built stand-alone wind systems to generate electricity (Figures 1.8 and 1.9), based on a propeller type rotor with two or three blades. Most of the wind chargers had direct current generators (6 to 32 V) and some later models generated 110 V. The electricity was stored in wet-cell lead–acid batteries that required careful maintenance for long life.

FIGURE 1.8 Direct current 100-W Windcharger with flap air brakes at U.S. Department of Agriculture's ARS Wind Station at Bushland, Texas. Notice the 4- and 100-kW Darrieus wind turbines in the background.

These systems with two or three propeller blades are quite different from farm windmills that utilized several blades covering most of the rotor-swept area. Farm windmills were well engineered for pumping low volumes of water, but too inefficient for generating electricity because the blade design and large numbers of blades meant slow rotor speed.

Wind chargers became obsolete in the United States when inexpensive (subsidized) electricity became available from rural electric cooperatives in the 1940s and 1950s. After the energy crisis of 1973, a number of these units were repaired for personal use or to sell. Small companies also imported wind machines from Australia and Europe to sell in the United States during the 1970s.

1.1.4 Generation of Electricity for Utilities

A number of attempts were made to design and construct large wind turbines for utility use [5–10]. These designs centered on different concepts for capturing wind energy (Figure 1.10): airfoil-shaped blades with horizontal or vertical rotor axes, Magnus effect, and Savonius designs. A vertical axis presents no rotor orientation problems arising from different wind directions.

A rotating cylinder in an airstream will experience a force or thrust perpendicular to the wind. This is known as the Magnus effect. In 1926, Flettner built a horizontal axis wind turbine with four blades. Each blade was a tapered cylinder driven by an electric motor. The cylinders (blades) were 5 m long and 0.8 m in diameter at midpoint. The rotor (on a 33-m tower) was 20 m in diameter and had a rated power of 30 kW at a wind speed of 10 m/sec.

Introduction

FIGURE 1.9 Jacobs, 4 kW, direct current generator. It was still in use in the '70s on a farm near Vega, Texas, USA.

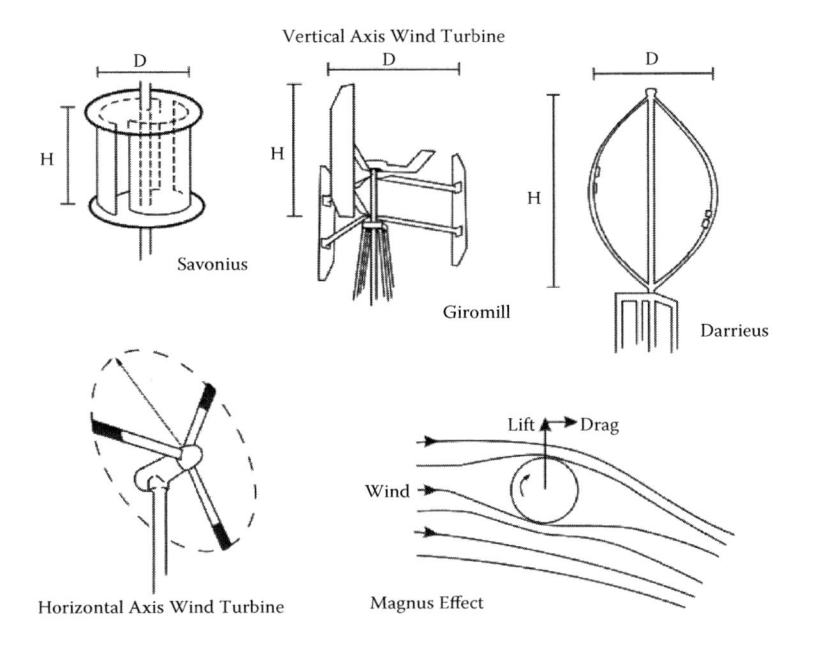

FIGURE 1.10 Various rotors.

Madaras proposed mounting vertical rotating cylinders on railroad cars propelled by the Magnus effect around a circular track. The generators were to be connected to the axles of the cars. In 1933, a prototype installation consisting of a 29-m tall cylinder 8.5 m in diameter mounted on a concrete base was spun when the wind was blowing and the force was measured. Results were inconclusive and the concept was abandoned.

The Magnus effect was used in Flettner rotors used to propel ships [11,12], and one ship operated using rotors for fuel savings from 1926 to 1933. In 1984, the Cousteau Society built a sailing ship called the *Alcyone* that used two fixed cylinders with an aspirated turbosail [13].

In Finland, Savonius built S-shaped rotors that resembled two halves of a cylinder separated by a distance smaller than the diameter. In 1927, Darrieus invented a wind machine whose blade was shaped like a jumping rope. His patent for a "giromill" also covered straight vertical blades. The Darrieus design was later reinvented by researchers in Canada [14].

In 1931, the Russians built a 100-kW wind turbine near Yalta on the Black Sea. The rotor was 30 m in diameter and sat on a 30-m rotating tower. The rotor was kept facing into the wind by moving the inclined supporting strut that connected the back of the turbine to a carriage on a circular track. The blade covering was galvanized steel and the gears were made of wood. The adjustable angle (pitch) of the blades to the rotor plane controlled the rotational speed and power. Annual output was around 280,000 kWh.

The Smith-Putnam wind turbine (Figure 1.11) was developed, fabricated, and erected in between 1939 and 1941 [15]. The turbine was placed on a 38-m tower located on Grandpa's Knob in Vermont and connected to the grid of Central Vermont Public Service. The rotor was 53 m in diameter. Its blades were stainless steel with a 3.4-m chord, and each weighed 8,700 kg. The generator was synchronized with the line frequency by adjusting the pitches of the blades.

At wind speeds above 35 m/sec the blades were changed to the feathered position (parallel to the wind) to shut the unit down. Rated power output was 1,250 kW at 14 m/sec. The rotor was on

FIGURE 1.11 Smith-Putnam 1250-kW wind turbine. (Photo courtesy of archive of Carl Wilcox.)

Introduction

the downwind side of the tower and the blades were free to move independently (teeter perpendicular to the wind) due to wind loading.

Testing of the wind turbine started in October 1941, and in May 1942, after 360 h of operation, cracks were discovered in the blades near the root. The root sections were strengthened and the cracks were repaired by arc welding. A main bearing failed in February 1943 and was not replaced until March 1945 because of a material shortage of materials due to World War II. After the bearing was replaced, the unit operated as a generating station for three weeks when a blade failed due to stress at the root. Total running time was only around 1,100 h. Even though the prototype project showed that a wind turbine could be connected to a utility grid, the project was not pursued because of economics. Photos of the construction of the Smith-Putnam wind turbine are available online [15].

Percy Thomas, an engineer with the Federal Power Commission, pursued the feasibility of wind machines. He compiled the first map for wind power in the United States and published reports on design and feasibility of wind turbines [6].

After World War II, research and development efforts on wind turbines were centered in Europe. E.W. Golding summarized the efforts in Great Britain [7], and further efforts were reported in the conference proceedings of the United Nations [8]. The British built two large wind turbines. One was built by the John Brown Company on Costa Hill, Orkney, in 1955. The unit was rated at 100 kW at 16 m/sec, with a rotor diameter of 15 m on a 24-m tower. The wind turbine was connected to a diesel-powered grid and ran only intermittently in 1955 due to operational problems.

The other unit was built by Enfield, based on a design by Edouard Andreau, a French engineer, and erected at St. Albans in 1952. The Enfield-Andreau wind turbine rotor was 24 m in diameter on a 30-m tower, with a rated power of 100 kW at 13 m/sec. The unit was different in that the blades were hollow. When they rotated, the air flowed through an air turbine connected to an alternator at ground level and exited from the tips of the blades (Figure 1.12). The unit was moved to Grand Vent, Algeria, for further testing in 1957. Frictional losses were too large for it to be successful.

The French built several prototype wind turbines from 1958 to 1966. An 800-kW wind turbine located at Nogent Le Roi had a rotor diameter of 31 m and was operated at constant speed by a rotor connected to a synchronous generator. The top weighed 162 metric tons and was mounted on a 32-m tower. The unit fed electricity into the national grid from 1958 to 1963. Two other units were located at St. Remy-Des-Landes. The smaller Neyrpic machine had a rotor diameter of 21 m on a 17-m tower, and the asynchronous generator produced 130 kW at 12 m/sec. The larger unit had a rated power of 1,000 kW at 17 m/sec and operated for seven months until operation ceased in June 1964 due to a broken turbine shaft. Although the prototypes clearly showed the feasibility of connecting wind turbines to electric grids, the French decided in 1964 to discontinue further wind energy research and development.

During the 1950s, Hütter of Germany designed and tested wind turbines that remained the most technologically advanced for the next two decades. The downwind rotors had lightweight fiberglass blades (Figure 1.13) mounted on a teetered hub with pitch control and coning. A 10-kW unit was developed and tested and led to a larger unit, 34 m in diameter that produced 100 kW at 8 m/sec [16]. This unit operated around 4,000 h from 1957 to 1968. However, the experiments proceeded slowly due to lack of funds and blade vibration problems.

In Denmark, several hundred systems based on the design by La Cour [17] were built, with rated power from 5 to 35 kW. The units had rotor diameters around 20 m and four blades connected mechanically to a generator on the ground. By 1900, around 30,000 wind turbines operated at farms and homes, and by 1918, some 120 local utilities operated wind turbines, typically 20 to 35 kW for a total of 3 MW and produced about 3% of the Danish electricity.

Danish interest waned in subsequent years, until a crisis in electricity production occurred during World War II. Since the Danes had no fossil fuel resources, they looked at connecting wind turbines into their national grid, and the Danish government started a program to develop large-scale wind turbines to produce electricity. During World War II, a series of wind turbines

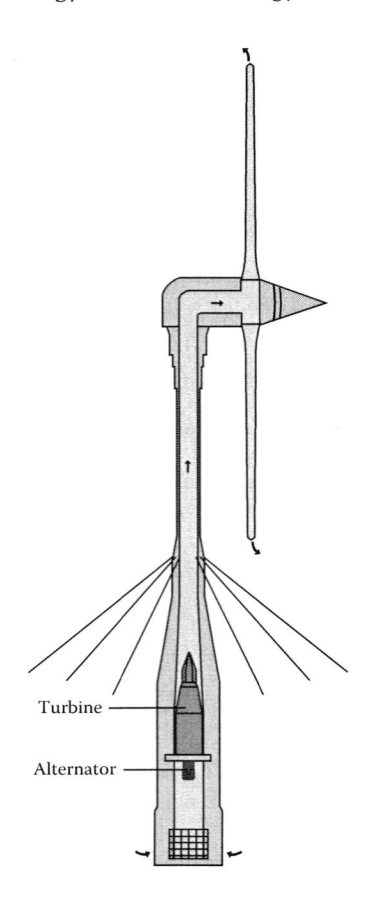

FIGURE 1.12 Enfield-Andreau wind turbine (100 kW).

in the 45-kW range were developed with direct current (DC) generators and produced around 4 million kWh per year.

The Danes had the only successful program that began in 1947 with a series of investigations of the feasibility of using wind power, and continued until 1968 [8, pp. 229–240]. A prototype wind turbine of 7.5 m diameter was built and remained in operation until 1960 when it was dismantled. A wind turbine at Bogo, originally constructed for DC power in 1942, was reconstructed for alternating current (AC) in 1952. Rotor diameter was 13.5 m and the system used a 45-kW generator. The results of the two experimental turbines were encouraging and culminated in the Gedser wind turbine (Figure 1.14).

The unit was erected in 1957, and from 1958 through 1967 it produced 2,242 MWh. It was shut down in 1967 when maintenance costs became too high. The rotor was 35 m in diameter and the 26-m high tower was made of prestressed concrete. The rotor was upwind of the tower, and the blades were fixed pitch types with tip brakes for overspeed control. The wind turbine had an asynchronous generator (rated for 200 kW at 15 m/sec) that provided stall control and also had an electromechanical yaw mechanism. Denmark and the United States furnished funds to place the Gedser wind turbine in operation for a short period in 1977 and 1978 for research that involved tests for aerodynamic performance and structural load limits.

The successful program of the Danes was overshadowed by the failures of other large machines due to technical problems, mainly stresses from vibration and control issues at high wind speeds. Some were economic failures despite agreement that no scientific barriers prevented the use of wind

FIGURE 1.13 German wind turbines. Left: 100 kW. Right: 10 kW. (Photo courtesy of NASA-Lewis.)

turbines tied to utility grids. In the 1960s, development of wind machines was abandoned because petroleum was available and inexpensive.

1.2 WIND FARMS

Wind farms appeared in California in 1982 as a result of federal laws and incentives along with mandates to avoid costs set by the California Energy Commission. The early years of wind farm history were dominated by installations in the U.S. and Denmark. By the 1990s, Europe surpassed the United States after large numbers of units were constructed in Germany (Figure 1.15a). By the end of 2012, the worldwide installed capacity (Figure 1.15b) was estimated at 282 GW with the largest capacity installed in China.

In addition, about 1 GW is generated by small wind turbines. In the first edition of this book based on 2007 data, I estimated that global installation would reach 240 GW by 2012—an underestimate of over 40 GW. Turbine sizes increased from 10- to 20-m diameter units generating 25 to 100 kW to megawatt units of 60- to 100-m diameters installed on towers exceeding 100 m in height. Over 5.5 GW capacity has been installed in offshore wind farms.

Electricity from wind farms is the cheapest renewable energy and less expensive than that produced by new coal and nuclear power plants. However, the increase of natural gas production in the United States decreased the prices of combined cycle gas turbines.

FIGURE 1.14 Gedser 200-kW wind turbine. (Photo courtesy of Danish Wind Industry Association.)

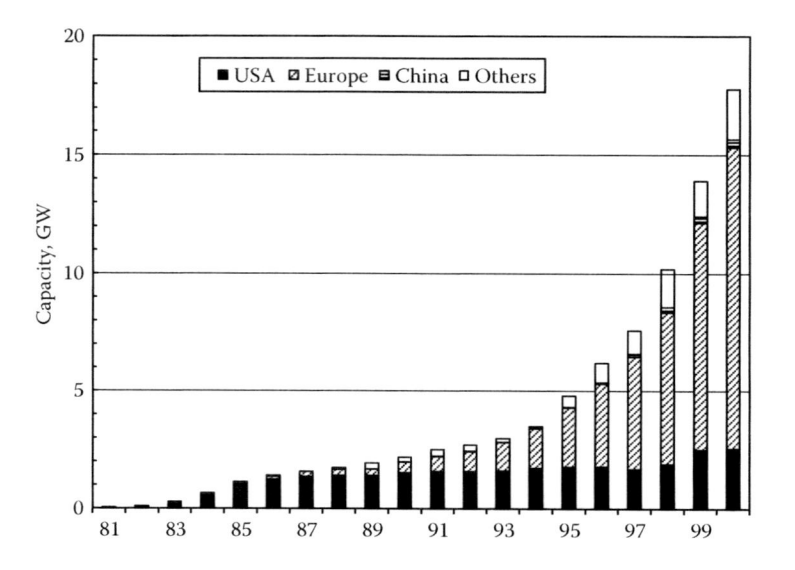

FIGURE 1.15 (a) World installed capacity of wind turbines, 1981–2000.

Introduction

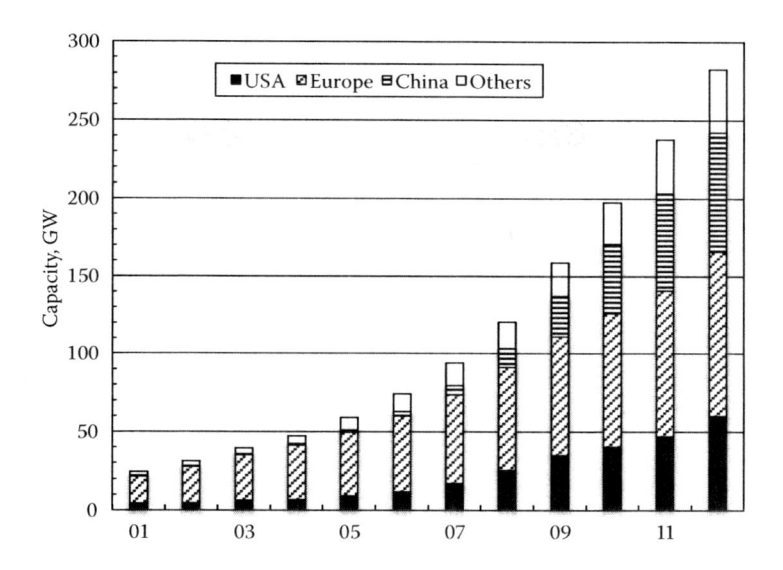

FIGURE 1.15 (*Continued*) (b) World installed capacity of wind turbines, 2001–2012.

Wind power has grown at an average rate of 28% per year from 1995 through 2012. However, the averages were 21% in 2011 and 19% in 2012. These numbers demonstrate the problem of exponential growth. At some point, a linear increase will become a norm and eventually the number of new installations will decrease.

My global prediction is that wind power will reach 600 GW by 2020 if a linear addition of 40 GW per year is used and capacity will be around 1,000 GW by 2030 to meet the goals for wind power set by China, Europe, and the United States. In any case, the numbers for wind power to date and for the future are astounding.

1.3 SMALL SYSTEMS

Small systems, in general, are wind turbines rated to 100 kW. They are designated micro (0 to 1.5 kW), small (1.5 to 50 kW), and mid (50 to 500 kW). As of 2012, the number installed (Table 1.1) was around 940,000 units (about 1,060,000 produced) with a capacity around 860 MW.

Estimating the number installed and capacity as of 2012 is difficult because of the number of manufacturers and also because China produced and installed the most units since the 1980s. However, around 100,000 of those earlier units (50 or 100 W) in China have been retired and/or replaced by larger units. I estimate that another 20,000 units in other parts of the world are no longer operational.

Most small systems are not connected to grids and utilize battery storage. Most fall within the size range of 50 to 300 kW. However, in the United States and other parts of the developed world, a fairly large market has developed for small (1 to 10 kW) wind systems connected to grids through inverters. Telecommunications systems need high reliability and some are hybrid combinations of wind, photovoltaic (PV), and diesel energy and battery storage. Some of their locations are accessible only by helicopter.

As one fourth of the world's population does not have electrical power and costs of diesel generation have increased, a number of hybrid installations now power villages. Most are hybrid systems powered by wind and PV cells or wind only, both utilizing battery storage. Another system combines wind and diesel power. Some systems include storage and others have wind turbines added to existing diesel power plants [17]. The wind–diesel systems range in size from less than 100 kW with

TABLE 1.1
Small Wind Systems Worldwide

Application	Number of Systems
Total, electric generation	940,000
Village power; wind, wind hybrid, wind diesel	2,500?
Telecommunications, military	3,500?
Farm windmill	300,000

one or more wind turbines to hundreds of kilowatts and multiple wind turbines, and even megawatt wind turbines.

1.4 COMMUNITY WIND

Community wind is another term for wind projects that involve local financial participation and control. These projects cover a wide range of sizes and types: small wind systems for homes and farms; mid-size wind turbines (up to megawatt size) for schools and businesses; multi-megawatt wind farms owned by cooperatives and municipalities; and wind farms involving independent power producers that generate tens of megawatts.

Denmark and Germany built their early markets on local or cooperative owners. Distributed wind is a term for a wind project inside the utility meter; energy is used onsite or there is the possibility of net metering. Small wind connected to a grid would be classified as distributed wind.

LINKS

American Wind Power Center and Museum. www.windmill.com (1/5/2013).
Danish Wind Industry Association. History of wind energy. http://wiki.windpower.org/index.php/Category:Historyof_wind_energy (2/21/2013).
Darrel Dodge. http://telosnet.com/wind (1/5/2013). Good overview of history of wind power development.
Erik Grove-Nielsen. Winds of change: 25 years of wind power development on planet earth. www.windsofchange.dk (2/21/2013). Photo history of developments from 1975 to 2000; site also has brochures of wind turbines.
European Wind Energy Association, 2007. *Wind directions:* 25th anniversary; *The road to maturity.* www.ewea.org/fileadmin/ewea_documents/documents/publications/WD/2007_september/wd-sept-focus.pdf (1/5/2013).
Farm windmills, http://windmillersgazette.com/index.html (1/5/2013).
International Molinological Society. www.timsmills.info (1/5/2013).

REFERENCES

1. V. Smil and W.E. Knowland. 1983. *Energy in the Developing World: Biomass Energies.* Plenum: New York.
2. D.G. Sheppard. 1994. Historical development of the windmill. In D.A. Spera, Ed., *Wind Turbine Technology: Fundamental Concepts of Wind Turbine Engineering.* New York: ASME Press.
3. T.L. Baker. 1984. *A Field Guide to American Windmills.* University of Oklahoma Press: Norman.
4. V. Nelson, R.N. Clark, and R. Foster. 2004. *Wind Water Pumping* CD. *Bombeo de Agua con Energía in Spanish.* West Texas A&M University Bookstore. www.wtbookstore.com
5. P.C. Putnam. 1948. *Power from the Wind.* New York: D. Van Nostrand.
6. P.H. Thomas. *Electric power from the wind,* 1946. *Wind power aerogenerator,* 1949. *Aerodynamics of the wind turbine,* 1954. *Fitting wind power to the utility network.* Federal Power Commission Reports.
7. E.W. Golding. 1955. *The Generation of Electricity by Wind Power.* New York: Halsted Press.

8. United Nations. 1994. *Proceedings of the United Nations Conference on New Sources of Energy: Wind Power,* Vol. 7.
9. D.J. Vargo. 1975. *Wind energy developments in the 20th century.* NASA Technical Reports Server, http://ntrs.nasa.gov/(1/13/2013).
10. F.R. Eldridge. 1980. *Wind Machines,* 2nd ed. New York: Van Nostrand Reinhold.
11. S.D. Orsini. 1983. Rotorships: sailwing ships without sails. *Oceans,* 16, January/February.
12. C.P. Gilmore. 1984. Spin sail. *Popular Science,* 224, 70.
13. J.A. Constants et al. 1985. *Alcyone: daughter of the wind and ship of the future.* Paper presented at Asian Development Bank Regional Conference on Sail–Motor Propulsion.
14. R.J. Templeton and R.S. Rangi. 1983. Vertical axis wind turbine development in Canada. *IEEE Proceedings,* 130A, 555.
15. P. Gipe. www.wind-works.org/photos/index.html (1/5/2013).
16. U. Hütter. 1964. *Operating experience obtained with a 100-kW wind power plant N73-29008/2.* Kanner & Associates and National Technical Information Service.
17. V.C. Nelson et al. 2001. *Wind hybrid systems technology characterization: report for National Renewable Energy Laboratory.* West Texas A&M University and New Mexico State University. http://solar.nmsu.edu/publications/wind_hybrid_nrel.pdf

2 Energy

2.1 GENERAL

Scientists have been very successful in understanding and finding unifying principles. Many people take the resulting technology for granted and do not understand the limitations of humans within the physical world. There are moral laws (or principles), civil laws, and physical laws. Moral laws have been broken by acts such as murder and adultery, and everyone has broken some civil law, for example, by driving over the speed limit. However, no one breaks a physical law. Therefore, we can only work with nature, and we cannot do anything that violates the physical world. Another way of stating this: you cannot fool Mother Nature.

The universe allows only four generalized interactions (forces between particles): nuclear, electromagnetic, weak, and gravitational [1]. In other words all the different types of energy in the universe can be traced back to one of these four interactions. This interaction or force is transmitted by an exchange particle.

The exchange particles for electromagnetic and gravitational interactions have zero rest masses; thus the transfer of energy and information occurs at the speed of light (3×10^8 m/sec or 186,000 miles per second). Although the gravitational interaction is very, very weak, it is noticeable when masses are large. The four interactions constitute a great example of how a scientific principle covers an immense number of phenomena.

Energy can be classified into many different types. Kinetic energy arises from the motions of particles, for example, wind or moving water. Potential energy results from the positions of particles, for example, water stored in a dam, the energy trapped in a coiled spring, and energy stored in molecules (e.g., gasoline). Other types of energy are mechanical, electrical, thermal (heat), chemical, magnetic, nuclear, biological, tidal, and geothermal.

The sources of solar energy are the nuclear interactions at the core of the sun. The energy comes from the conversion of hydrogen nuclei into helium nuclei. This energy is transmitted to the earth primarily by electromagnetic waves that may also be represented by particles (photons). Of the solar energy (3.85×10^6 exajoules/year) on the earth, a fraction (2.25×10^3 exajoules/year) is then transformed into wind energy, and a very small amount of wind energy is used by humans. A satellite photo of the night sky of the earth [2] illustrates the tremendous amount of energy consumed by humans.

Industrialized societies run on energy—a tautological statement in the sense that it is obvious. Population, gross domestic product (GDP), consumption and production of energy, and production of pollution for the world are interrelated (Figure 2.1). The world numbers for the latest year of data available are:

Population: 7.0×10^9 (2012)
Gross domestic product: $\$69 \times 10^{12}$ (2012)
Energy consumption: 548 exajoules (2011)
Carbon dioxide (CO_2) emissions: 32×10^9 tons (2010)

A large change in energy consumption between 2005 and 2011 resulted from growth in China and India (Figure 2.2). Eighty-six percent of the energy consumed in the world in 2011 (Figure 2.3) was from fossil fuels, petroleum, coal, and natural gas; the quantity in the United States was 82%. Notice that renewable energy for production of electricity was 7% (larger than nuclear power);

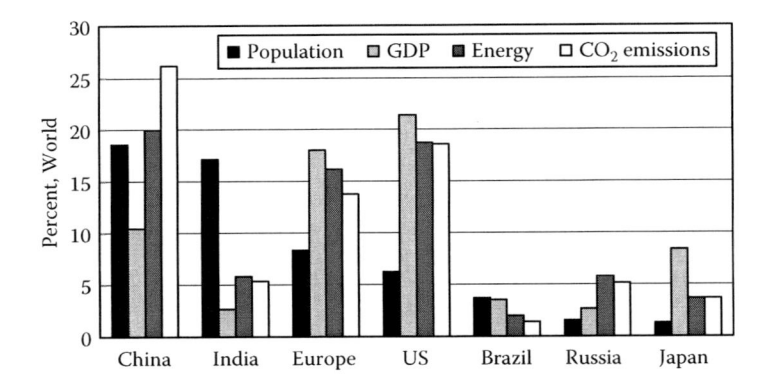

FIGURE 2.1 Comparisons of population (rank in world), gross domestic product, energy consumption, and carbon dioxide emissions for 2011.

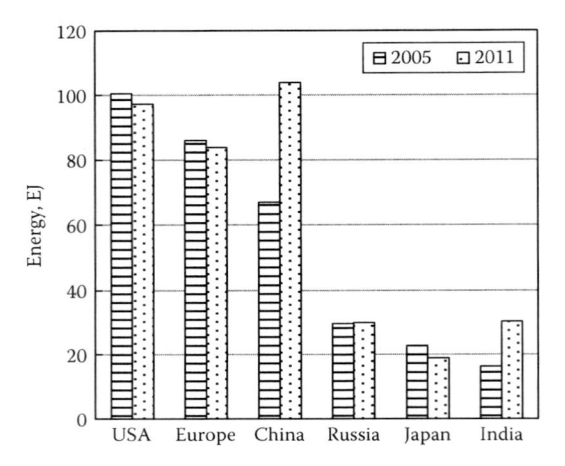

FIGURE 2.2 Comparison of energy consumption from 2005 to 2011.

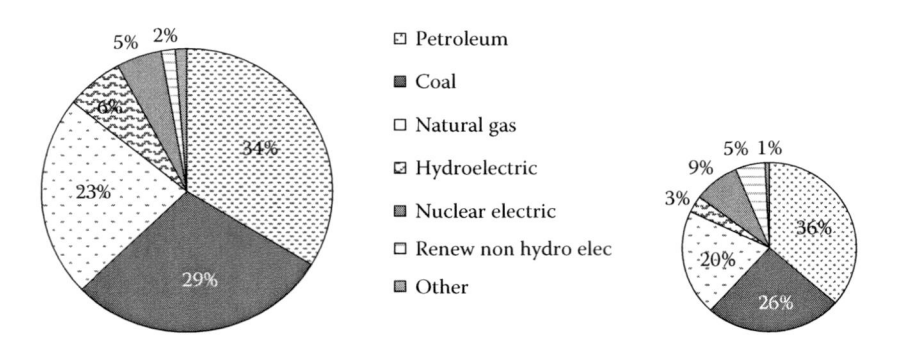

FIGURE 2.3 World (left) and United States (right) energy consumption by source in 2011.

non-hydroelectric energy was only 2% (Figure 2.4). Wind and solar energy for production of electricity have increased rapidly in recent years, and the mandate for ethanol has increased the use of biofuels in the United States (Figure 2.5).

The United States has 6% of the world population (2012). It also generates around 21% of gross production and 19% of CO_2 emissions worldwide and consumes 19% of the world's energy.

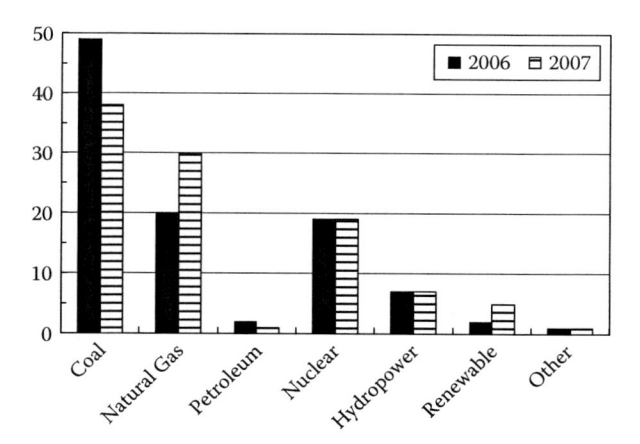

FIGURE 2.4 Change in U.S. electricity use from 2006 to 2007.

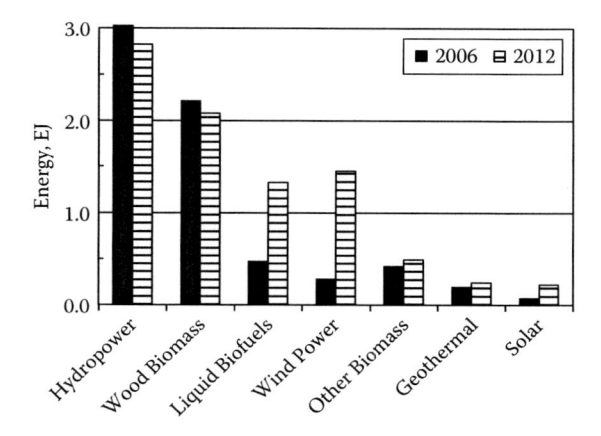

FIGURE 2.5 Change in U.S. renewable energy supply from 2006 to 2012.

Europe and the United States combined consume around 40% of the world's energy. Note that the countries listed in Figure 2.1 consume 72% of the energy and produce 67% of the world's GDP and 72% of its CO_2 emissions.

For the United States and other developed countries, energy consumption is smaller now than it was in 2005 due to the 2008 recession, the reduction of CO_2 emissions through the Kyoto treaty,[*] and the displacement of coal by natural gas for generation of electricity in the U.S. The developed countries consume the most energy and produce the most pollution, based on larger amounts of energy per person. On a per-person basis, the U.S. ranks worst for energy consumption and carbon dioxide emissions.

Formerly underdeveloped areas, primarily China and India—the two largest countries in terms of population—are beginning to emulate the developed countries in the areas of consumption of energy and material resources and emissions of greenhouse gases. China now leads the world in CO_2 emissions due to widespread use of coal and expanding vehicle use (more vehicles are sold annually in China than in the U.S.). One dilemma in the developing world is that despite expanding energy use, large numbers of rural areas do not have electricity.

---

[*] The U.S. was not a signatory.

The energy consumption in the United States increased from 32 quads in 1950 to 101 quads in 2005. One quad = 10^{15} British thermal units = 1.055 exajoules. Primarily due to the shock of the oil crisis of 1973, the industrial sector increased energy efficiency use, but you must remember that correlation between GDP and energy consumption does not indicate cause and effect. The oil crisis of 1973 showed that efficiency is a major component in the use of energy and in gross national product.

It is enlightening to consider how the U.S. energy use has changed since World War II. Ask your grandparents about their lives in the 1950s and then compare their recollections with today's use of:

Family residence: space heating and cooling, number of lights, amount of space per person
Transportation: number and types of family vehicles
Commercial activities: space heating and cooling for buildings, lighting
Industrial production: efficiency

2.1.1 Renewable Energy

Solar energy is described as renewable and/or sustainable because it will be available as long as the sun continues to shine. Estimates for the remaining life of the main stage of the sun are another 4 to 5 billion years. The energy from the sun (electromagnetic radiation) is known as insolation. The other main renewable energy types are wind, bioenergy, geothermal, and hydro (generated by water movement).

Wind energy is derived from the uneven heating of the earth's surface because more heat input at the equator triggers the accompanying transfer of water and thermal energy by evaporation and precipitation. In this sense, rivers and dams that produce hydro energy are really solar energy storage sites. Another major aspect of solar energy is the conversion of solar energy into biomass by photosynthesis.

Animal products such as oil from fat and biogas from manure are derived from solar energy. Another type of renewable energy is geothermal. It consists of heat from the earth caused by the decay of radioactive particles and residual heat from gravitation during formation of the earth. Volcanoes are fiery examples of the movement of geothermal energy from the extremely hot interior of the earth to the cooler surface. Tidal energy is generated by the gravitational interactions of the earth and the moon.

It is difficult to estimate the total renewable energy supply. An educated estimate indicates that 14% of the world's energy comes from bioenergy, primarily wood and charcoal, but also crop residues and even animal dung used for cooking and heating. Bioenergy use contributes to deforestation and the loss of topsoil in developing countries. Production of ethanol from biomass is now used in liquid fuels for transportation, especially in Brazil and the U.S.

In contrast, fossil fuels contain stored solar energy from past geological ages. Although the quantities of oil, natural gas, and coal are large, they are finite and will not be sustainable over hundreds of years.

2.1.2 Advantages and Disadvantages of Renewable Energy

Renewable energy provides considerable advantages. It is sustainable (nondepletable), ubiquitous (found everywhere in the world unlike fossil fuels and minerals), and essentially nonpolluting. Wind turbines and photovoltaic (PV) cells do not use water in the production of electricity. This is a major advantage in dry areas of the world like most of the western U.S. Conversely, thermal electric plants including nuclear powered facilities use large quantities of water.

The disadvantages of renewable energy are low density and variability that result in higher initial cost because of the need for large capture area and storage or backup power. For different forms of renewable energy, other disadvantages or perceived problems are visual pollution, odor from biomass, avian and bats at wind farms, and brine produced by geothermal energy.

In addition, any installation of a large facility will present perceived and real problems to the local people. For conventional power plants using fossil fuels, for power plants using nuclear energy, and even for renewable energy, the not-in-my-backyard problem arises. In the U.S., considerable opposition to a wind farm offshore of Cape Cod surfaced. Several areas such as the coasts of Florida have been declared off limits for drilling for oil and natural gas. Many infrastructure problems associated with transmission lines for electricity and pipelines for oil and gas have occurred.

2.1.3 ECONOMICS

Business entities always couch their concerns in terms of economics. The following statements are common:

> We cannot have a cleaner environment because it is uneconomical.
> Renewable energy is not economical.
> We must be allowed to continue our operations as in the past. If we have to install new equipment for emission reduction, we cannot compete with other energy sources.
> We will have to reduce employment; jobs will go overseas.

The types of economics to consider are pecuniary, social, and physical. The common view of economics is pecuniary (dollars). Social economics (sometimes called externalities) are borne by all members of society and may be negative or positive.

Many businesses want the general public to pay for their environmental costs. A good example is the use of coal in China where all cities of all sizes face major problems with air pollution. Governments have laws (social economics) for clean air, but they are not enforced. The cost will be paid in the future by widespread health problems, especially for today's children. If environmental problems affect someone else today or in the future, who pays? Estimates of the pollution costs for generation of electricity by coal range from $0.005 to 0.10/kWh.

Physical economics (energetics) covers energy costs and the efficiencies of the processes. Some consider energetics the energy balance or energy returned on energy invested. A system for producing energy must be a net energy gainer. What is the energy content at the end use versus how much energy is used for production, transport, and transmission? The energetics of a process must be calculated over the life of a system, and the energetics must be positive.

Industries face fundamental limitations imposed by the physical laws of nature. In the end, Mother Nature always wins based on the corollary of paying now or probably paying more in the future. On that note, we should be looking at life cycle costs rather than our ordinary way of doing business—low initial costs followed by payments over time.

Finally, we have to look at incentives and penalties for energy entities. Each energy entity wants incentives (subsidies) for itself and penalties for its competitors. Incentives come in the forms of reduced or no taxes, not having to pay social and/or environmental costs on a product, and government investments in research and development.

Penalties may be imposed as taxes and fines imposed by environmental and other regulations. It is estimated that we use energy sources in direct proportion to incentives received in the past. We can note many examples of incentives for fossil fuels and nuclear power. At one time, the U.S. allowed a huge incentive for the production of oil: a 27.5% depreciation allowance taken off the bottom lines of taxes.

A thought on energy and GDP: a solar clothes drying device (clothes line) does not add to the GDP, but every electric and gas dryer contributes to it. Clothes lines and dryers both perform the same function. We may need to think in terms of results, efficient ways to accomplish a function or process, and actual life cycle cost. Why do we need heavy cars or sport utility vehicles with big motors that accelerate rapidly to transport people?

2.2 DEFINITIONS OF ENERGY AND POWER

To understand renewable energy and the environment, *energy* and *power* must be defined. *Work* (W) is the force (F) on an object moved through distance (D). Work is equal to force times distance:

$$W = F \times D$$

$$\text{Joule (J)} = \text{Newton (N)} \times \text{meter (m)} \tag{2.1}$$

Some units of energy are the Joule (J), calorie, British thermal unit (BTU), kilowatt hour (kWh), and even barrel (bbl). A newton is a unit of force. A number of symbols will be used in this chapter, and problems can be solved using personal computers, spreadsheets, and calculators. Examples are supplied for illustration and understanding.

Many people have mental blocks as soon as they see mathematical symbols, but everyone uses symbols. Ask any person what *piano* means and he or she understands the symbol. However, a South Seas Islander understands a piano as "a big black box, you hit him in teeth, and he cries." By the same principle, Equation 2.1 can be understood as a shorthand notation for the concepts described above it.

Energy is required to move objects, do work, and change positions of objects, so energy and work are measured by the same units. A very useful converter is the *Unit Juggler* (www.unitjuggler.com). A few common conversion definitions are:

calorie = amount of energy required to raise 1 g of water 1°C (Celsius)
BTU = amount of energy required to raise 1 lb of water 1°F (Fahrenheit)

Some conversion factors for energy are:

1 calorie = 4.12 J
1 kilocalorie = 1,000 calories (used in nutrition)
1 BTU = 1,055 J
1 barrel of oil (42 gallons) = 6.12×10^9 J = 1.7×10^3 kWh
1 gallon (U.S.) = 3.78 liters (L) = 33 kWh
1 metric ton (t) of coal = 2.5×10^7 BTU = 2.2×10^{10} J
1 cubic foot (ft^3) of natural gas = 1,000 BTU
1 therm = 10^5 BTU = 100 ft^3 of natural gas
1 quad = 10^{15} BTU = 1.055 EJ
1 kWh = 3.6×10^6 J = 3.4×10^3 BTU
1 kW = 1.33 horsepower (hp)

Natural gas is sold by the mcf (1,000 cubic feet) and has an energy content around 10^6 BTU. You should be careful when comparing energy from coal with other sources because 1 metric ton = 1,000 kg = 2,200 lb, 1 ton or long ton = 2,400 lb, and 1 short ton = 2,000 lb. Metric tons will be used in this chapter unless noted. Also, different types of coal have different energy contents. A barrel of oil (160 L, 42 gal) is refined to around 166 L (44 gal) of components, of which 72 L (19 gal) is gasoline.

Objects in motion can do work and therefore possess kinetic energy (KE):

$$\text{KE} = 0.5 \, m \, v^2 \tag{2.2}$$

where *m* is the mass of the object and *v* is its speed.

Energy

Example 2.1

A car with a mass of 1,000 kg moving at 15 m/sec has kinetic energy calculated as:

$$KE = 0.5 \times 1{,}000 \times 15 \times 15 = 112{,}500 \text{ J} = 1.1 \times 10^5 \text{ (to two significant figures)}$$

Because objects interact, for example, by gravity or electromagnetics, they can do work or have potential energy (*PE*) due to their relative positions. Raising a 1-kg mass 2 m high requires 20 J of energy. At that upper level, the object has 20 J of potential energy. Energy from fossil fuels is chemical and also potential due to electromagnetic interactions. Power is the rate of energy use or production and equals energy divided by time:

$$P = E/t, \text{ watt} = \text{J/sec} \qquad (2.3)$$

If either power or energy is known, the other quantity can be calculated for any period. Always remember that a kilowatt (kW) is a measure of power and a kilowatt hour (kWh) is a measure of energy:

$$E = P \times t \qquad (2.4)$$

Example 2.2

A 5 kW electric motor that runs for 2 hr consumes 10 kWh of energy.

Example 2.3

Ten 100-watt light bulbs that are left on all day will consume 24 kWh of energy.

Heat is a thermal form of energy. Heat is simply the internal kinetic energy (random motions of atoms). Rub your hands together and they get warmer. As you heat your home, you are increasing the speeds of particles of air and other materials inside the structure. Heat and temperature (T) are different. Heat is energy; temperature is the potential for transfer of heat from a hot place to a cold place. Do not equate temperature to heat (energy). To illustrate the difference between heat (energy) and temperature, would you rather stick your finger in a cup of hot coffee at a temperature of 90°C or get hit by a high-speed proton at a temperature of 1,000,000°C? One has much more energy than the other one.

2.3 ENERGY FUNDAMENTALS

A major unifying concept in a discussion of energy is how energy is transferred. The area of physics dealing with heat is called thermodynamics. Most of our understanding of energy can be embodied in the following laws or principles of thermodynamics.

First, energy is conserved. Energy is not created or destroyed; it only transforms from one form to another. In lay terms, this means that all you can do is break even. A number of patents have been issued for perpetual motion machines [3] intended to produce more energy than the energy needed to run them. A number of people have invested money in such machines and lost their money because their devices contradicted the first law of thermodynamics.

Second, thermal energy (heat) cannot be transformed totally into work. In lay terms, you cannot even break even. Every transformation involves energy efficiency below 100%. Energy is required to move heat from a cold place to a hot place. Examples are a refrigerator or heat pump.

Another way of looking at the concept is that systems tend toward disorder and transformations of energy increase disorder. In succinct terms, entropy is increasing.

Some forms of energy are more useful than other. For example, the energy in a liter of gasoline is not lost; it is transformed into heat by a car engine. However, after the transformation, the energy is dispersed into a low-grade form (more entropy) and cannot be used to move the car. The efficiency from energy input to end product (energetics) must be considered and calculated. Fuel cells have much higher efficiency than internal combustion engines, so why aren't highways filled with cars powered by fuel cells?

As an aside for scientists and students of science, the following most famous equation says that mass is just a very concentrated form of energy. Conversion of a small amount of mass gives a lot of energy such as that produced by an atomic or hydrogen bomb or a nuclear reactor: $E = mc^2$ where c is the speed of light.

2.4 ENERGY DILEMMA AND LAWS OF THERMODYNAMICS

There is no real energy *crisis* because energy cannot be created or destroyed. It can only transform to another form. We have an energy *dilemma* caused by our use of finite energy resources and their effects on the environment, primarily from the burning of fossil fuels.

The first and primary objective of any energy policy must be conservation and efficiency. Conservation and efficiency represent the most economical solutions for alleviating our energy problems. The other major aspect of the energy dilemma is the high energy content of liquid fuels for transportation that makes them difficult or costly to replace.

2.4.1 CONSERVATION

Conservation simply means "do not turn it on or use it if you do not need it." Admonitions to reduce thermostat settings and lower speed limits are conservation measures. High prices and energy shortages increase conservation; for example, in the California electrical crisis of 2000–2001, consumption of electricity was reduced.

In general, utility and energy companies like to sell more electricity and energy rather than have customers reduce use. Texas increased the speed limit on some highways to 129 kilometers per hour (kph)—80 miles per hour (mph). In reality, to save energy the U.S. should have a national speed limit of 105 kph (65 mph) on some interstate highways and a reduction to 96 kph (60 mph) on interstate highways in the East.

2.4.2 EFFICIENCY

Efficiency is the measure of energy of a function or product divided by the energy input:

$$\text{Efficiency} = \text{energy out/energy in} \quad (2.5)$$

Energy can be used to do work (mechanical energy), heat an object or space (thermal energy), can be transformed to electrical energy, or be stored as potential or chemical energy. Physical principles determine an upper limit on efficiency for each transformation. In thermal processes, the temperatures of the hot and cold reservoirs determine this efficiency:

$$\text{Eff} = \frac{T_H - T_C}{T_H} \quad (2.6)$$

where T_H and T_C are the temperatures of the hot and cold reservoirs, respectively. Temperatures must be expressed in Kelvin. The conversion equation is $T_K = T_C + 273$. Thermal electric power plants have efficiencies of 35 to 40%. In other words, 40% of the chemical (nuclear) energy is converted into electricity and 60% of the chemical (or nuclear) energy is rejected as waste heat.

Example 2.4

An electrical generating plant uses steam at 700°C (973K). On the downside, the steam is cooled by water to 300°C (573K). The maximum efficiency possible is around 0.41 or 41%.

Since efficiency is always less than 1, energy must be obtained from outside the system for a system or device to continue to operate. For a series of energy transformations, the total efficiency is the multiplication product of the individual efficiencies.

Example 2.5

Efficiency of Home Lighting Powered by Coal-Fired Plant

Transformation	Efficiency (%)
Mining of coal	96
Transportation of coal	97
Generation of electricity	38
Transmission of electricity	93
Electricity to light (incandescent, CFL, LED)	5, 20, 30
Overall efficiency, coal to light	1.6, 6.6, 9.9

You can see why fluorescence lights (15 to 25% efficiency) for commercial buildings, compact fluorescence lights (CFLs), and light-emitting diodes (LEDs; 25 to 50% efficiency) for homes are so important. Although CFLs and LEDs cost more, their higher efficiencies reduce the needs for new power plants. Countries, states, and even cities now have regulations requiring increased efficiency of lights. Incandescent lighting will be phased out eventually, and savings in air conditioning costs will be realized because other types of lighting emit less heat.

In the physical world, subsidies and economics (dollars) will not change the final outcome. All they do is tilt consumption or use in favor of different energy resources. For example, at some future point, it will take more energy to drill for oil than the energy in the oil produced. It will then become foolish to subsidize the drilling for oil as an energy source. It is possible that the product may be so useful as a liquid fuel or feedstock for other products that it may be subsidized by other energy sources. As another example, a glass of orange juice is a net energy loser in temperate climates. What is the energetics of producing ethanol from corn, especially corn irrigated by wells that need power to operate?

Prior to the oil crisis of 1973, industry and business maintained that efficiency was not cost-effective and that the gross domestic product (GDP) was tied directly to the amount of energy used. However, industry changed and the U.S. saved billions of dollars since 1973 through increased efficiency in industry and transportation. Now more efficient appliances, lights, and other devices for homes are saving energy, but much more conservation and efficiency must be achieved in the coming decades.

Every U.S. president since 1973 has called for energy independence, primarily in reaction to the importation of foreign oil. In 2006, President George W. Bush's energy policy maintained that we had to drill for more oil and gas, and as in the past, the automobile industry fought against increasing fuel efficiency. The industry's argument was couched in terms of economics—we cannot compete with foreign manufacturers of small cars; consumers will not buy fuel-efficient cars (while advertising pushes large motors, acceleration and power, and SUVs)—and safety.

In discussions, students said gas prices in the U.S. would have to reach $1 to $1.40/L ($4 to $5/gallon) before they would buy fuel-efficient vehicles. Of course, Europeans have been paying those prices for many years. It is thus not surprising that when crude oil prices exceeded $100/bbl in 2008, the sale of fuel-efficient vehicles increased, then as gasoline became cheaper in the U.S., fuel-efficient vehicle sales declined. Total safety means everyone should drive a truck or an M1 tank, or at least a huge car—to heck with fuel efficiency.

Another way to enhance efficiency is cogeneration—a technique that combines heat and power. During electricity production, the low-grade (lower-temperature) energy can be used for other processes. In most cases, 60% of the heat from electricity generation by steam (coal, oil, gas, and even nuclear) is not used. In Europe, some electric power plants have associated heating districts.

As an example of efficiency, in 1975 the U.S. Congress passed laws specifying corporate average fuel economy (CAFE) for vehicles weighing less than 3,886 kg. Pickup trucks and large vans were not covered by the CAFE. This law saved the U.S. millions of dollars for imported oil. The problem is that sport utility vehicles (SUVs) were counted as light trucks and their fuel consumption is around 5.5 km/L or 12 miles per gallon (mpg). As a result overall fuel efficiency declined as SUVs gained market share.

Despite continued objections by the automobile industry, finally in 2007 the CAFE was increased to 15 km/L (34 mpg) by 2020. Under President Obama, an agreement reached in 2011 with major automotive manufacturers (except Volkswagen) is that the CAFE should be raised to 23 km/L (54.5 mpg) for cars and light trucks by 2025. The European Union and Japan fuel economy standards for 2012 are around 19 L/km (45 mpg), and their proposed standards for 2025 will still be greater than those of the U.S.

Note that the Big Three U.S. automobile manufacturers received over $2 billion from the government for research and development through the Partnership for New Generation of Vehicles [4]. The goal was to design a sedan for five people that would achieve 34 km/L (80 mpg) fuel consumption. The automotive manufacturers later said that goal could not be reached. President George W. Bush promoted government incentives for using fuel cells and ethanol.

Amory Lovins, who was emphatically right about the soft energy path in response to the first energy crisis [5], strongly advocated hybrid and light-weight cars. Guess what? Hybrid cars entered the market in 2000. By 2013, a lot of hybrid models and even a number of plug-in hybrids and electric vehicles were available. [A personal note: I have owned hybrid models for years and bought an electric vehicle (average 7.5 km/kWh) in 2012.] Consider what large numbers of hybrid and electric cars will do to help alleviate the present energy dilemma that leads excessive oil imports in the U.S.

Again, we must ask where the federal government should award incentives. It may be cheaper to subsidize more efficient cars than subsidize drilling for oil. Consider the cost for oil if the costs for the Gulf War (Oil War I) and the Iraq War (Oil War II) are included.

In the past, the Organization of Petroleum Exporting Countries (OPEC) wanted to keep the price of oil in a range that made the countries a lot of money but not high enough to encourage conservation and efficiency. However, at some point the demand for oil across the world will exceed the supply. When world oil production starts to decline, we will face even higher prices.

2.5 EXPONENTIAL GROWTH

Our energy dilemma can be analyzed by using on fundamental principles. A corollary of the first law of thermodynamics is: it is a physical impossibility to have continued exponential growth of any product or exponential consumption of any resource in a finite system.

The present rate of consumption and the size of the system allow people to perceive resources as either infinite or finite. The total energy output of the sun and the amount of mass in the solar system

Energy

are infinite sources at our present rates of energy and material use even though the solar system is finite. The amount of solar energy received by the earth is a huge resource. The energy dilemma is defined within the context of the system and arises from the earth's finite amounts of fossil fuels.

An easy way to understand exponential growth (Figure 2.6) is to use the example of money. Suppose Sheri receives a beginning salary of $1/year with the stipulation that it will double every year (100% growth rate). It is easy to calculate the salary by year (Table 2.1). After thirty years, her salary is $1,000 million per year. Notice that for any year, the amount needed for the next year is equal to the total sum for all the previous years plus 1. If a small growth rate is used, the doubling time ($T2$) can be estimated by:

$$T2 = 69/R \qquad (2.7)$$

where R = percent of growth per unit time, generally years. Doubling times for different yearly rates are given in Table 2.2. Many factors serve as indicators of growth:

Population: 2 to 3% per year
Gasoline consumption: 3% per year
World oil production: 5 to 7% per year
Electricity consumption: 7% per year

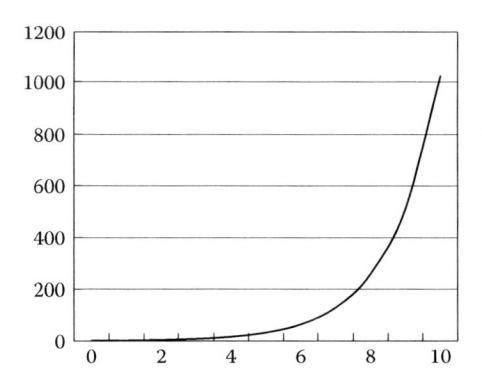

FIGURE 2.6 Exponential growth at growth rate of 100% per year.

TABLE 2.1
Salary at Growth Rate of 100% Per Year

Year	Salary, $	Amount = 2^t	Cumulative, $
0	1	2^0	1
1	2	2^1	3
2	4	2^2	7
3	8	2^3	15
4	16	2^4	31
5	32	2^5	63
6	64	2^6	127
7	128	2^7	255
8	256	2^8	511
t		2^t	$2^{t+1} - 1$
30	$1*10^9$	2^{30}	$2^{31} - 1$

TABLE 2.2
Doubling Times for Different Rates of Growth

Growth, %/year	Doubling Time, years
1	69
2	35
3	23
4	17
5	14
6	12
7	10
8	9
9	8
10	7
15	5

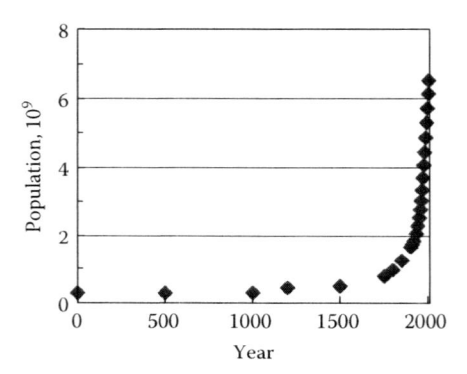

FIGURE 2.7 World population growth from year 0 to 2005.

If we plotted the values per year for smaller rates of growth, the curve would be the same as shown in Figure 2.6. Only the time scale along the bottom would be different Figure 2.7. The projection of future population growth (Figure 2.8) assumes a rate decrease to 0.5% in 2050. The United Nations projects a leveling of population at 9×10^9 to 11×10^9 people by 2200. However, even with smaller growth rates, the final result remains the same. When consumption grows exponentially, enormous resources do not last very long. Order of magnitude calculations make the analysis very clear.

2.6 USE OF FOSSIL FUELS

Although the 2008 recession reduced demand for energy in developed countries, the world demand increased from 510 EJ in 2007 to 549 EJ in 2011 due to the increase in demands in China and India. Consumption of fossil fuels (Figure 2.9) also increased because fossil fuels supplied around 90% of the energy.

However, it is physically impossible to continue to consume fossil fuels at exponential growth rates or even consume at the present rate of demand over the long term due to finite amounts of resources. Two excellent sources of world, regional, and national energy data are the U.S. Energy Information Administration (EIA; www.eia.gov) and British Petroleum's Statistical Review of World Energy (www.bp.com).

Energy

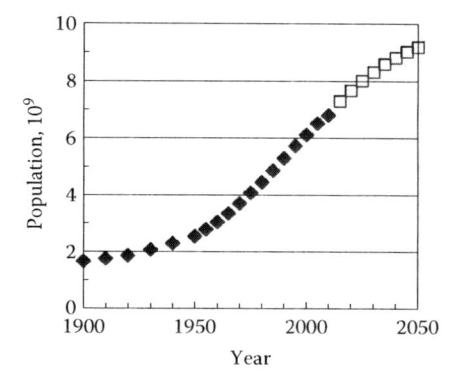

FIGURE 2.8 World population from 1900 to 2050 (United Nations projections for 2015 through 2050 under median variant).

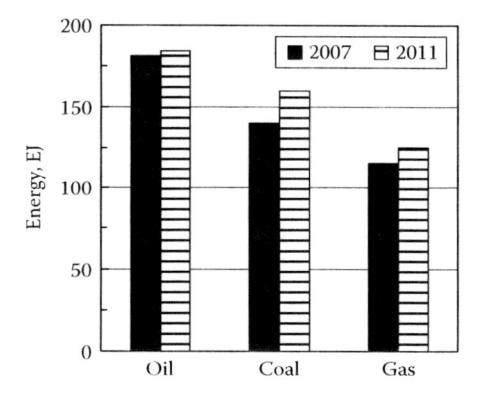

FIGURE 2.9 World fossil fuel consumption by type for 2007 and 2011 (gas = dry natural gas).

TABLE 2.3
Estimated Yearly Cost of Imported Oil for United States

Year	Demand (Gbbl)	Produced (Gbbl)	Imported (Gbbl)	$/bbl	Cost (Billions of Dollars)
1973	6.2	4	2.2	42	92
1990	6.1	3.52	2.61	37	97
2007	6.9	2.51	4.39	72	316
2012	6.0	3.17	2.78	102	284

2.6.1 Petroleum

The magnitude of the problem is demonstrated by the cost of imported oil in the U.S. (Table 2.3). In 1973, when petroleum consumption was 5.8 Gbbl/year and approximately 37% was imported, the cost of oil was around $92 billion per year at $42/bbl (the cost would be higher—around $100/bbl in 2012 dollars—if adjusted for inflation).

Although oil imports were reduced in the 1980s (Figure 2.10), imported energy was still very expensive. In the 1990s, U.S. oil consumption and imports increased again. The cost for imported oil reached $316 billion by 2007. Oil imports reached 50% by the mid 1900s and demand was at a

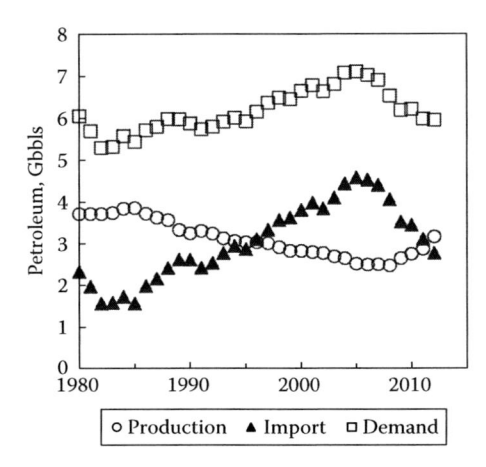

FIGURE 2.10 U.S. petroleum production, imports, and demand from 1980 to 2012.

high of 7.2 Gbbl in 2005. As demand decreased and U.S. production increased, imports dropped below 50% of demand in 2012. Demand and imports decreased due to the 2008 recession, increased U.S. production of crude oil, and increased efficiency. As a result the cost of imported oil was reduced to $284 billion by 2012.

The advent of horizontal drilling and fracking to obtain oil and gas from shale formations stopped the decline in production of oil and gas in the U.S. This technology will be used in other parts of the world to extend the duration and amount of future oil and gas production. Notice that crude oil production and oil supply and consumption are different; petroleum supplies include crude oil, natural gas liquids, plant liquids, and other liquids.

The important concept is that crude estimates of resources give fairly good answers as to when production for finite resources will peak. Also, predictions on the future use of resources can be made from past production as production and consumption of a finite resource will probably follow the bell curve.

Hubbert began his analysis of the U.S. oil production [6] in the early 1950s when he was with Shell Research. In 1956, Hubbert predicted that the U.S. oil production would peak mid 1970s. He was very close, as the actual peak occurred in 1970. The prediction (logistic curve) of U.S. oil production in Figure 2.11 is based on actual oil production through 2012, and the prediction was calculated on a spreadsheet using the method of Deffeyes [7, chap. 3]. Data include production from Alaskan oil fields and offshore facilities. The U.S. Energy Information Administration (EIA) predicts U.S. petroleum production in 2030 at 7.0 Gbbl for the reference case compared with 3.2 Gbbl in 2012—a 100% increase in production.

Even if a larger resource base is assumed, exponential growth means the larger resource will be used up around the same time. Also, as a resource is used, it becomes more difficult, i.e., production requires more energy, which also means more money. The amounts of oil and natural gas discovered per foot of hole drilled decrease exponentially. The same type of analysis and predictions can be made for natural gas, coal, and nuclear ore.

The bell curve, also called the normal or Gaussian curve, will not be exact for predicting future production because advanced technology will allow us to recover more fossil fuels and extend the time they will be available. However, the end result will be the same.

The actual production of oil (Figure 2.12) and natural gas (Figure 2.13) in Texas [8] corroborates the above analysis. Notice the difference between predictions made in 1992 and the actual oil and natural gas production in Texas since then. The predictions were based on existing and advanced technology for oil at $20 to $25/bbl and the state comptroller based its prediction on the continuation of past production (bell curve).

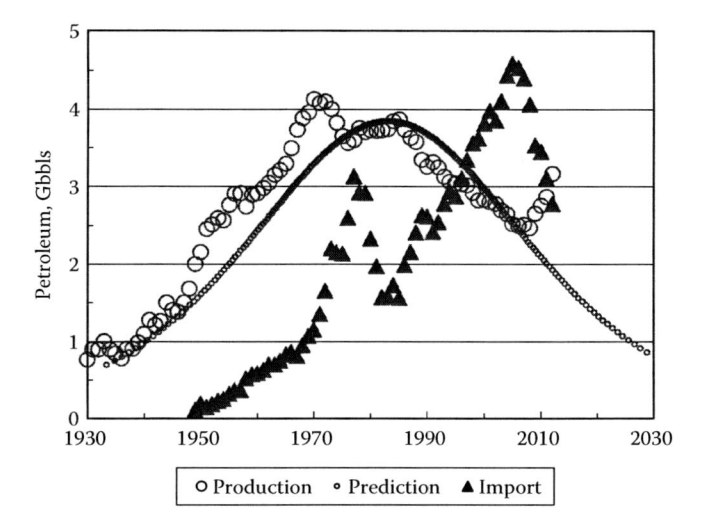

FIGURE 2.11 U.S. petroleum production, net imports, and production predictions using logistic curve.

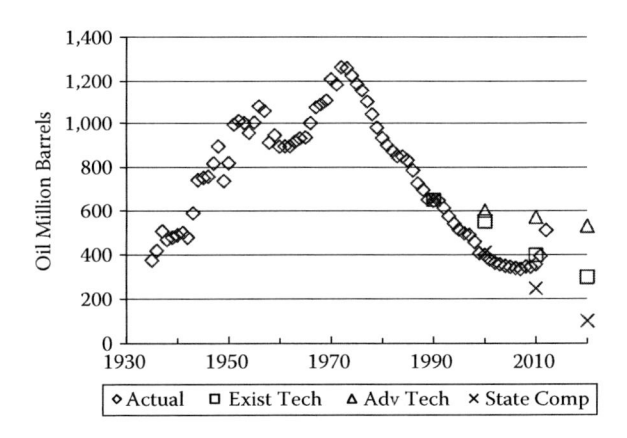

FIGURE 2.12 Texas crude oil production and future predictions. (*Source:* Texas Railroad Commission.)

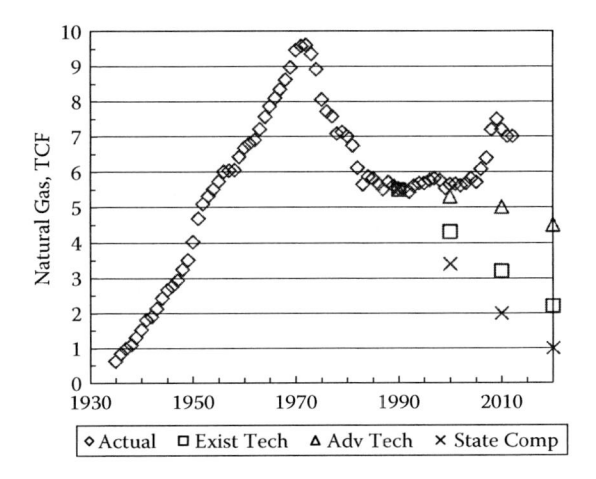

FIGURE 2.13 Texas natural gas production and future predictions. (*Source:* Texas Railroad Commission.)

Oil production again increased in Texas starting in 2008—the first increase since 1970—again due to advanced technology. The number of new wells exceeded the number shut down (plugged). The prediction for natural gas was based on $3 per thousand cubic feet. Oil production in Texas followed the low prediction curve. Natural gas production leveled off, and then a large increase occurred in 2008 as a result of more drilling and advanced technology. However, low gas prices in 2010 led to reduced drilling. Many more wells were drilled for natural gas than for oil from 1990 through 2009. While Texas is the major U.S. producer of oil and natural gas, it became a net importer of energy in 1994 and 1995.

World oil production [9] will follow the same pattern as U.S. production. Notice that the bell curve predicts a peak of world oil production (Figure 2.14) around 2016. There are a number of websites focusing on peak oil data. The oil poster (www.oilposter.org, 2/21/2013) is very well done and shows a world oil peak in 2010. Future production is stretched out because it includes heavy oil, deep water oil, polar oil, shale oil, and natural gas liquids, all of which will be more expensive. Note the cost of the BP *Horizon* oil leak in the Gulf of Mexico in 2010.

The reaction to the oil crises of 1973 and 1980 was increased efficiency, demonstrated by a dip in production. However, as developing countries demand more energy, demand and production will be approximated generally by a bell curve. In the past the U.S. EIA predicted cheap energy ($20/bbl) for 2030. Even in 2012, the agency predicted for 2030 that conventional liquid (petroleum) production would be at 35.2 Gbbl and price at $125/bbl for the reference case. It also predicted 4.2 Gbbl of biofuels for 2030. Its long-term predictions are probably low because petroleum prices will remain very volatile. To review EIA predictions for U.S. and international production, check the forecast and analysis section of its website (www.eia.doe.gov/analysis).

2.6.2 Natural Gas

People and organizations tout compressed natural gas for vehicles because of the cost of imported oil, the future decline of oil production, and fewer emissions from generating electricity. However, over the long term, the same problem persists: a finite resource will be used fairly quickly [10] because of increasing demand.

The production of natural gas (Figure 2.15) is increasing across the world. U.S. production climbed in 2008 because of increased drilling and advanced technology, especially for shale formations [11]. Production of natural gas in Russia exceeded that of the U.S. The two countries produced 50% of the world production in 1995 and 38% in 2011.

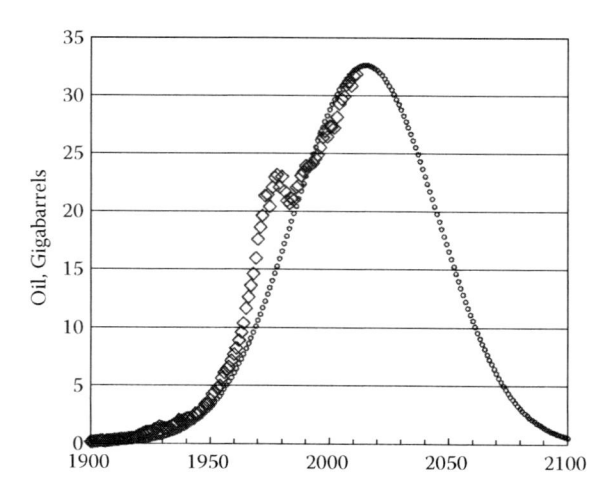

FIGURE 2.14 World oil production per year and production prediction (peak in 2016) using bell curve.

Energy

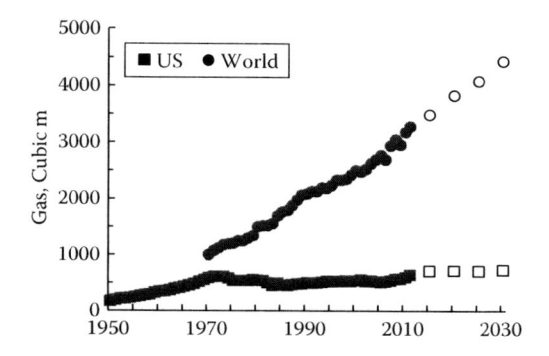

FIGURE 2.15 World and U.S. natural gas production and predictions.

Total production in the U.S. will decrease because reserves are around 8.5 Tm^3 compared to Russia's 74 Tm^3. Present reserves in the U.S. will last fewer than 100 years at the 2012 rate of consumption. The U.S. EIA predicts increased consumption to 2030. The peak of natural gas production will probably occur in the 2030s decade although some predict peak production by 2020 [12].

The production of natural gas from shale formations in the U.S. increased to 30% in 2011 from 8% in 2007 along with a corresponding reduction from a high price of $11/mcf in 2008 to a low of $1.79/mcf in May 2012. Although the price rebounded to $3.00/mcf in October 2012, low prices mean fewer new gas wells.

Natural gas is an important feedstock for fertilizer and has also been promoted as the feedstock for a future hydrogen economy. Both developments will require enormous amounts of natural gas. Carbon dioxide emissions in the U.S. have decreased since 2008 due to decreased demand and its replacement by coal in electricity production.

It is important to remember that new wells are needed for both oil and natural gas to replace decreased production from previous wells. For 2011, wells drilled in the U.S. accounted for about 30% of the world total. Since 1949, 2,581,782 oil and gas wells (exploratory, developmental, dry) have been drilled in the U.S., and around 850,000 are still in production. It is obvious that the U.S. accounted for around 30% of the world demand. More wells have been drilled in the U.S. than in any other region. For example, Saudi Arabia has the world's largest oil reserves and drilled only around 10,000 wells [13, chap. 5].

2.6.3 COAL

Each fossil fuel industry touts the use of its product. The coal industry promotes the sustainable development of coal and the conversion of coal to liquid fuels. Clean coal—which is really stretching its total environmental impact—is a concept intended to promote coal plants that sequester carbon dioxide. In 2012, coal provided 26% of the world's primary energy and 43% of global electricity. Production of coal has increased by 47% in the last 25 years; production in 2011 was 7.5 Gt.

Production of coal has increased, especially since 2000 (Figure 2.16) because 80% of the electricity in China is provided by coal, and China constructs several coal-power plants every year. Coal also fuels major portions of heating and cooking in China.

The U.S. has the largest coal reserves (Figure 2.17) that are estimated to last 200 years. Does that 200 years include increased production of coal since coal producers want to increase their share of the energy market? Of course, the use of coal produces pollution and carbon dioxide emissions. For more information, go to the U.S. EIA site; www.americancoalcouncil.org provides the industry view.

In the long term, the use of fossil fuels may become the "fickle finger of fate" (Figure 2.18). The world is approaching the midpoint of the 400-year reign of fossil fuels as the major energy source.

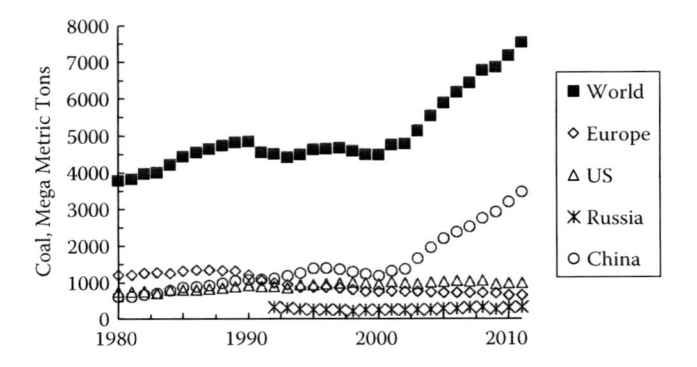

FIGURE 2.16 World and major producing nations' coal production in 2011.

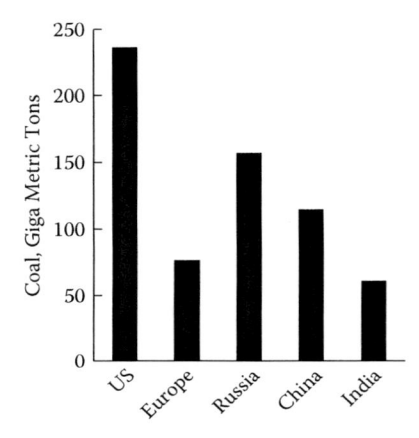

FIGURE 2.17 Major world coal reserves.

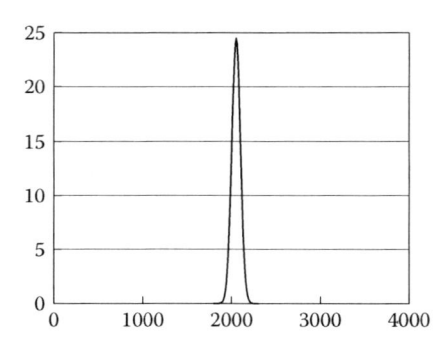

FIGURE 2.18 Fossil fuel exploration and use in human history from year 0 (projected to 4000). Compare this graph with Figure 2.14.

Also, global climatic change caused by the consumption of fossil fuels will produce a major impact on civilization.

2.7 NUCLEAR ENERGY

The first commercial nuclear plant was built in 1957. As of 2012 [14], the installed capacity of 435 plants around the world was 356 GW (production of 2,507 TWH according to 2011 data).

Energy

The 104 plants in the U.S. had installed capacity of 115 GW and production of 790 TWH. In 2013, one U.S. nuclear plant was closed permanently due to cost overruns for major repairs.

The capacity was 365 GW in 2006. Some countries in Europe are phasing out nuclear power; 6.3 GW was decommissioned in 2011. Worldwide, sixty-five reactors with a capacity of 65 GW were under construction in 2013, but operation dates are some years away because of long construction periods. The entire world knows about the failures of the nuclear power plants in Japan due to the May 2011 tsunami that reduced the nation's capacity by 12 GW and then resulted in all nuclear plants being shutdown in Japan.

In Example 2.6 What was in first edition was left out of second edition. Reinsert

If you do not use the equation, a spreadsheet is very useful for calculations, as you can play with different scenarios of growth and size of the resource.

year	Consumption	Cumulative
0	3.00E+10	
1	3.09E+10	3.09E+10
2	3.18E+10	6.27E+10
3	3.28E+10	9.55E+10
.	. . . .	. . . .
23	5.92E+10	1.00E+12
24	6.10E+10	1.06E+12
25	6.28E+10	1.13E+12
26	6.47E+10	1.19E+12
27	6.66E+10	1.26E+12

So at around 25 years all the conventional oil is gone. In the real world there is not the abrupt drop-off, as supply cannot meet demand. However, the example reinforces a previous statement: Exponential growth means large resources do not last very long.

Nuclear power provides about 13% of global electricity, with the largest percentage (75% of that) generated in France. The U.S. has not built any new nuclear plants since 1996, and the production of U.S. electricity by nuclear power declined from 23 to 20%. The reason is that new electric plants are primarily fired by natural gas. The U.S. nuclear plants achieved about a 90% capacity factor—a large improvement over the 66% in 1990. U.S. nuclear power received large amounts of funding for research and development and continues to receive substantial federal funding.

2.8 MATHEMATICS OF EXPONENTIAL GROWTH

Values of future consumption r can be calculated (Example 2.6) from a present rate r_0 and fractional growth per time period k:

$$r = r_0 e^{kt} \tag{2.8}$$

where e is the base of the natural log and t is time.

Example 2.6

Present consumption is 100 units/year and growth rate is 7% per year. Therefore, $r_0 = 100$ units/year, $k = 0.07$/year. Suppose $t = 100$ years. We then calculate:

$$r = 100\ e^{0.07*100} = 100\ e^7 = 100 \times 1{,}097 = 1 \times 10^5 \text{ per year}$$

The consumption per year after 100 years is 1,000 times larger than the present rate of consumption. Note that exponents never have associated units.

2.8.1 Doubling Time

The doubling time $T2$ in years for any growth rate can be calculated from Equation 2.8:

$$r = 2\,r_0,\ 2r_0 = r_0 e^{kT2}\ \text{or}\ 2 = e^{kT2}$$

Take the natural log ln of both sides of the equation:

$$\ln 2 = k * T2,\ T2 = 0.69/k$$

If right side values are multiplied by 100, $T2 = 69/R$ [Equation (2.7)] where R is the percentage growth rate per year.

2.8.2 Resource Consumption

The total sum of the resource consumed from an initial time to any time T can be estimated by summing consumption per year. This can be done via a spreadsheet on a personal computer or calculated. If r is known as a function of time, the total consumption can be found by integration. For exponential growth, the total consumption is given by:

$$C = \int r\,dt = \int_0^T r_0 e^{kt}\,dt$$

$$C = \frac{r_0}{k}\left(e^{kT} - 1\right) \qquad (2.9)$$

2.9 LIFETIME OF FINITE RESOURCE

If the magnitude of a resource is known or can be estimated, the end time T_E (when the resource is used up) can be calculated for different growth rates. The size of resource S is inserted in Equation (2.9) and the resulting equation is solved for T_E:

$$S = \frac{r_0}{k}\left(e^{kT_E} - 1\right)$$

$$T_E = \frac{1}{k}\ln\left(k\frac{S}{r_0} + 1\right) \qquad (2.10)$$

If the demand is small enough, reduced exponentially, or reduced at the depletion rate, a resource can essentially last a very long time. However, with increased growth, T_E can be calculated for various resources (Table 2.4 and Example 2.7), and the time before the resource is used up is generally short. Remember, these are only estimates of resources, and other estimates will be higher or lower.

TABLE 2.4
Estimated 2011 Resources and Reserves

Resource	Amount
U.S. crude oil	30.9×10^9 bbl
U.S. oil	80×10^9 bbl
U.S. natural gas	8.5×10^{12} m^3
U.S. coal	237×10^9 t
U.S. uranium oxide (2008)	2.5×10^5 t @ \$110/kg
	5.6×10^5 t @ \$220/kg
World crude oil (conventional)	1.6×10^{12} bbl
World oil (heavy, sand, shale, deep sea, polar)	2.1×10^{12} bbl
World natural gas	196×10^{12} m^3
World coal	861×10^9 t
World uranium oxide	5.4×10^6 t @ \$130/kg
	6.3×10^6 t @ \$260/kg

Notes: t = metric tons. bbl = barrels. m^3 = cubic meters.
Sources: Energy Information Administration. *Statistical Review of World Energy 2012* (www.eia.doe.gov/cneaf/nuclear/page/reserves/ures.html) and European Nuclear Organization (www.euronuclear.org/info/encyclopedia/u/uranium-reserves.htm)

Example 2.7

How long will world oil last if consumption grows at 3%/year? First, we designate $r_0 = 30 \times 10^9$ barrels/year, $S = 1{,}100 \times 10^9$ barrels, and $k = 0.03$. We then place the values In Equation (2.10):

$$T_E = \frac{1}{0.03}\ln\left(0.03\frac{1100*10^9}{30*10^9}+1\right) = 33\ln(2.1) = 33 \times 0.74 = 24 \text{ years}$$

If you do not use the equation, a spreadsheet is very useful for calculations, as you can try different scenarios of growth and size of a resource.

Based on the calculations, all conventional oil will be gone by the twenty-fifth year. In the real world, however, no abrupt drop-off occurs when supply cannot meet demand. However, the example reinforces the previous statement that exponential growth means large resources do not last very long.

Year	Consumption	Cumulative
0	3.00E + 10	
1	3.09E + 10	3.09E + 10
2	3.18E + 10	6.27E + 10
3	3.28E + 10	9.55E + 10
...	...	...
23	5.92E + 10	1.00E + 12
24	6.10E + 10	1.06E + 12
25	**6.28E + 10**	**1.13E + 12**
26	6.47E + 10	1.19E + 12
27	6.66E + 10	1.26E + 12

According to the energy companies, the continued growth of energy use in the U.S. is to be fueled by our largest fossil fuel resource (coal) and nuclear energy. With the increased natural gas from shale formations, natural gas would displace some of the coal and serve as the bridge to large renewable energy use in the future, especially in the electricity sector. How long can coal last if we continue to increase production to offset decline in production of oil and reduce the need for oil imports? The preceding analysis will allow you to make order of magnitude estimates. Also, increased or even current production rates of fossil fuels may produce major environmental effects. Global warming has become an international political issue.

2.10 GLOBAL WARMING

Global warming clearly demonstrates that physical phenomena do not react to political or economic statements. Global warming is primarily the result of human activity. For the first time in human history, our activities have produced an impact on a global scale.

> "Global atmospheric concentrations of carbon dioxide, methane and nitrous oxide have increased markedly as a result of human activities since 1750 and now far exceed pre-industrial values determined from ice cores spanning many thousands of years …. The global increases in carbon dioxide concentration are due primarily to fossil fuel use and land use change, while those of methane and nitrous oxide are primarily due to agriculture." [15, p. 2]

Concentrations of carbon dioxide in the atmosphere (Figure 2.19) are projected to double due to future energy use based on today's trend [16]. The Kyoto Protocol of 1996 to reduce greenhouse gas emissions became effective in 2005 as Russia became the fifty-fifth country to ratify the agreement. The goal was for the participants collectively to reduce emissions of greenhouse gases 5.2% below 1990 emission levels by 2012. While the 5.2% figure was collective, individual countries were assigned higher or lower targets and some were permitted increases. For example, the U.S. was expected to reduce emissions by 7%. However this did not happen. The U.S. failed to ratify the treaty because the perceived economic costs would be too high and the provisions for reducing future emissions for developing countries, especially China, were inadequate.

If participant countries allow emissions above target levels, they will be required to engage in emissions trading. Notably, participating countries in Europe are using different methods for carbon dioxide trading with other countries including wind farm installations and forest planting. Carbon dioxide emissions will still increase, even if nations reduce their emissions to 1990 levels because of population growth and increased energy use in the underdeveloped world. As the Arctic thaws, methane—a more potent greenhouse gas than CO_2—will further increase global warming [18].

Increased temperatures, effects on weather, and sea level rises are the major consequences. Overall, the increased temperature will have negative effects compared to the climate conditions from

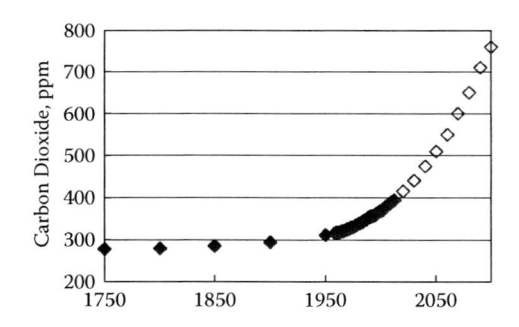

FIGURE 2.19 Carbon dioxide in atmosphere and projected growth without emission reductions.

1900 to 2000. By 2100, sea levels are projected to increase by 0.2 to 1 m. An increase of 2 m is unlikely but physically possible. As a result of decreased sea ice and continued increases in carbon dioxide emissions, melting of the Greenland ice sheets would increase sea level by over 7 m and the West Antarctic ice sheet would add another 5 m. Large seacoast cities will have to be relocated or build massive infrastructures to keep out the ocean. Who will pay for this? The national or local governments?

2.11 SUMMARY

Continued exponential growth is a physical impossibility in a finite (closed) system. Calculations of future conditions are only estimations. Possible solutions to our energy dilemma are:

1. Reducing fossil fuel demands to depletion rates.
2. Using renewable energy at sustainable rates and establish a steady-state society.
3. Redefining the size of the system by colonizing space and other planets. This measure will not, however, solve the energy dilemma on earth. From our present view, the resources of the solar system are infinite and our galaxy contains over 100×10^9 stars.

Because the earth is finite, it imposes limits on population, fresh water supplies, fossil fuels, minerals [19], and even on food production and fish catches. Therefore, a change to a sustainable society that depends primarily on renewable energy will be imperative within this century. The world will have to take certain steps during the transition period (2007 through 2020). In order of priority, the steps are:

1. Implement conservation and energy efficiency. Since the first energy crisis, this has proven the most cost-effective mode of operation. It is much cheaper to save a barrel of oil than to discover new oil or import it.
2. Increase the use of renewable energy.
3. Reduce dependence on oil and natural gas during the transition period.
4. Use clean coal to control carbon dioxide and include all social costs (externalities).
5. Implement incentives and penalties in line with items 1 and 2.

Efficiency can be improved in all major sectors: residential, commercial, industrial, transportation, and electrical. The greatest gains can be accomplished in the transportation, residential, and commercial sectors. National, state, and even local building codes can improve energy efficiency in buildings. Finally, individuals can participate in conservation and energy efficiency activities. They can also serve as advocates for conservation, efficiency, renewable energy, and environmental protection.

Possible future predictions for human society are conservation and efficiency, transition to sustainable energy, steady state with no growth, catastrophe, or catastrophe with some revival (Figure 2.20). As overpopulation and overconsumption exert effects on the earth via an uncontrolled

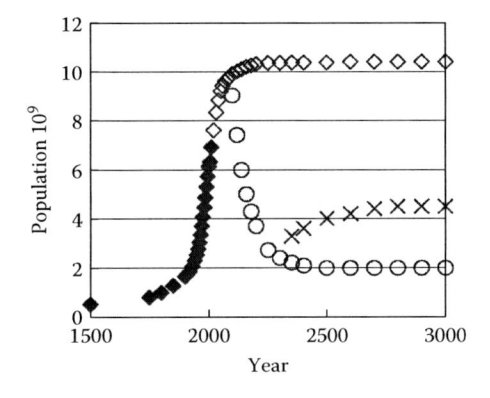

FIGURE 2.20 Possible future population paths.

experiment, the most probable future for the human population is catastrophe or catastrophe with some revival, especially because politicians and other decision makers respond mostly to short-term economic and other concerns.

LINKS

Energy Information Administration of the U.S. Department of Energy (data on U.S. and international energy resources and production; reports and data files can be downloaded along with PDFs and spreadsheets). www.eia.doe.gov

International Energy Agency, www.eia.org

Texas Railroad Commission (regulates the state's oil and gas production). www.rrc.state.tx.us

Peak Oil, www.peakoil.com

United Nations (population data and growth projections). www.un.org/esa/population/unpop.htm

U.S. Census (world population Information). www.census.gov

Worldmapper (morphed maps showing topical data such as population, oil exports, oil imports, and other resource information). www.worldmapper.org

U.S. Department of Energy (excellent source of information). www.nrel.gov/docs/fy13osti/54909.pdf

British Petroleum (includes downloadable Excel workbook). http://www.bp.com/sectionbodycopy.do?categoryId=7500&contentId=7068481

REFERENCES

1. Interactions. http://hyperphysics.phy-astr.gsu.edu/hbase/forces/funfor.html
2. Night sky of earth. http://antwrp.gsfc.nasa.gov/apod/ap010827.html
3. K. Adler. 1986. The perpetual search for perpetual motion. *American Heritage of Invention and Technology*, Summer, 58.
4. G. Zorpette. 1999. Waiting for the supercar. *Scientific American*, April, 46.
5. A. Lovins. 1977. *Soft Energy Paths: Toward a Durable Peace*. San Francisco: Ballinger.
6. K. Hubbert. 1969. Energy resources. *Resources and Man: National Academy of Sciences*, 157–242;1971. Energy resources of the earth, *Scientific American*, 60.
7. K.S. Deffeyes. 2005. *Beyond Oil: The View from Hubbert's Peak*. New York: Hill and Wang.
8. State of Texas Energy Policy Partnership (STEPP). 1993. Report. Texas' energy resources: oil, p. 9. Texas' energy resources: natural gas, p. 12. Vol. 1. Austin: Texas Railroad Commission. www.seco.cpa.state.tx.us/seco_links.htm
9. K.S. Deffeyes. 2001. *Hubbert's Peak: The Impending World Oil Shortage*. Princeton, NJ: Princeton University Press.
10. J. Darley. 2004. *High Noon for Natural Gas*. White River Junction, VT: Chelsea Green.
11. B. Powers. 2012. *Cold, Hungry and in the Dark: Exploding the Natural Gas Supply Myth*. Gabriola Island, Canada: New Society Publishers.
12. R.W. Bentley. 2002. Global oil and gas depletion: an overview. *Energy Policy*, 189. http://www.oilcrisis.com/bentley/depletionoverview.pdf
13. M.R. Simmons. 2005. *Twilight in the Desert: The Coming Saudi Oil Shock and the World Economy*. New York: John Wiley & Sons.
14. U.S. Energy Information Administration. World Nuclear Reactors. International Energy Statistics. www.eia.gov; Nuclear Energy Institute. www.nei.org/resourcesandstats/; World Nuclear Organization. www.world-nuclear.org
15. IPCC Working Group I. 2007. Summary for policymakers. In S. Solomon, D. Qin, M. Manning et al., Eds., *Climate Change 2007: The Physical Science Basis*. New York: Cambridge University Press.
16. Intergovernmental Panel on Climate Change. Carbon dioxide: projected emissions and concentrations. www.ipcc-data.org/ddc_co2.html
17. U.S. Climate Change Science Program. 2007. Final report: scenarios of greenhouse gas emissions and atmospheric concentrations and review of integrated scenario development and application, www.climatescience.gov/Library/sap/sap2-1/finalreport/sap2-1a-final-technical-summary.pdf
18. S. Simpson, 2009. The peril below the ice. *Scientific American: Earth 3.0*, 18, 30.
19. M.C. Klare. 2001. *Resource Wars: The New Landscape of Global Conflict*. New York: Metropolitan Books.

SUGGESTED READINGS

R.N. Anderson. 1998. Oil production in the 21st century. *Scientific American*, March, 86.
T. Appenzeller. 2004. The end of cheap oil. *National Geographic*, June, 80.
A.A. Bartlett. 2000. Analysis of U.S. and world oil production patterns using Hubbert-style curves. *Mathematical Geology, 12,* http://jclahr.com/bartlett/20000100,%20Mathematical%20Geology.pdf
L.R. Brown. 2009. *Plan B 4.0*. New York: Norton.
C.J. Campbell. 2005. *Oil Crisis*. Essex, U.K. Multi-Science Publishing.
C.J. Campbell and Jean H. Laherrere. 1998. The end of cheap oil. *Scientific American*, March, 78.
W. Clark. 1974. *Energy for Survival: The Alternative to Extinction*. Garden City, NY: Anchor Press.
G. Daily and K. Ellison. 2002. *The New Economy of Nature*. Washington: Island Press.
J. Darley. 2004. *High Noon for Natural Gas*. White River Junction, VT: Chelsea Green.
T. Flannery. 2006. *The Weather Makers*. New York: Atlantic Monthly Press.
S.A. Fouda. 1998. Liquid fuels from natural gas. *Scientific American,* March, 92.
R.L. George. 1998. Mining for oil. *Scientific American,* March, 84.
R. Heinberg. 2007. *Peak Everything*. New York: New Society.
M.T. Klare. 2004. *Blood and Oil*. New York: Metropolitan Books.
T.H. Lee, B.C. Ball, and R.D. Tabors. 1990. *Energy Aftermath*. Boston: Harvard Business School Press.
N. Lenssen. 1993. Providing energy in developing countries. In *State of the World 1993*, New York: Norton, p. 101.
A. Mckillop, Ed. 2005. *The Final Energy Crisis*. London: Pluto Press.
H.T. Odum. 1975. *Environment, Power, and Society*. Ann Arbor, MI: Wiley Interscience.
M. Parfit. 2005. Future power. *National Geographic*, August, 2.
R.H. Romer. 1976. *Energy, an Introduction to Physics*. San Francisco: W.H. Freeman.
Scientific American. 2005. Crossroads for Planet Earth: Special Issue, September.
Scientific American. 1990. Energy for Planet Earth: Special Issue, September.
M.R. Simmons. 2005. *Twilight in the Desert*. New York: John Wiley & Sons.
W. Youngquist. 1997. *GeoDestinies: The Inevitable Control of Earth Resources over Nations and Individuals*. Portland, OR: National Book.

QUESTIONS AND ACTIVITIES

1. Go to a U.S. Census site and look at the population clock at upper right. What is the population of the United States? The world?
2. List three ways you are going to save energy this year.
3. Go to the Energy Information Administration website and access data for latest year available. What is world oil production? What is world coal production?
4. Would you rather stick your finger in a cup of hot coffee (temperature = 80°C) or be hit by a high-speed proton that has a temperature of 1,000,000°C. Justify your answer.
5. Place your hand near a 100-W incandescent light bulb and a 20- to 40-W fluorescent light bulb. Qualitatively describe the light and heat outputs of the two bulbs.

ORDER OF MAGNITUDE (OM) ESTIMATES

Estimates of energy consumption, production, supply, and demand are needed and order of magnitude calculations will suffice. Order of magnitude means expressing a quantity as one or two significant digits to a power of 10. For example, determining the number of seconds in a year is easy if a calculator is used:

365 days × 24 hr/day × 60 min/hr × 60 sec/hr = 31,536,000 sec (answer to five significant digits)

Rounding 31,536,000 to one significant digit, the answer is 3×10^7 sec. For two significant digits, the answer is 3.2×10^7 sec. To calculate an order of magnitude estimate, round all inputs to one number with a power of 10, then multiply the numbers and add the powers of 10. Without a calculator, the above problem becomes:

$$4 \times 10^2 \times 2 \times 10^1 \times 6 \times 10^1 \times 6 \times 10^1 = 4 \times 2 \times 6 \times 6 \times 10^5 = 288 \times 10^5 = 3 \times 10^2 \times 10^5 = 3 \times 10^7$$

Order of magnitude problems below are preceded by "OM." If you have trouble with powers of 10, please consult your instructor.

PROBLEMS

1. A snowball with a mass of 0.5 kg is thrown at 10 m/sec. How much kinetic energy does it possess? What happens to that energy after you are hit with that snowball?
2. OM: The Chamber of Commerce and the Board of Development are always promoting their city as a place for new industry. If the city has a population of 100,000 and a growth rate of 10% per year, what is the population after five doubling times? How many years is that?
3. What is the doubling time if the growth rate is 0.5%? The world population in 2013 was around 7.1×10^9.
4. OM: If world population is 7×10^9, estimate how many years before the population reaches 28×10^9.
5. OM: How many people will populate the earth by the year 2100? Assume present rate of growth of world population.
6. OM: If the growth rate of population could be reduced to 0.5% per year, how many years would it take to reach 24×10^9 people?
7. OM: The population of the world is predicted to reach 10×10^9. Mexico City is one of the largest cities in the world at 2×10^7 people. How many new cities the size of Mexico City must be built to accommodate this increase in population?
8. OM: The most economical size of a nuclear power plant is around 1,000 MW. How many nuclear power plants would have to be built in the U.S. over the next fifty years to meet its long-term historical growth of 7% per year in demand for electricity?
9. OM: Assume world electricity demand increases by 10% per year over the next thirty years. To meet all that increased demand, how many 1,000-MW nuclear plants must be installed by the end of thirty years? What is the total cost for those nuclear plants at $5,000/kW?
10. OM: If China's electricity demand increases by 50% over the next thirty years, how many 300-MW coal plants will be needed at the end of thirty years? What is the total cost for those coal plants at $2,000/kW?
11. OM: For problem 10, how many metric tons of coal will be needed for the thirtieth year? Assume plants operate at 90% capacity and 40% efficiency.
12. What is the efficiency at a nuclear power plant if the incoming steam is at 700°C and the outgoing steam is at 320°C? Remember, you have to use Kelvin.
13. The Hawaii Natural Energy Institute tested a 100-kW ocean thermal energy conversion (OTEC) system. The surface temperature is 30°C, and at a depth of 1 km the temperature is 10°C. Calculate the maximum theoretical efficiency of an OTEC engine. Remember, you have to use Kelvin.
14. For a binary cycle geothermal power plant, the incoming temperature is 110°C and the outgoing temperature is 71°C. Calculate the maximum theoretical efficiency of the steam turbine.
15. OM: Using data on U.S. coal reserves, at today's rate of consumption, how long will the nation's reserves last?
16. OM: Assume a growth rate of coal consumption for the U.S. of 10% per year because of the use of coal for liquid fuels. How long will the U.S. coal last?

17. OM: Assume a growth rate of coal consumption for China of 15% per year. How long will China's coal last?
18. Estimate the power installed for lighting your home. Estimate the energy used for lighting for one year.
19. Estimate the energy saving if you convert your home lighting to LED lights because they are more efficient (produce more light per watt).
20. What is the maximum power (electrical) used by your residence (assume all your appliances, lights, and electronics are on at the same time)?
21. OM: Estimated world oil reserves are 2.1×10^{12} bbl. How long will the reserves last at the present rate of consumption?
22. OM: Same as previous problem, but assume a demand increase of 2.5% per year. How long will the oil last?
23. OM: Calculate how long world coal reserves will last if world demand increases at a rate of 5% per year.
24. OM: The U.S. now has 250 million cars that consume 10 million bbl of gasoline per day. Suppose the Chinese government goal is the same ratio of people to cars within thirty years. How many cars will China have and how many barrels of oil will It consume per year?
25. OM: How long will U.S. uranium last for the present installed nuclear power plants? Assume a nuclear power plant uses around 3×10^4 kg of uranium oxide to generate 1 TWH of electricity.
26. OM: Same as Problem 25, but assume a 2% per year growth in nuclear power plants.
27. OM: How long will world uranium last for the present world nuclear power plants?
28. OM: Same as Problem 27, but assume a 4% per year growth in nuclear power plants.

3 Wind Characteristics

3.1 GLOBAL CIRCULATION

The motion of the atmosphere can vary in distance and time from the very small to the very large (Table 3.1). There is an interaction between each of these scales and the flow of air is complex. Global circulation encloses eddies that enclose smaller eddies and the enclosures continue until finally the microscale is reached.

The two main factors in global circulation are the solar radiation and the rotation of the earth and the atmosphere. The seasonal variation is due to the tilt of the earth's axis to the plane of the earth's movement around the sun. Since solar radiation is greater per unit area when the sun is directly overhead, heat is transported from the regions near the equator toward the poles.

Because the earth rotates on its axis and there is conservation of angular momentum, the wind will shift as it moves along a longitudinal direction. The three-cell model explains the predominant surface winds (Figure 3.1). Those regions in the trade winds are generally good locations for the utilization of wind power; however, there are exceptions, as Jamaica is not nearly as windy as Hawaii.

Superimposed on this circulation is the migration of cyclones and anticyclones across the midlatitudes that disrupt the general flow. Also, the jet streams, the fast cores of the central westerlies at the upper levels, influence the surface winds.

Local winds are due to local pressure differences and are influenced by the topography, friction of the surface due to mountains, valleys, and other features. The diurnal (24 hr) variation is due to temperature differences between day and night. The temperature differences between the land and sea also cause breezes but they do not penetrate very far inland (Figure 3.2).

3.2 EXTRACTABLE LIMITS OF WIND POWER

Solar energy drives the wind, which is then dissipated due to turbulence and friction at the earth's surface. The earth's atmosphere can be considered a giant duct, and if energy is used at one location, it is not available elsewhere. Therefore, it is important to distinguish between the kinetic energy in the wind and the rates and limits of the extraction of that energy, the power in the wind, and the maximum power extractable.

A comparison can be made on the basis of the kinetic energy of the winds per unit area of the earth's surface. Of the solar input, only 2% is converted into wind power, and 35% of that is dissipated within 1 km of the earth's surface. This is the wind power available for conversion to other forms of energy.

The amount extracted is limited by the criterion of not changing the climate but the uncertainties surrounding such criteria are huge. Humans would have to substitute specific types of wind turbines for naturally occurring frictional features such as trees, mountains, and other natural features.

Gustavson [1] assumed the extractable limit as 10% of the available wind power within 1 km of the surface. When these values are applied to the contiguous forty-eight states of the U.S., the limit would be 2×10^{12} W (2 TW), or 62 quads/year. A similar analysis can be made for the entire world. The calculation shows that wind energy represents a very large energy source.

On a global scale, wind can be compared to other renewable sources (Table 3.2). In locations with high wind speeds, wind power is comparable to or better than the amount of solar power available. The wind energy available represents approximately 20 times the rate of global energy consumption.

TABLE 3.1
Time and Space Scale for Atmospheric Motion

Name	Time	Length	Example
General circulation	Weeks to years	1,000 to 40,000 km	Trade winds, Jet stream
Synoptic scale	Days to weeks	100 to 5,000 km	Cyclones, Hurricanes, Typhoons
Mesoscale	Minutes to days	1 to 100 km	Thunderstorms, Land-sea breezes, Tornadoes
Microscale	Seconds to minutes	<1 km	Turbulence

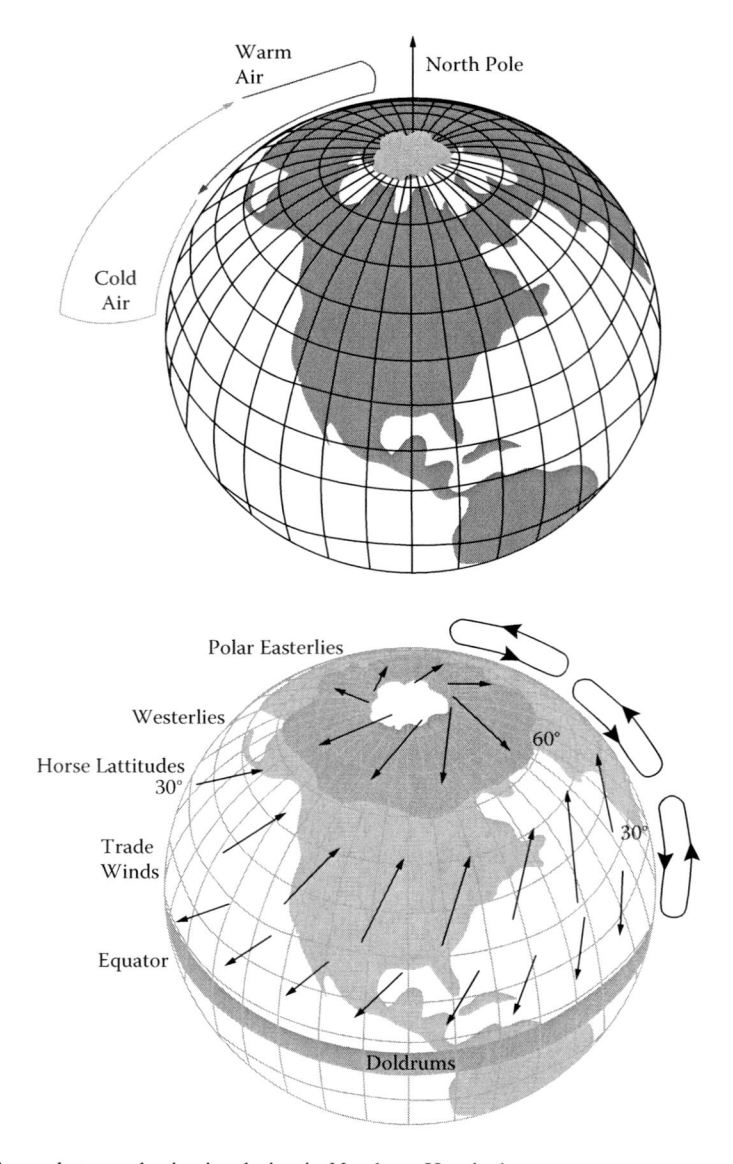

FIGURE 3.1 General atmospheric circulation in Northern Hemisphere.

Wind Characteristics

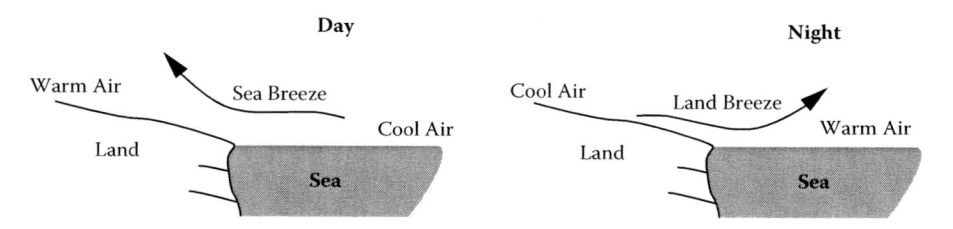

FIGURE 3.2 Sea breezes (day) and land breezes (night).

TABLE 3.2
Summary of Global Values for Renewable Sources

	Power, W	Extractable Power, W	Energy, quads/year
Solar	$1.8*10^{17}$		
Wind	$3.6*10^{15}$	$1.3*10^{14}$	3,900
Hydro	$9.0*10^{12}$	$2.9*10^{12}$	86
Geothermal	$2.7*10^{13}$	$1.3*10^{11}$	4
Tides	$3.0*10^{12}$	$6.0*10^{11}$	1.9

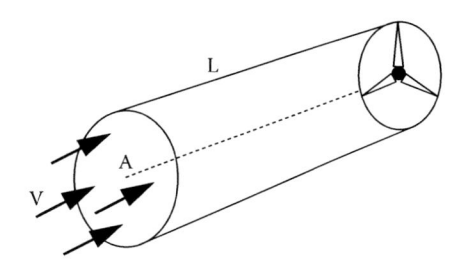

FIGURE 3.3 Flow of wind through a cylinder of area A.

3.3 WIND POWER

The moving molecules of air have kinetic energy, so the amount of air molecules moving across some area during some time period determines the power locally (Figure 3.3). This area is not the surface area of the earth, which was referred to in the estimation of extractable power and energy, but the area perpendicular to the wind flow. The mass m in the volume of the cylinder that will pass across the area A in time t can be determined from the density of the air ρ and the volume of the cylinder V. The power is the kinetic energy (KE) of the air molecules divided by the time:

$$P = KE/t = 0.5 \, m \, v^2/t \qquad (3.1)$$

$$\rho = m/V$$

$$V = \text{area} \times \text{length} = A \times L$$

$$m = \rho \times V = \rho \times A \times L$$

Substitute the value of mass into Equation (3.1). Only those molecules with a velocity $v = L/t$ will cross the area in time t and those further to the left will not, so the power is given by

$$P = 0.5\ \rho\ A\ L\ v^2/t = 0.5\ \rho\ A\ L/t\ v^2 = 0.5\ \rho\ A\ v\ v^2 = 0.5\ \rho\ Av^3$$

The power/area, referred to as wind power density, is

$$P/A = 0.5\ \rho\ v^3 \qquad (3.2)$$

From Equation (3.2) the power/area in the wind can be calculated for different wind speeds (Table 3.3). However, not all the power in the wind can be extracted, as the maximum theoretical efficiency for wind turbines is 59%.

Note that if the wind speed is doubled, the power is increased 8 times, and the power at 25 m/sec is 125 times the power at 5 m/sec. Because there is so much power and energy in the wind at high speeds, some damage to structures and trees occurs during severe storms and major damage is caused by tornadoes and hurricanes of class 3 and above. This is also the reason wind turbines do not extract all the available energy at high wind speeds. All wind turbines have some means of control or they would be destroyed in high winds.

Example 3.1

A wind turbine with a radius of 2 m, area = 12.6 m^2, would have approximately 100 kW of wind power across that area due to a 25 m/sec wind speed.

A first estimation of wind power/area can be calculated using the annual mean wind speed that can be estimated from the mean hourly speeds or other wind speed measurements. However, the use of average or mean wind speeds will underestimate the wind power because of the cubic relationship. For example, Culebra, Puerto Rico; Tiana Beach, New York; and San Gorgonio, California have annual average wind speeds of 6.3 m/sec, but their annual average power potentials are 220, 285, and 365 W/m^2, respectively [2]. For a better estimate of the wind power potential for any extended time period, you must know the frequency distribution of the wind speeds; the amount of time for each wind speed value, or a wind speed histogram; and the number of observations within each wind speed range.

TABLE 3.3
Estimated Wind Power Per Area, Perpendicular to the Wind

Wind Speed, m/s	Power, kW/m^2
0	0
5	0.06
10	0.50
15	1.68
20	4.00
25	7.81
30	13.50

Wind Characteristics

Example 3.2

Suppose the wind blows at 5 m/sec for 1r h and 15 m/sec for another hour. During the 2-hr period, the average wind speed is (5 + 15)/2 = 10 m/sec. Power/area calculated from the average wind speed is 500 watts/m². However, the power/area for the first hour is 62.5, and for the second hour is 1687.5; the average for the 2 hr is 875 W/m², which is 375 W/m² larger than the value calculated by using the average wind speed.

Wind power also depends on the air density:

$$\rho = 1.2929 \frac{Pr - VP}{760} \frac{273}{T}, \frac{kg}{m^3} \tag{3.3}$$

where Pr = atmospheric pressure and VP = vapor pressure, both expressed as millimeters of mercury, and T = temperature in Kelvin.

The vapor pressure term is a small correction, around 1%, and can be neglected. High temperatures and low pressures reduce the density of air and thus reduce the power per area. A major factor for change in density is the change in pressure with elevation. A 1,000 m increase in elevation will reduce the pressure by 10%, and thus reduce the power by 10%. If only elevation is known, air density can be estimated by

$$\rho = 1.226 - (1.194 \times 10^{-4})z \tag{3.4}$$

The standard density for comparing output of wind turbines is 1.226 kg/m³, which corresponds to a temperature of 15°C and an air pressure of sea level. For example, the average density for Amarillo, Texas, is around 1.1 kg/m³. When this value is compared to standard pressure, sea level, and 15°C (288K), Amarillo would have 10% less power for the same wind speeds. With the measurement of wind speed, pressure, and temperature, wind power/area can be calculated from Equation (3.2).

The energy per area for a time period of the same wind speed is

$$\frac{E}{A} = \frac{P}{A} t \text{ kWh/m}^2 \tag{3.5}$$

3.4 WIND SHEAR

Wind shear is a change in wind speed or direction over some distance (Figure 3.4) and it can even be vertical (Figure 3.5). The change in wind speed with height (horizontal wind shear) is an important factor in estimating wind turbine energy production. The change in wind speed with height has been measured for different atmospheric conditions [3, chap. 4].

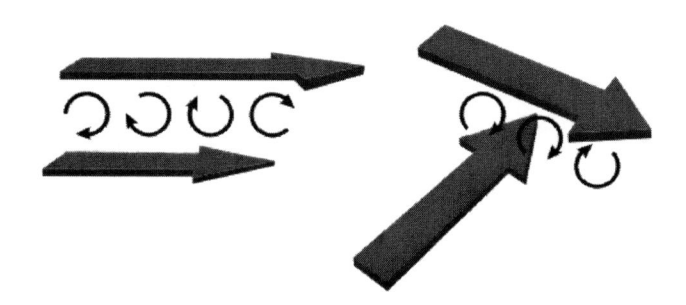

FIGURE 3.4 Left: wind shear caused by difference in wind speed with height. Right: wind shear caused by difference in wind direction.

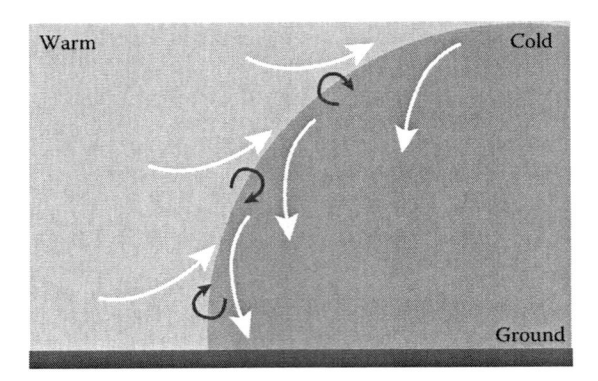

FIGURE 3.5 Example of vertical wind shear.

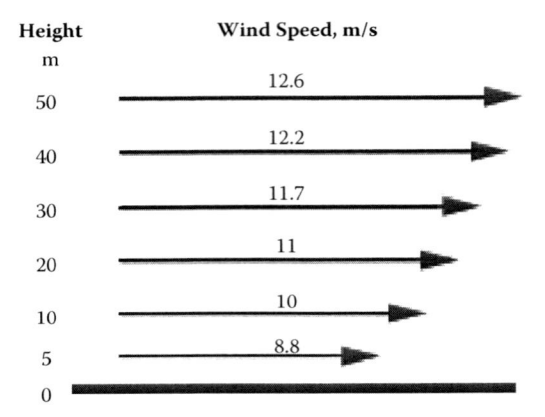

FIGURE 3.6 Wind shear change in wind speed with height. Calculations are for known wind speed of 10 m/sec at 10 m, $\alpha = 1/7$.

The general methods of estimating wind speeds at higher heights from known wind speeds at lower heights are power law, logarithm with surface roughness, and logarithm with surface roughness that has zero wind velocity at ground level. The power law for wind shear is

$$v = v_0 \left(\frac{H}{H_0}\right)^\alpha \qquad (3.6)$$

where v_0 = measured wind speed, H_0 = height of known wind speed v_0, and H = height.

The wind shear exponent α is around 1/7 (0.14) for a stable atmosphere (decrease in temperature with height); however, it will vary, depending on terrain and atmospheric conditions. From Equation (3.6) the change in wind speed with height can be estimated (Figure 3.6). Notice that for $\alpha = 0.14$, the wind power at 50 m is double the value at 10 m. This is a convenient way to estimate power, so many wind maps show wind speed and power classes for 10 and 50 m heights. However, for wind farms, wind power potential is determined for heights from 50 m to hub heights and maps of wind power/area and average wind speed are available for 80 and 100 m heights.

The wind shear exponent values in continental areas will be closer to 0.20 for heights of 10 to 40 m and above, with large differences from low values during the day to high values at night.

Wind Characteristics

Measurements taken at heights of 10, 20, and 50 m for the northwest Texas region [3] for 12-hr periods (6 to 18 hr spanning day and night) showed a large difference between 10, 20 m and 50 m levels. Data for sixteen sites in Texas and a site in New Mexico showed the same result: a change in diurnal wind speed pattern around 40 m [4]. Wind speeds were sampled at 1 Hz and averaging time was 1 hr.

The data were averaged by hour over a month, and the results averaged over a year to obtain an annual average day (Figure 3.7). This same pattern is noted for data taken at heights above 50 m (Figure 3.8). Other sites in the central plains of the U.S. show the same pattern [5].

Since wind speed is still increasing with height for these areas, the issue for wind farms is the trade-off between increased output with wind turbine height and increased cost for taller towers. These results clearly show that wind speed data need to be taken at least at a height of 40 m or higher to find shifts in pattern between day and night wind speeds. Once data at 10 m and 40 to 50 m are compiled, wind shear can be used to predict wind speeds at higher levels. The higher night wind speeds mean more power; however, those hours also present less demand, so if a wind farm is selling at the market price, that energy may be worth less.

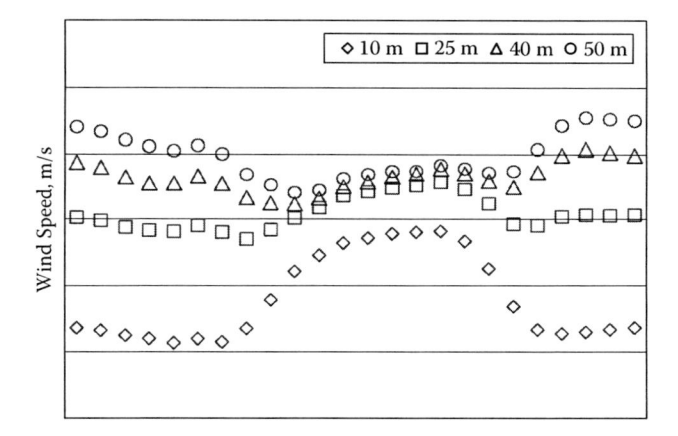

FIGURE 3.7 Annual average wind speed by time of day at 10, 25, 40, and 50 m heights, Dalhart, Texas, April 1996 to April 2000.

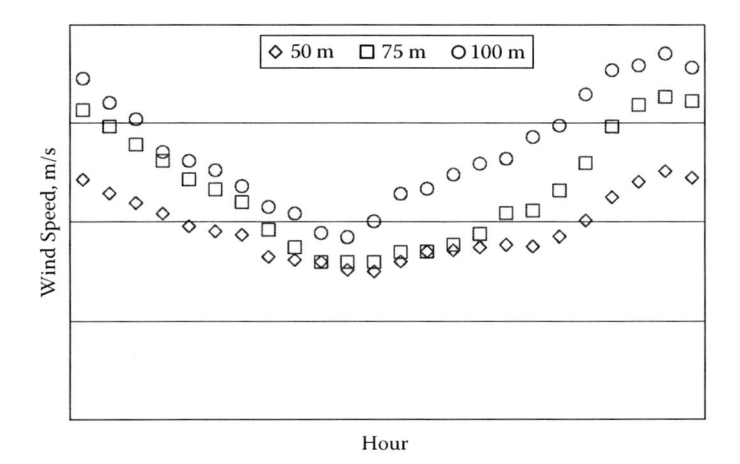

FIGURE 3.8 Annual average wind speed by time of day at 50, 75, and 100 m heights, Washburn, Texas, September 2003 to September 2006.

The wind shear exponent changes from low values during the day to high values at night over a 2-hr period (Figure 3.9). Time of day data were averaged over each month. The low values occur for more hours in the summer. Some locations, primarily mountain passes, demonstrate little wind shear (Figure 3.10). In this case, taller towers for wind turbines would not be needed.

The world standard height is 10 m for meteorology measurements for weather; however, using 10-m data and the 0.14 wind shear exponent to estimate wind power potential for 50 m for many locations will vastly underestimate the power potential for wind farms. The other formulas for estimating wind speed with height are

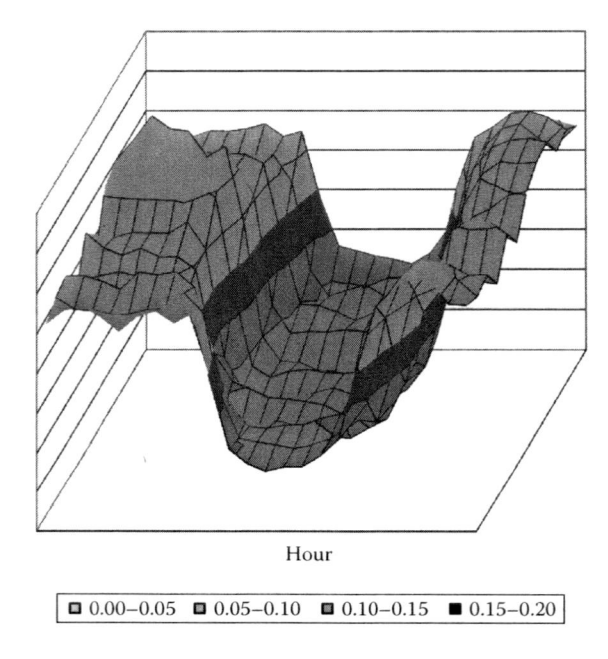

FIGURE 3.9 Wind shear exponent between 10 and 50 m for average month by time of day, Dalhart, Texas, April 1996 to April 2000.

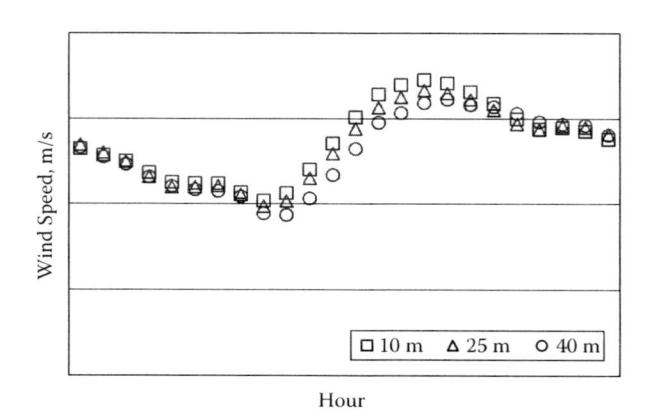

FIGURE 3.10 Annual average wind speed by time of day at 10, 25, and 40 m height, Guadalupe Pass, Texas, 1995 to 1999.

TABLE 3.4
Typical Values of the Roughness Parameter, z_0

Terrain Description	z_0, m
Snow, flat ground	0.0001
Calm open sea	0.0001
Blown sea	0.001
Snow, cultivated farmland	0.002
Grass	0.02–0.05
Crops	0.05
Farmland and grassy plains	0.002–0.3
Few trees	0.06
Many trees, hedges, few buildings	0.3
Forest and woodlands	0.4–1.2
Cities and large towns	1.2
Centers of cities with tall buildings	3.0

$$v = v \frac{\ln\left(\frac{H}{z_0}\right)}{\ln\left(\frac{H_0}{z_0}\right)} \tag{3.7}$$

$$v = v \frac{\ln\left(1 + \frac{H}{z_0}\right)}{\ln\left(1 + \frac{H_0}{z_0}\right)} \tag{3.8}$$

where z_0 is the roughness parameter. Equation (3.8) allows a zero wind speed at the surface. The roughness parameter ranges from 0.01 to 0.03 m for flat open terrain with short grass to larger than 1 m for rough terrain (Table 3.4).

Example 3.3

A meteorology (met) tower is located close to the edge of town. If the wind speed is 8 m/sec at 10 m height, what is the wind speed at 50 m? Use Equation (3.8) and select $z_0 = 1.2$.

$$v = 8 \frac{\ln\left(\frac{50}{1.2}\right)}{\ln\left(\frac{10}{1.2}\right)} = 8 \frac{\ln(41.7)}{\ln(8.3)} = 8 \frac{3.7}{2.1} = 14.1 \, m/s$$

This compares to 10 m/sec using the power law with a shear exponent = 0.14

3.5 WIND DIRECTION

Changes in wind direction are due to the general circulation of atmosphere, again on an annual basis (seasonal) to the mesoscale (4 to 5 days). The seasonal changes of prevailing wind direction could be as little as 30° in trade wind regions to as high as 180° in temperate regions. In the plains of the U.S., the predominant directions of the winds are from the south to southwest in the spring and summer

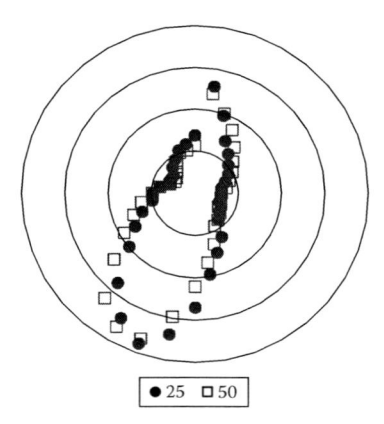

FIGURE 3.11 Annual average wind direction, 10-degree sectors at 25 and 50 m height, Dalhart, Texas, April 1996 to April 2000.

and from the north in the winter. Traditionally, wind direction changes are illustrated by a graph indicating percent of winds from a direction or a rose diagram (Figure 3.11).

Wind direction changes can also occur on a diurnal basis. However, a wind shear of change in wind direction with height is generally nonexistent or small, except for very short periods as weather fronts move through. Wind direction data (hour average wind speeds) from sixteen stations in Texas and one in New Mexico [6] did not reveal any significant wind shear of change in direction. Even on Padre Island, Texas, the land–sea breeze was not significant. Pivot tables were used to check the relationship of wind speed, wind direction, and time of day for the seventeen met stations, plus two tall-tower stations.

3.6 WIND POWER POTENTIAL

The most comprehensive, long-term source of data on wind speeds, pressures, and temperatures is the network of national weather stations in the U.S. Other sources include the National Climatic Center in Asheville, North Carolina, Federal Aviation Administration stations, and U.S. military bases and Coast Guard installations. In the early 1960s anemometers at national weather stations were moved from their previous locations (20 to 30 m heights) on airport control towers, hangers, and other structures to towers (around 6 m height) near runways and at least 1 km from buildings.

Wind speed data at U.S. National Weather Service (NWS) stations were recorded on strip charts and an observer estimated a wind speed over 1 to 2 min each hour. Hourly wind speed data along with pressure, temperature, and other climatological data were recorded on magnetic tape. The National Weather Service converted to automated surface observation systems in 1993 and 1994. Wind speed and direction are sampled at 1 Hz, averaged over 5 sec, and rounded. Then a 2-min running average is calculated from 24 samples of 5-sec each. Data on CD-ROMs, downloaded to a computer through the Internet, and monthly summary sheets can be purchased (http://lwf.ncdc.noaa.gov/oa/ncdc.html).

If the wind speeds are known, the average wind power or average wind energy per unit area can be estimated for any convenient period, usually months, seasons, or years. When more than a year of data are available, the year or month data are averaged to obtain annual values by year or month. The wind power per area is known as wind power potential or wind power density:

$$\frac{P_{avg}}{A} = \frac{\sum_{j=1}^{N} \frac{P_j}{A}}{N} = \frac{\sum_{j=1}^{N} \frac{0.5\rho_j v_j^3 A}{A}}{N} = \frac{\sum_{j=1}^{N} 0.5\rho_j v_j^3}{N} \tag{3.9}$$

where N is the number of observations.

Wind Characteristics

Average values of temperature and pressure can be used to calculate average density, and the average power/area can be calculated for the available wind speed data. The result will be fairly accurate since the pressure and temperature will not vary over a month or year nearly as much as wind speeds vary.

$$\frac{P_{avg}}{A} = \frac{0.5\rho_{avg}}{N} \sum_{j=1}^{N} v_j^3 \qquad (3.10)$$

If the observations of wind speeds are compiled into a histogram, the number of observations n_j in each wind speed bin may be converted to a frequency or probability by dividing the number of observations in a bin by the total number of observations:

$$N = \sum_{j=1}^{c} n_j, \quad f_j = \frac{n_j}{N}, \text{ and } \sum_{j=1}^{c} f_j = 1 \qquad (3.11)$$

where c is the number of classes or bins. If the wind speed units are changed or the wind speed changes due to height, the resulting histogram or frequency distribution should be normalized to contain the same number of observations.

Of course, for a large number of observations, a computer program or spreadsheet alleviates a lot of drudgery. Notice that the average wind speed (mean wind speed) is only a summation of the probability times the wind speed for each class in a frequency distribution:

$$v_a = \sum_{j=1}^{c} f_j v_j \qquad (3.12)$$

The average power/area can be calculated from a selected wind speed histogram or wind speed frequency distribution by

$$\frac{P_{avg}}{A} = \frac{0.5\rho_{avg}}{N} \sum_{j=1}^{c} n_j v_j^3 = 0.5\rho_{avg} \sum_{j=1}^{c} f_j v_j^3 \qquad (3.13)$$

Note the wind power/area is calculated from the sum. In one sense, the individual power/area values are in energy/time for each class (bin). If the energy in each bin is calculated and summed, the average wind power potential can also be calculated from this total energy divided by the number of hours.

3.7 TURBULENCE

Winds vary by locations and times and are influenced by terrain, vegetation, and obstacles. In addition to mean wind speed, the variability of a set of data is represented by the standard deviation. For more detail, see Rohatgi and Nelson [7, chap. 9, 10]. The standard deviation for a set of wind speed data is

$$\sigma = \left[\frac{1}{1-N} \sum_{j=1}^{N} (v_j - \bar{v})^2 \right]^{0.5} \qquad (3.14)$$

where $\bar{v}$ is the mean wind speed. Because $N-1$ is close to N for a large sample, the standard deviation for data loggers and spreadsheets is calculated from

$$\sigma^2 = \frac{\sum_{j=1}^{N} v_j^2}{N} - \overline{v}, \quad \overline{v} = \frac{\sum_{j=1}^{N} v_j}{N}$$

In general, two different calculations are used: (1) the standard deviation of the average values and (2) the standard deviation of a set of data. If the average 1-hr wind speeds are placed in 1 m/sec bins for a month or a year, a standard deviation can be calculated for each bin. This is different from the standard deviation calculation of 1-Hz data that are averaged over 10 min or 1 hr.

Turbulence intensity is usually calculated for short periods (minutes to an hour) and is calculated as mean wind speed divided by the standard deviation:

$$I = \frac{\overline{v}}{\sigma} \tag{3.15}$$

3.8 WIND SPEED HISTOGRAMS

A wind speed histogram shows the number of hours (or other time period used) the wind blew at each wind speed class (Table 3.5). Wind speeds were sampled at 1 Hz and averaged for 1 hr. Year wind speed histograms for 1996 through 1999 were averaged to obtain a representative annual value. An average density of 1.1 kg/m³ was used to calculate the average wind power potential. The average wind speed was 8.2 m/sec, and the average wind power potential was 467 W/m² for a height of 50 m.

TABLE 3.5
Annual Average: Wind Speed Histogram, Frequency and Calculation of Mean Wind Speed and Wind Power Potential at 50 m for White Deer, Texas, 1996–1999

Bin Class	Wind Speed, m/s	Hours	Frequency	$f_j v_j$	$f_j v_j^3$	Duration %	kWh/m²
1	0.5	54	0.01	0.00	0.0	100	0
2	1.5	146	0.02	0.03	0.1	99	0
3	2.5	353	0.04	0.10	0.6	98	3
4	3.5	487	0.06	0.19	2.4	94	11
5	4.5	617	0.07	0.32	6.4	88	31
6	5.5	747	0.09	0.47	14.2	81	68
7	6.5	844	0.10	0.63	26.4	73	127
8	7.5	950	0.11	0.81	45.7	63	220
9	8.5	949	0.11	0.92	66.5	52	320
10	9.5	940	0.11	1.02	92.0	41	443
11	10.5	801	0.09	0.96	105.9	31	510
12	11.5	702	0.08	0.92	122.0	21	588
13	12.5	486	0.06	0.69	108.4	13	522
14	13.5	302	0.03	0.47	84.8	8	409
15	14.5	175	0.02	0.29	60.9	4	293
16	15.5	85	0.01	0.15	35.9	2	173
17	16.5	52	0.01	0.10	26.9	1	130
18	17.5	32	0.00	0.06	19.6	1	94
19	18.5	22	0.00	0.05	15.7	0	76
20	19.5	12	0.00	0.03	10.5	0	51
21	20.5	4	0.00	0.01	3.6	0	17
	Sum	8,760	1	8.2	849		4,088
			Power/area		467		

Wind Characteristics

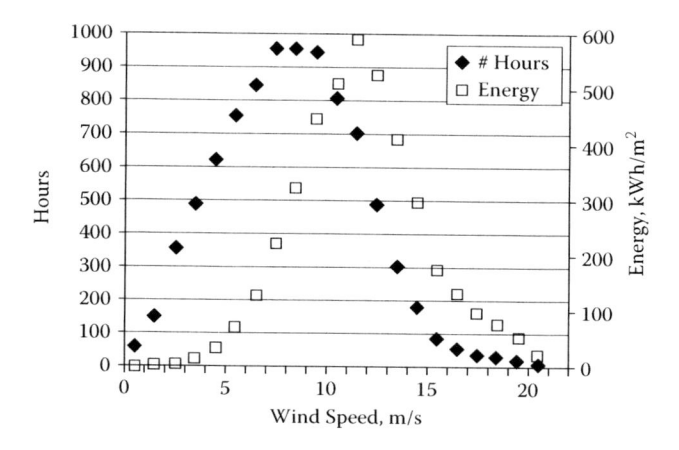

FIGURE 3.12 Annual average, comparison of wind speed and energy histograms at 50 m height, White Deer, Texas, 1996 to 1999.

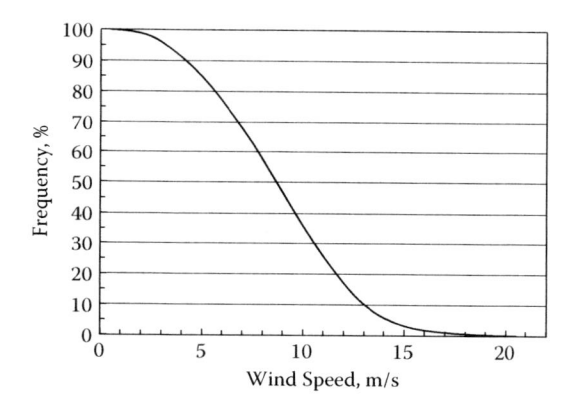

FIGURE 3.13 Wind speed duration curve at 50 m height, White Deer, Texas, 1996 to 1999.

The plots of the wind speed and energy histograms (Figure 3.12) show the relationship of wind and energy. There is little energy in low wind speeds because of the low speed, and little energy at high wind speeds because of the short durations of high wind speeds.

3.9 DURATION CURVE

Wind data can also be represented by a speed–duration curve (Figure 3.13) that plots cumulative frequency starting at the largest wind speed (subtract 100 from percent frequencies of cumulative frequencies if starting at the lowest wind speed). The percent duration is usually converted to the number of hours in a year by multiplying by 8,760. From wind speed–duration curves, estimates of the time the wind speed is above a given value can be obtained. The data in Table 3.5 and the curve in Figure 3.12 show, for example, that a wind of 3 m/sec or greater blows 95% of the time or 8,300 hours in a year for that location.

In general, whatever the wind speed is at any point in time, the behavior over the next hour should be similar. This is called persistence, calculated as $v(t + t_0) \sim v(t_0)$, where t is variable. However, a histogram does not give a time sequence of data, nor does a wind speed–duration curve indicate the lengths of calm periods. As more wind turbines are installed, wind farm operators and utilities will be interested in predicting wind speeds, average variations by season and time of day, durations of low wind speeds, and values for the 1 to 36 hr ahead.

3.10 VARIATIONS IN WIND POWER POTENTIAL

Since the motion of the atmosphere varies on a scale from seconds to years, wind power and wind energy will also vary on the same time scale. The annual average wind power (6 m height) for Amarillo, Texas, was 220 watts/m^2 for 1962 through 1977 [8]; however, variations from one year to the next can be quite large. A minimum of two years of data is required to obtain an estimate for annual wind power potential, and five years of data are needed to obtain a mean value within 6% of the long term mean. Many researchers assume that two or three years of data suffice when combined with longer term regional data for comparison to determine wind power potential.

The annual wind power potential (Figure 3.14) for White Deer and Dalhart, Texas, shows the correlation between sites that are 140 km apart in the same region. Data were sampled at 1 Hz and averaged over 1 hr. Therefore, for a region where long term base data are available for comparison, one to two years of data would suffice for determining wind power potential at a specific location.

The seasonal variation for most of the U.S. constitutes high wind speeds in the spring and low wind speeds in the summer (Figure 3.15). Notice that the standard deviations at 10 and 50 m are

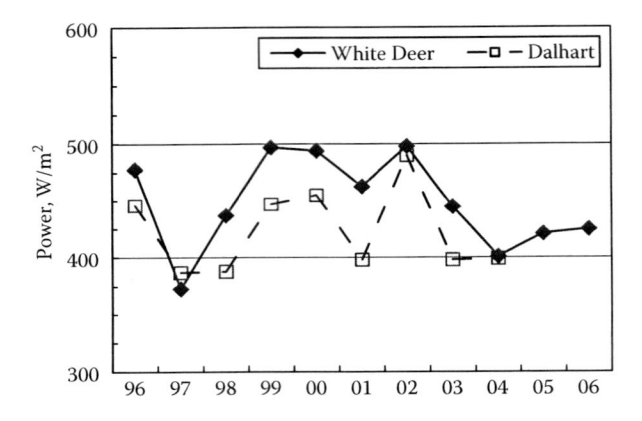

FIGURE 3.14 Annual variation of wind power potential at 50 m for White Deer and Dalhart, Texas.

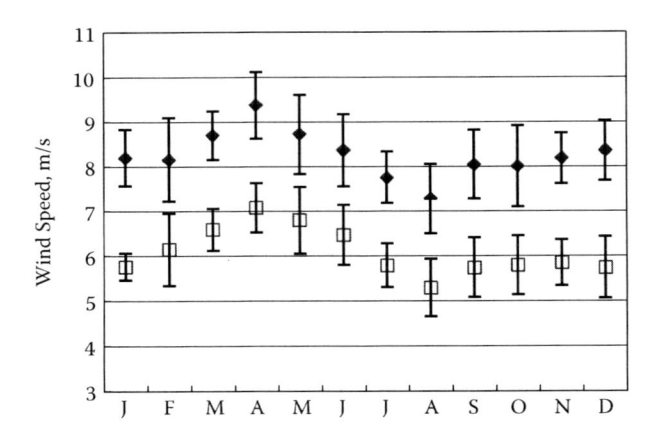

FIGURE 3.15 Annual wind speed and standard deviation by month at 10 m and 50 m for White Deer, Texas, 1996 to 2006.

Wind Characteristics

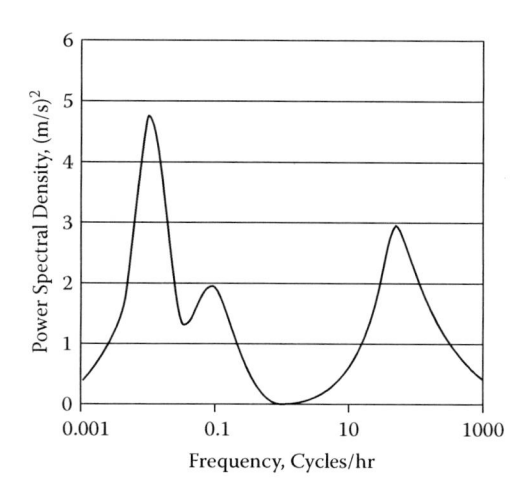

FIGURE 3.16 Example of power spectrum for wind speed. (*Source:* Van der Hoven, I. 1957. *Journal of Meteorology* 14, 160. With permission.).

comparable and the average value for both is 0.6 m/sec. Also, the standard deviation of the wind speed by month is close to the standard deviation of wind speeds for an individual month (744 data points). The most notable exception to general seasonal variation occurs in the mountain passes in California between the coast and inland deserts. The windy season corresponds to heating of the deserts in the summer when hot air rises and is replaced with cooler air flowing in from the ocean.

There are also variations with the movement of synoptic weather patterns represented by a four- to five-day variation. The diurnal (daily) variation is due to heating during the day. These frequency representations (Figure 3.16) are common to many locations [9]. The peak at 0.01 cycle/hr corresponds to a period of 100 hr that represents the four- to five-day variation. The peak near 0.1 cycle/hr corresponds to the diurnal variation.

During the investigation of power storage for a wind–diesel system, an appropriate wind speed power spectrum became a significant issue [10]. A power spectrum was developed from thirteen years of hourly average data, one year of 5 min average and gusty day data. and 1 sec data, all at 10 m height. The general shape is similar to the Van der Hoven spectrum; however, few of his peaks were found in the power spectrum at the U.S. Department of Agriculture, Agricultural Research Service (USDA-ARS) in Bushland, Texas.

While higher average wind speeds tend to suggest higher amplitudes in the high frequency end of the spectrum, this is not always true. Similar results were found for a power spectrum from three years of 15 min average data (sample rate, 1 Hz) at a 50 m height near Dalhart, Texas (Alternative Energy Institute met site). For wind speed data around the 40 m height, there be no diurnal peaks in the continental areas of the U.S. The Van der Hoven spectrum is not really useful for the wind turbine industry.

3.11 WIND SPEED DISTRIBUTIONS

If data are not available, wind speeds can be predicted from one or two parameters. A number of distributions have been tried, but the only two in general use are the Rayleigh and Weibull distributions. Both distributions give poor estimates of power for low mean wind speed situations. At higher wind speeds, both yield adequate estimates for many locations; however, for regions with steady winds such as trade winds, the Weibull distribution is better. The Rayleigh distribution is simpler because it depends only on mean wind speed. It is calculated as

$$F(v) = \Delta v \frac{\pi}{2} \frac{v}{v_a^2} \exp\left[-\frac{\pi}{4}\left(\frac{v}{v_a}\right)^2\right] \qquad (3.16)$$

where $F(v)$ = frequency of occurrence associated with each wind speed v at the center of Δv; Δv = width of class or bin; and v_a = average wind speed (same as mean wind speed). The wind speed histogram for one year can be calculated from $8{,}760 \times F(v)$. The Rayleigh frequency is calculated for two different values, $v = 3$ and 9 m/sec; $v_a = 8$ m/sec and $\Delta v = 2$ m/sec:

$$F(3) = 2\frac{\pi}{2}\frac{3}{8^2}\exp\left[-\frac{\pi}{4}\left(\frac{3}{8}\right)^2\right] = 0.147 e^{-0.11} = 0.132$$

$$F(9) = 2\frac{\pi}{2}\frac{9}{8^2}\exp\left[-\frac{\pi}{4}\left(\frac{9}{8}\right)^2\right] = 0.44 e^{-0.994} = 0.164$$

As a check, the sum of the frequencies (probabilities) should be close to 1. If not, you have made a mistake. Also, the curve will be smoother for smaller bin widths; however, 1 m/sec will suffice. For large bin widths, the wind speed histogram may have to be renormalized by bin value × 8,760/ (sum of observations).

The Weibull distribution is characterized by the shape parameter k (dimensionless) and the scale parameter c (m/sec). The Rayleigh distribution is a special case of the Weibull distribution where $k = 2$. For regions of the trade winds where the winds are fairly steady, the shape factor may be as high as 4 to 5. For most sites in Europe and the U.S., k varies between 1.8 and 2.4.

$$F(v) = \Delta v \frac{k}{c}\left(\frac{v}{c}\right)^{k-1}\exp\left[-\left(\frac{v}{c}\right)^k\right] \qquad (3.17)$$

In many parts of the world, wind speed data are sparse. If only the average wind speed by day or month is known, the average values and their deviations are used to estimate the two parameters. Rohatgi and Nelson [7, chap. 9] estimated the Weibull parameters by three methods: (1) a plot of c and k from log–log paper and analysis of standard deviations, and (2) analysis of the energy pattern factor.

A higher k value means wind speeds are peaked around the average wind speed (Figure 3.17). The values in the graph were calculated for a mean wind speed of 6 m/sec for

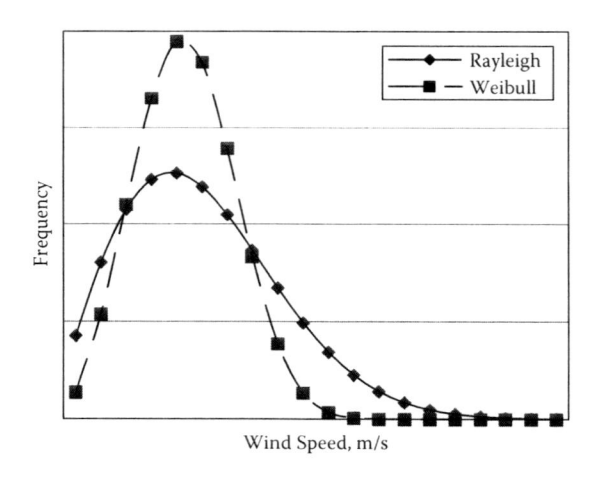

FIGURE 3.17 Example wind speed frequencies calculated using Rayleigh and Weibull distributions.

the Rayleigh distribution and $c = 6$ m/sec and $k = 3$ for the Weibull distribution; both used a bin width of 1 m/sec.

The energy pattern factor is rarely used. It is an estimate of the variability of wind speed calculated as the relationship between the mean of the cubes of each data point divided by the cube of the mean for a series of data (see Example 3.2 for a series of two points). The energy pattern factor is always greater than 1, and in the Southern High Plains, varies from 1.6 to 3.4.

3.12 GENERAL COMMENTS

Previous studies of wind behavior were performed by meteorologists who were mainly interested in weather and wanted to research turbulence and momentum transfer. Since 1975, numerous studies of wind characteristics have been funded because they pertain to wind energy potential and effects on wind turbines.

Most U.S. research was conducted by the Battelle Pacific Northwest Laboratory (PNL) and then transferred to the National Wind Technology Center (NWTC) of the National Renewable Energy Laboratory (NREL). A list of publications on wind characteristics is available from NREL. States and universities have also funded projects for estimating wind energy potential. For more data after using the national atlas, contact your state energy office or the American Wind Energy Association in the U.S.

National laboratories in many countries in the European Union took the same steps, for example, the Danish laboratory in Risoe. To obtain information from other countries, the procedures are the same: contact national entities, universities, institutes, and state, national, and international wind energy associations.

Now that wind farms affect the grid, power system operators and wind farm operators require wind forecasting data [11] to increase their economics. Private companies and national laboratories have and are developing models to forecast wind resources over periods from one hour to a day ahead. Grid operators use temperature forecasts to predict demand and now use wind forecasts to anticipate wind generation levels and adjust their generation units accordingly.

Current day and next day graphs of forecast and actual wind power production [12] are available online for the transmission system of the Electric Reliability Council of Texas. Improved short-term forecasts allow wind farm operators make better day-ahead market decisions. Also forecasting systems will help warn of extreme wind events and most U.S. wind farms now receive such forecasts.

LINKS

National Climatic Center wind data. http://lwf.ncdc.noaa.gov/oa/ncdc.html
National Wind Technology Center. www.nrel.gov/wind/resource_assessment.html

REFERENCES

1. M.R. Gustavson. 1978. Wind power extraction limits. In Nelson, V., Ed., *Proceedings of National Conference of the American Wind Energy Association*, p. 101.
2. D.L. Elliott et al. 1986. *Wind Energy Resource Atlas of the United States*. DOE/CH 10093-4. http://rredc.nrel.gov/wind/pubs/atlas/
3. E. Gilmore. 1987. *Wind characteristics: northwest Texas region, May 1978–December 1985*. Report 87-1., Alternative Energy Institute, West Texas A&M University.
4. T. Han. 2004. Wind shear and wind speed variation analysis for wind farm projects for Texas. Master's thesis, West Texas A&M University.
5. M. Schwarts and D. Elliott. 2006, Wind shear characteristics at central plains tall towers. NREL/PR-500-39989. http://www.nrel.gov/docs/fy06osti/39989.pdf
6. K. Herrera. 2006. Wind direction analysis for Texas. Master's thesis, West Texas A&M University.

7. J. Rohatgi and V. Nelson. 1994. *Wind characteristics: an analysis for the generation of wind power*. Alternative Energy Institute, West Texas A&M University.
8. V. Nelson and E. Gilmore. 1974. *Potential for wind generated power in Texas*. Report NT/8. Austin: Governor's Energy Advisory Council of Texas.
9. I. Van der Hoven. 1957. Power spectrum of horizontal wind speed in the frequency range from 0.0007 to 900 cycles per hour. *Journal of Meteorology*, 14, 160.
10. E.D. Eggleston. 1996. Wind speed power spectrum analysis for Bushland, Texas. In *Proceedings of Windpower Conference*, p. 429.
11. D. Lew, M. Milligan, G. Jordan et al. 2011. The value of wind power forecasting. NREL/CP-5500-50814. www.nrel.gov/docs/fy11osti/50814.pdf
12. ERCOT. Forecasted and actual wind power production. http://www.ercot.com/gridinfo/csc/

QUESTIONS AND ACTIVITIES

1. What is the wind power class where you live? In the U.S., go to the NREL site on wind data (http://rredc.nrel.gov/wind/pubs/atlas/) or consult your state map. In other countries, try to find wind data values for the area close to your home.
2. Note day and time. Go outside and estimate the wind speed. Now go to the weather information channel on your TV and note the wind speed. If your estimate is far off, what could be the reason? Going out on a calm day does not count.

PROBLEMS

Use a spreadsheet if applicable and available.

1. Calculate the power in kilowatts across the following areas for wind speeds of 5, 15, and 25 m/sec. Use area diameters of 5, 10, 50, and 100 m. Air density = 1.0 kg/m^3.
2. Solar power potential is around 1 kW/m^2. What wind speed gives the same power potential?
3. Calculate the factor for the increase in wind speed if the original wind speed was taken at a height of 10 m. New heights are 20 and 50 m. Use the power law with exponent $\alpha = 0.14$.
4. Calculate the factor for the increase in wind speed if the original wind speed was taken at a height of 10 m. New heights are at 50 and 100 m. Use the power law with exponent $\alpha = 0.20$.
5. Calculate the factor for the increase in wind speed if the original wind speed was taken at a height of 50 m. New heights are at 80 and 100 m. Use the power law with exponent $\alpha = 0.20$.
6. Houston Intercontinental Airport is surrounded by trees (20 m tall). Calculate the factor for increase in wind speed from 10 to 100 m. Use the ln relationship and an estimated z_0 from Table 3.4.
7. What is the air density difference between sea level and an altitude of 3,000 m?
8. In the Great Plains, temperature differ widely between summer (100°F) and winter (–20°F). What is the difference in air density? Assume you are at the same elevation and average pressure is the same.

Note: For problems with wind speed distributions, remember the wind speed must be the number in the middle of the bin. If you use a bin width of 1 m/sec, the numbers have to be 0.5, 1.5, etc. In general, bin widths of 1 m/sec are more than adequate. Smaller bin widths mean more calculations.

9. Calculate the wind speed distribution using the Rayleigh distribution for an average wind speed of 8 m/sec. Use 1 m/sec bin widths.
10. Calculate the wind speed distribution for a Weibull distribution for $c = 8$ m/sec and $k = 1.7$. Use 1 m/sec bin widths.
11. Calculate the wind speed distribution for a Weibull distribution for $c = 8$ m/sec and $k = 3$.
12. From Figure 3.13, what is the percent of the time the wind speed is 5 m/sec or more?
13. From Figure 3.13, what is the percent of the time the wind speed is 12 m/sec or more?

Wind Characteristics

14. For a 10-min period, the mean wind speed is 8 m/sec and the standard deviation is 1.5 m/sec. What is the turbulence intensity?
15. At the Delaware Mountains wind farm, very high winds with gusts over 60 m/sec were recorded. An average value for 15 min was 40 m/sec with a standard deviation of 8 m/sec. What was the turbulence intensity?

Use the following table to calculate answers for Problems 16 through 21. The most convenient way is to use a spreadsheet.

Bin j	Speed (m/sec)	No. Obs.
1	1	20
2	3	30
3	5	50
4	7	100

16. Calculate the frequency for each class (bin). Remember, sum of $f_j = 1$.
17. Calculate the wind power/area for $j = 5$ bin and 10 bin.
18. Calculate the average (mean) wind speed.
19. Calculate the wind power potential (power/area).
20. From the mean wind speed of Problem 18, calculate the power/area. How will that value compare (smaller, same, larger) to the value in Problem 19. Justify your answer.
21. From the answer to Problem 18, use the mean wind speed and calculate a Rayleigh distribution for an average wind speed = 10.2 m/sec. Use $\Delta v = 2$ m/sec.
22. Go to the ERCOT site for forecast wind power data. Using current date, note date, time, forecast, and actual wind power production.

4 Wind Resource Assessment

The two aspects of wind resource assessment are: (1) determination of general wind power and (2) determination of wind power potential and predicted energy production for wind farms. Wind resource assessment for wind farms will be covered in Chapter 9.

General wind power or area was determined from available wind speed data and then wind maps were developed. In general, the available wind speed data covered heights of 6 to 20 m; however, some anemometers were on tops of buildings or control towers at airports and their heights influenced the accuracy of the data. In many parts of the world, the amount of wind speed data was limited to daily or even monthly averages. Wind classes were developed for 10 m height, because that is the standard for world meteorological data. The wind power potential at 50 m was double that at 10 m due to the assumption that the wind shear exponent was 1/7 for all locations.

A world wind map that showed the wind classes at 50 m height for typical open, well-exposed sites was prepared by the Pacific Northwest Laboratory using data compiled in 1980 [1]. The assessment was made by critically analyzing all available wind data and previous assessments to estimate the broad-scale distribution of wind power potential. Many data were used cautiously because of the lack of information about anemometer height and exposure.

Global pressure and wind patterns, upper air wind data, and boundary layer meteorology were also used to obtain consistent estimates of wind energy resources. Actual wind speed frequency distributions were used when available or Weibull distributions were used to estimate the wind power potential. If only mean wind speeds were available, Rayleigh distributions were used.

Most general results were known, for example, the presence of strong trade winds—northeast in the northern hemisphere and southeast in the southern hemisphere. At mid-latitudes (about 40 to 60 degrees) the flow is westerly, and strong westerlies circle the world all year round in the southern hemisphere. The flow creates strong winds at the tip of South America, southwest coast of South Africa, southern coast of Australia, the island of Tasmania, and New Zealand. The flow of the westerlies in the northern hemisphere is broken up by large land masses.

The region off the northern coast of South America also shows high wind speeds. The wind around the poles is predominantly easterly. The world wind map is very broad and should be viewed with caution when estimating wind power potential. Country, state, and regional maps formulated from better data and at much higher resolution are now available for many parts of the world. The National Renewable Energy Laboratory (NREL) maintains a map search site [2].

Note: If you are searching the Internet for world winds, links referring to NASA World Wind are for open-source Windows software to view satellite images of the earth and have nothing to do with world wind maps or estimations of world wind power potential.

Archer and Jacobson [3] quantified the world's wind speeds in 2000 as indications of wind power potential at 80 m height. A least square extrapolation technique was used to estimate wind speeds at 80 m from observed wind speeds at 10 m and a network of sounding stations. Globally, ~13% of the stations revealed class 3 (mean wind speeds ≥ 6.9 m/s) and above winds at 80 m suitable for wind farms. This is a conservative estimate; for example, India does not show any winds above class 2, and has a number of wind farms.

Wind maps are presented for Europe, North America, Australia and New Zealand, Asia, and Africa. In general, the maps show the same regions of high winds as the previous world wind map. The major difference is that each met station is classified by a dot indicating wind class. Again, these maps should be used with caution as mean wind speeds are only indications of wind power potential and mean wind speeds are only shown for one year; however, they are considered representative of the five-year period from 1998 to 2002.

Wind Atlases of the World [4] show world wind maps that include values of mean wind speeds at 10 m height for 1976 through 1995. Again, the global westerlies in the southern hemisphere are prominent.

REmapping the World, a project for renewable energy resource assessment, will provide information to individuals and to governments. The interactive map provides global wind data at heights of 20, 50, and 80 m at 15-km resolution for a single year. It estimates that 40% of the world's land mass has wind speeds of 6 m/sec or more. The global wind map can be viewed at firstlook.3tiergroup.com.

As more data have been collected specifically on wind power potential for nations, states, and regions, digital wind maps are now available. They display better resolution than the older maps, and the values are more accurate as data above 10 m have become available. However, the data collected by private wind farm developers are not available to the public, so data at heights of 50 m and higher are still being collected to provide regional databases. Anemometer loan programs are available for private individuals in some states in the U.S., and after some period, the data generally become public.

Computer tools for modeling wind resources have been developed by a number of groups including the National Wind Technology Center (NWTC) and National Renewable Energy Laboratory (NREL) in the United States, WAsP developed by Risø, National Laboratory, Denmark, and private industries. Information from AWS True Wind about the Northwest Wind Mapping Project describes the mapping process.

The advanced MesoMap™ mesoscale modeling system simulates complex meteorological phenomena not adequately represented in standard wind flow models. It models sea breezes, offshore winds, mountain and valley winds, low-level nighttime jets, temperature inversions, surface roughness effects, flow separations in steep terrain, and channeling through mountain passes. This model utilizes historical upper air and surface meteorological data, thereby providing a consistent long-term, three-dimensional wind resource record. This record can later be used as a substitute for long-term surface wind measurements in the correlate–measure–predict (CMP) method that adjusts short-term site measurements to long-term climatological norms.

The modeling results can help identify where limited wind measurement resources should be applied. Based on prior model validations, the expected range of discrepancy between measured and predicted winds in complex terrain is 3 to 7%.

Now remember what a 5% error in wind speed does to the error in wind power. Therefore, siting for wind farms is still important and on-site data are imperative for financing a project.

4.1 UNITED STATES

A number of wind power and wind energy maps cover the U.S. However, the earlier maps did not account for the height differences of the anemometers. As part of the overall evaluation of wind energy, two major contracts were awarded to General Electric and Lockheed in 1975. Their estimates of the wind energy potential for a height of 50 m indicated that most of the U.S. has fairly large potential. The problem is that most of these values were estimated from data taken at a height of 6 to 10 m, with the value at 50 m estimated at double that at 10 m.

The Pacific Northwest Laboratory oversaw a comprehensive assessment of wind energy potential. The Wind Energy Resource Atlas covers the United States and its territories [5]. Wind power potential by year and season were also estimated for each state and region. The wind power classes (Table 4.1) were estimated for a grid of 20 min longitude by 15 min latitude (27 by 25 km, 16 by 15 miles). The atlas and the wind maps were updated in 1985 [6]. The different wind power maps show similar gross features. Regions of better wind power are in the Great Plains, along the coasts, in Hawaii, and at selected sites, such as ridges, mesas, and mountain passes.

TABLE 4.1
Classes of Wind Power Potential at 10 and 50 m Levels

	10 m		50 m	
Class	Power W/m²	Speed m/s	Power W/m²	Speed m/s
1	0	0	0	0
	100	4.4	200	5.6
2	100	4.4	200	5.6
	150	5.1	300	6.4
3	150	5.1	300	6.4
	200	5.6	400	7.0
4	200	5.6	400	7.0
	250	6.0	500	7.5
5	300	6.4	600	8.0
	400	7.0	800	8.8
6	400	7.0	800	8.8
	1,000	9.4	2,000	11.9
7	1,000	9.4	2,000	11.9

Note: Values at 50 m are based on 1/7 power law from data at 10 m. Wind speeds are the equivalent value based on a Rayleigh distribution to give that power.

Now all of this information has been placed online and the maps are in digital format (Figure 4.1). The NWTC and the NREL are updating the wind maps for states using terrain modeling, and maps have been completed for a number of states (Figure 4.2). The procedure uses actual data for verification. Information and data on wind resources are available from NWTC [7], and digital wind maps of the U.S. are available from Wind Powering America [8]. There are wind speed maps for 80-m utility scale and maps of wind energy resource potential for 80 and 100 m.

New computer tools and technical analyses that use satellite, weather balloon, and meteorological tower data are being used to create better maps for assessing wind power potential. Geographic information systems (GIS) provide wind maps with selected overlays, for example, transmission lines, roads, parks, and wildlife areas, to assist in wind resource assessment. The higher resolution of these maps (1 km) allows better assessments of wind farm locations also shows higher class winds in areas where none were thought to exist. The wind maps for the Northwest region [9] and Texas [10] have online interactive features for zooming in on local areas.

NWTC had a program of collecting data on tall towers, up to 100 m. The data from the thirteen tall towers in the Central Plains show that wind speeds and, of course, wind power potential continue to increase with height. Because wind speed increases with height, some regions with class 2 winds that were presumed to have little potential for wind farms have now become viable if they are near large load centers. Three years of meteorological data from the two tall tower sites in Texas and five years of met data from sixteen met sites in Texas and one met site in New Mexico are available to the public [11]. Some met data from the General Land Office in Texas, are also available.

4.2 EUROPEAN UNION

The Europeans made concerted efforts to assess wind resources beginning with the publication of the *European Wind Atlas* [12] in 1989. Part 1 provides an overall view of the wind resources. Part 2 provides information for determining wind resources and siting wind turbines. It also provides descriptions and statistics for the 220 met stations in the countries of the European Community

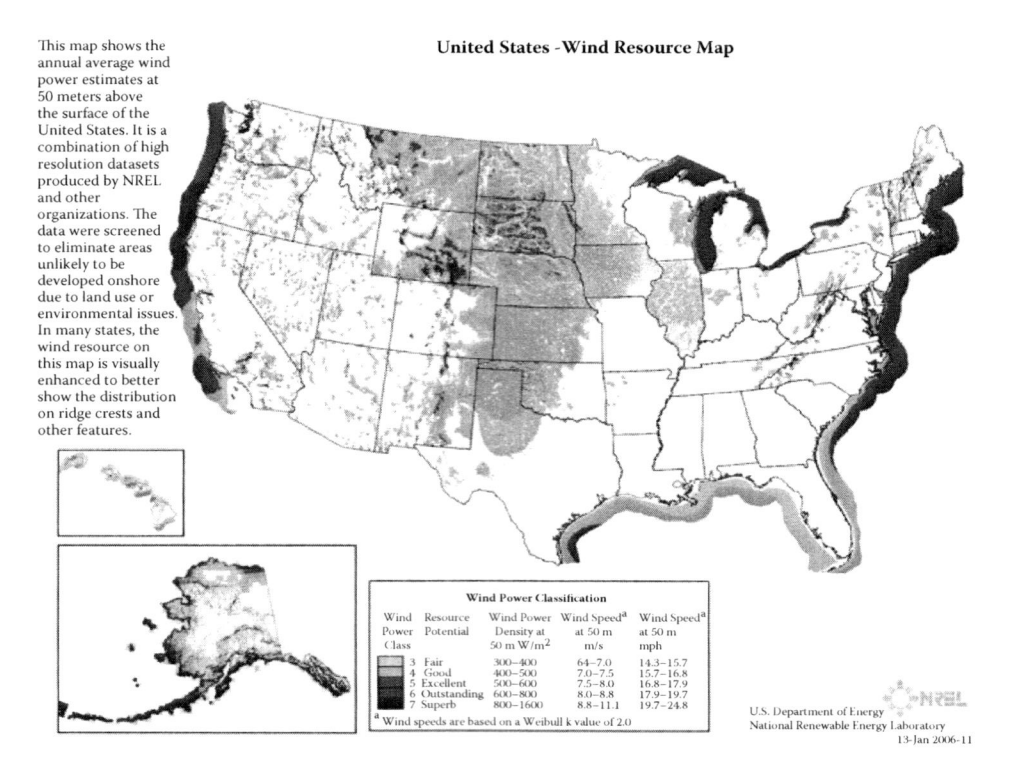

FIGURE 4.1 Wind power map of United States. (*Source:* National Renewable Energy Laboratory.)

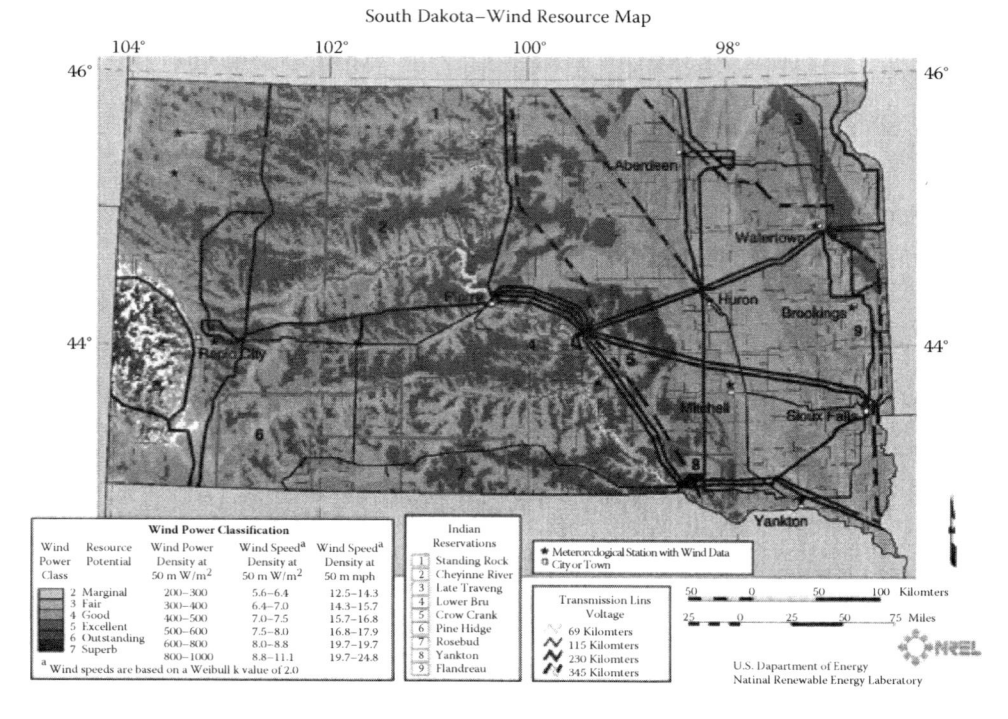

FIGURE 4.2 Wind power map with terrain enhancement for South Dakota. (Source: National Renewable Energy Laboratory.)

Wind Resource Assessment 69

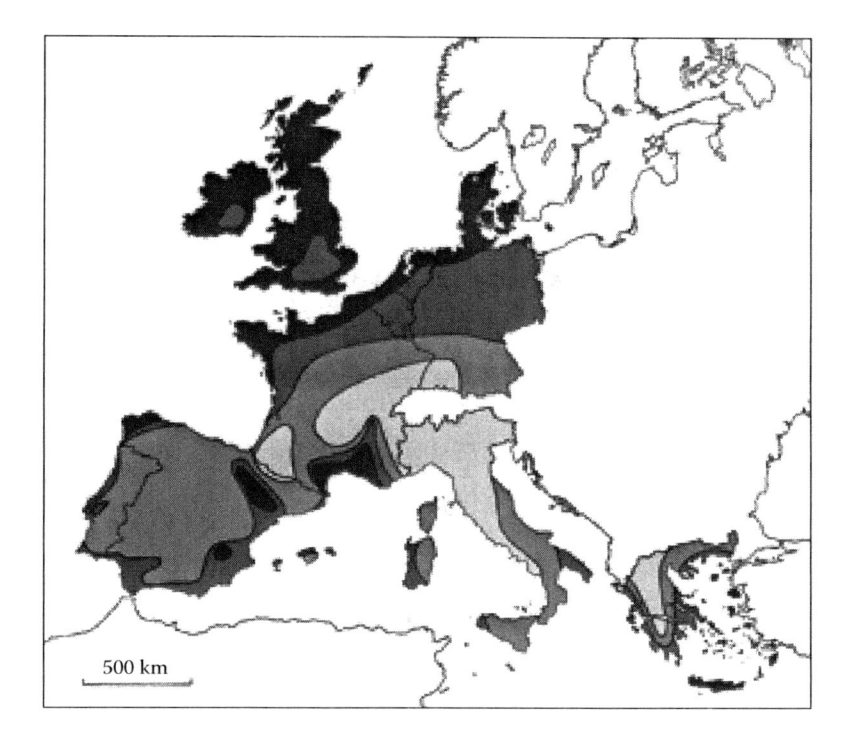

FIGURE 4.3 Europe wind resources at 50 m above ground level for five topographic conditions. (*Source:* 1989 European Wind Atlas. © Riso National Laboratory, Denmark.)

(EC) and includes methods for calculating the influence on the wind from landscape features such as coastlines, forests, hills, and buildings. Part 3 explains the meteorological background and analysis methods used to prepare the atlas and includes the physical and statistical bases for the models.

The wind map for the EC (Figure 4.3) shows high winds for northern United Kingdom and Denmark and across the northern coasts from Spain to Denmark. Wind maps were available for thirteen countries. The wind power classes are somewhat different from those of the U.S. and cover different terrains (Table 4.2). Since the original publication, the EC expanded into the European Union (EU) and wind maps are available for more countries.

4.3 OTHER COUNTRIES

Wind power maps and isovent (contour lines of wind speed) maps are available for countries and regions around the world since wind energy is now included in many national energy policies. Some countries do not make wind maps available. More countries are developing wind maps and NREL is helping to develop wind maps worldwide. As of 2012, NREL listed thirty countries [13]. Some maps are available through the Solar and Wind Energy Resource Assessment (SWERA) Program and the Asia Alternative Energy Program. SWERA provides online, high-quality renewable energy maps and other resource information at no cost for countries and regions around the world. Renewable energy maps, atlases, and assessments can be downloaded [14].

Wind Atlases of the World contains links for over fifty countries [15], many of which contain wind data and wind statistics on disk. At present, the WAsP wind atlas methodology has been employed in around 105 countries and territories.

A database of wind characteristics is compiled and maintained by the Technical University of Denmark and the Risø National Laboratory also in Denmark [16]. The database contains

TABLE 4.2
Wind Classes for Different Terrains, *European Wind Atlas*

Class	Shelter m/s	Terrain W/m²	Open Plain m/s	W/m²	Sea Coast m/s	W/m²	Open Sea m/s	W/m²	Hills and Ridges m/s	W/m²
5	>6.0	>250	>7.5	>500	>8.5	>700	>9.0	>800	>11.5	>1,800
4	5.0–6.0	150–200	6.5–7.5	300–500	7.0–8.5	400–700	8.0–9.0	600–800	10.0–11.5	1,200–1,800
3	4.5–5.0	100–150	5.5–6.5	200–300	6.0–7.0	250–400	7.0–8.0	400–600	8.5–10.0	700–1,200
2	3.5–4.5	50–100	4.5–5.5	100–200	5.0–6.0	150–250	5.5–7.0	200–400	7.0–8.5	400–700
1	<3.5	<50	<4.5	<100	<5.0	<150	<5.5	<200	<7.0	<400

four categories of wind data: time series, wind resource measurements, structural wind turbine response measurements, and wind farm measurements and also provides links for wind maps. RETScreen International has collected renewable energy resource maps [17] for the world, forty-seven countries, and various states and/or regions.

Examples of older maps are given in Rohatgi and Nelson [18, chaps. 5 and 6]. China and India have installed substantial wind turbine capacity and resource assessment obviously preceded the installations.

More detailed assessments are available from measurements to provide a database for wind power by state, region, and nation and delineate possible locations for wind farms. Micrositing for wind turbines within wind farms is important.

4.4 OCEAN WINDS

Ocean winds are and have been measured by ships and instruments on buoys. Now complete coverage of the oceans is available using reflected microwaves from satellites [19,20]. A physics-based algorithm calculates ocean wind speed and direction at 10 m from surface roughness measurements. Water vapor, cloud water, and rain rate are also calculated. This algorithm is a product of fifteen years of refinements and improvements. Data are compiled by orbital daily observations mapped to a 0.25-degree grid, and averages are calculated for three days, a week, and a month. Images of the data can be viewed on websites for the world and by region.

At the Remote Sensing Systems website, images can be viewed in the browse/download section. For SSM/I and TMI satellites, the wind speed images do not include direction and have a maximum value of 12 m/sec. Dynamic Data Imaging lets users select region, dates, and zoom factors, and also gives statistics.

The QSCAT satellite images include direction and higher wind speeds (Figure 4.4 and Figure 4.5). They do not provide dynamic imaging. The system divides the world into regions. Ocean wind data are not available within 25 km of the shore because radar reflections off the bottom of the ocean skew the data.

Notice that ocean winds indicate onshore winds for islands, coasts, and also some inland regions of higher winds. Two regions of average wind speeds of 10 m/sec due to the northeast trade winds are in the Teohuantepec isthmus in Mexico and the Arenal region of Costa Rica, where winds are funneled by the land topography.

Around 5,000 MW (2012) have been installed in offshore wind farms in Europe because offshore wind speeds are generally higher and onshore land has a high value. The U.S. is considering offshore wind farms near Cape Cod, Massachusetts, and along the Gulf Coast of Texas. Wind farm developers have expressed interest in offshore production in the Great Lakes and on both coasts. NREL has begun a program for offshore wind resource assessment [21]. The maps will extend from coastal areas to 90 km offshore and have a horizontal resolution of 200 m. The final maps of wind speed and power will be for 50 m height, and the model data will be modified from data taken at sites on the coast and offshore.

4.4.1 TEXAS GULF COAST

Ocean wind data were used to calculate wind speed and power (10 m height) for 0.25-degree pixels for a 5 × 5 degree area (longitude 25–30 N, latitude 93–98 W) from 1988 through 1994 [22]. The average wind speed was around 5 m/sec, and further out in the Gulf of Mexico exceeded 6 m/sec (Figure 4.6). December and January produce the highest winds, and June is the low wind month.

The NREL and the Texas State Energy Conservation Office are sharing the cost of a project that will produce high-quality and validated offshore wind resource maps [23]. These data include the near-shore region not covered by ocean wind data. Maps of mean annual wind speed at 10 to 300 m and mean annual wind power potential at 50 m are available. The project found class 3 winds along

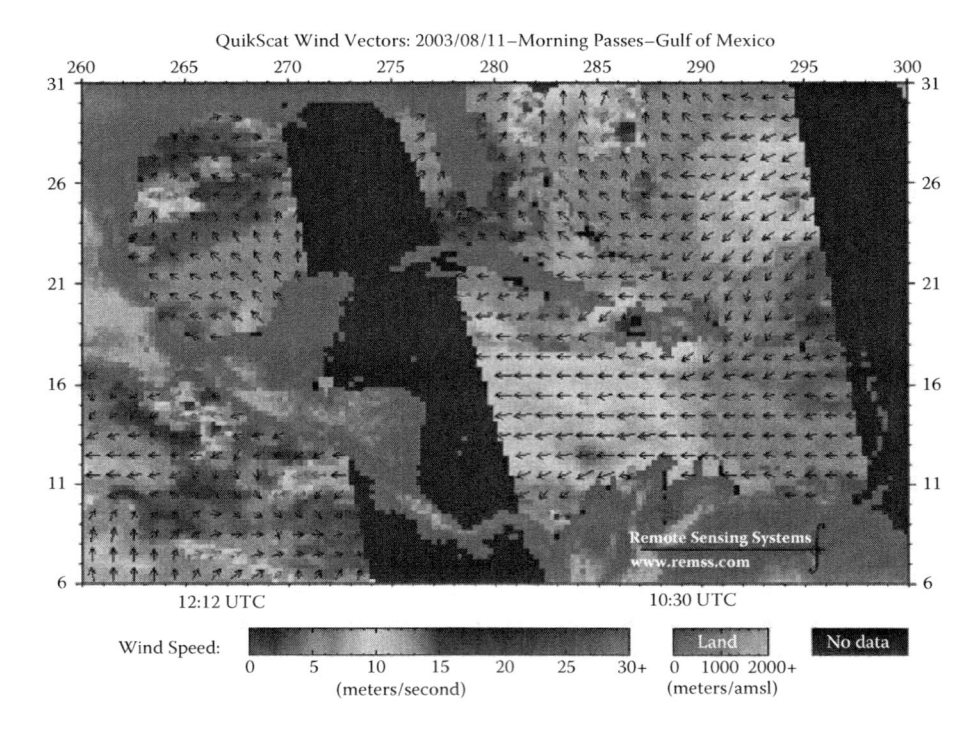

FIGURE 4.4 Daily satellite passes over Gulf of Mexico.

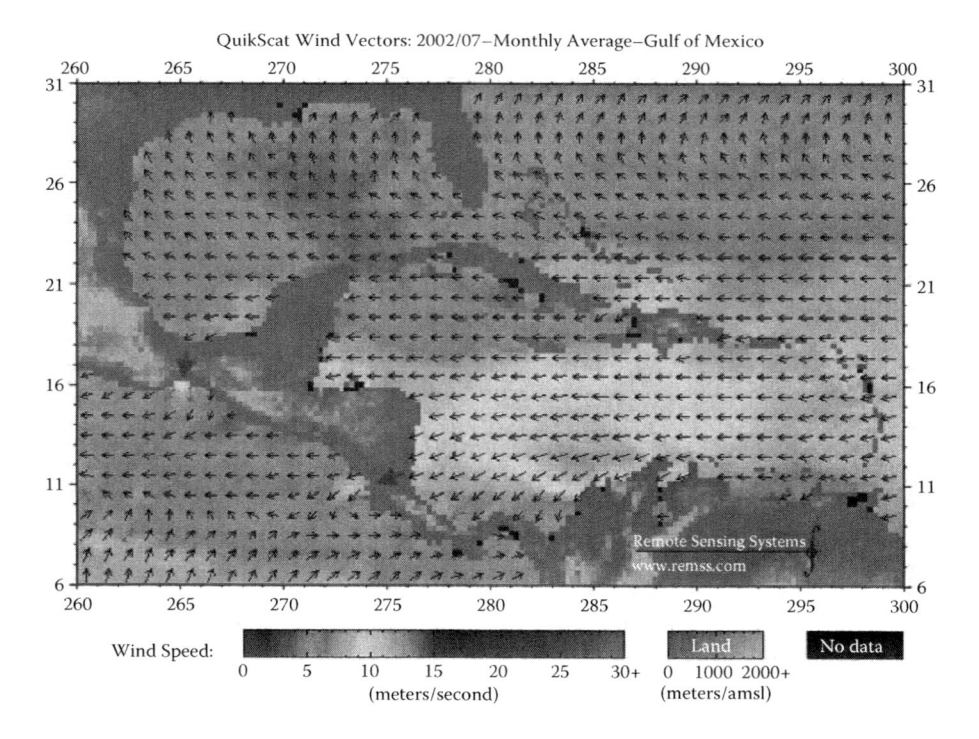

FIGURE 4.5 Average wind speeds for July 2002. Large arrows on land indicate excellent onshore wind regions. Average wind speed = 10 m/sec.

Wind Resource Assessment

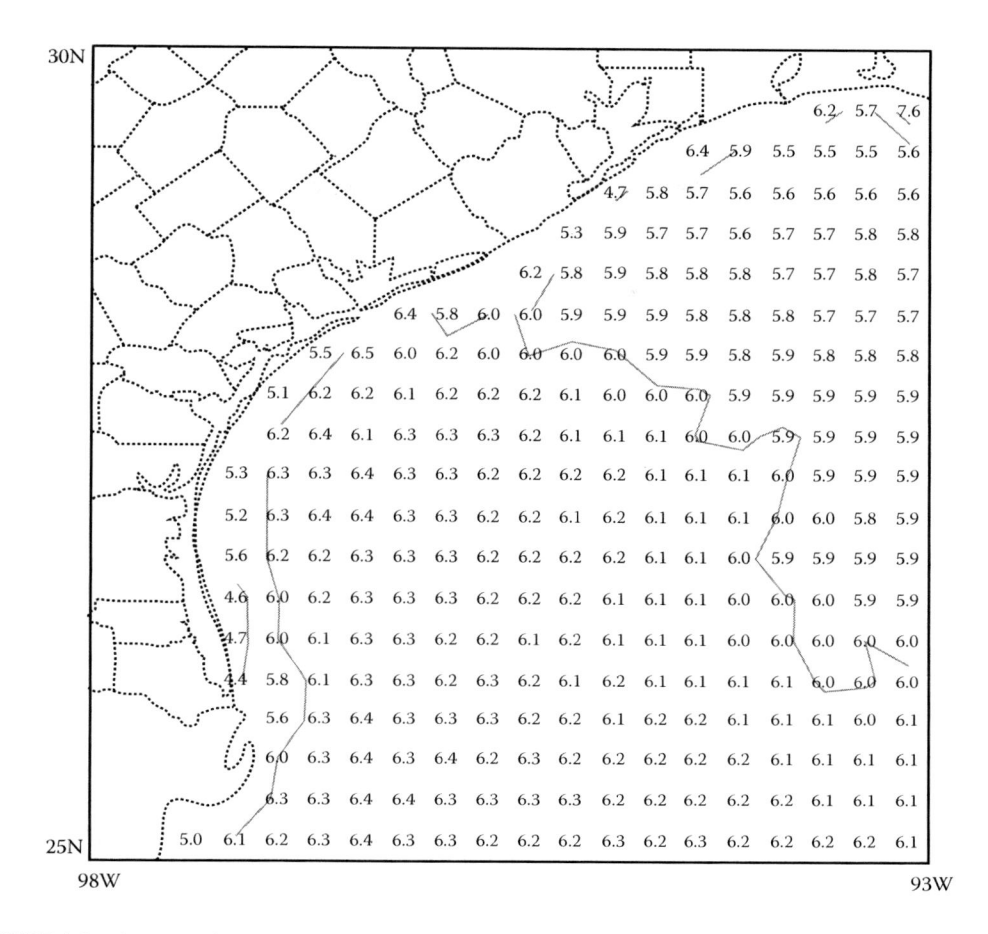

FIGURE 4.6 Average wind speeds (m/sec) at 10 m height for Texas coast of Gulf of Mexico, 1988 through 1994.

the northern third of the Gulf Coast and class 4 winds with a region of class 5 winds from Corpus Christi almost to the border with Mexico. NREL also plans to use the data to analyze the offshore wind shear and other wind characteristics for offshore wind turbine design and performance. The state has control of the land for a distance of 16 km from the coast and is interested in leasing areas for wind farms.

4.4.2 WORLD

The *European Wind Atlas* also has offshore wind data (Figure 4.7). The Predicting Offshore Wind Energy Resources project [24] aimed to assess offshore wind power potential in European Union waters, taking into account coastal effects and highlighting sea areas where hazardous wind or wave conditions exist. These estimates can then be used to pinpoint areas that are favorable for siting wind farms. More detailed monitoring can then be undertaken to improve the initial wind power estimates at selected sites.

4.5 INSTRUMENTATION

An anemometer is a device for measuring airflow. Devices for measure wind speed include pitot tube, cup, vane, propeller, hot wire, hot film, sonic, and laser Doppler anemometers. The common (and less expensive) devices are the cup and propeller anemometers. However, their response times

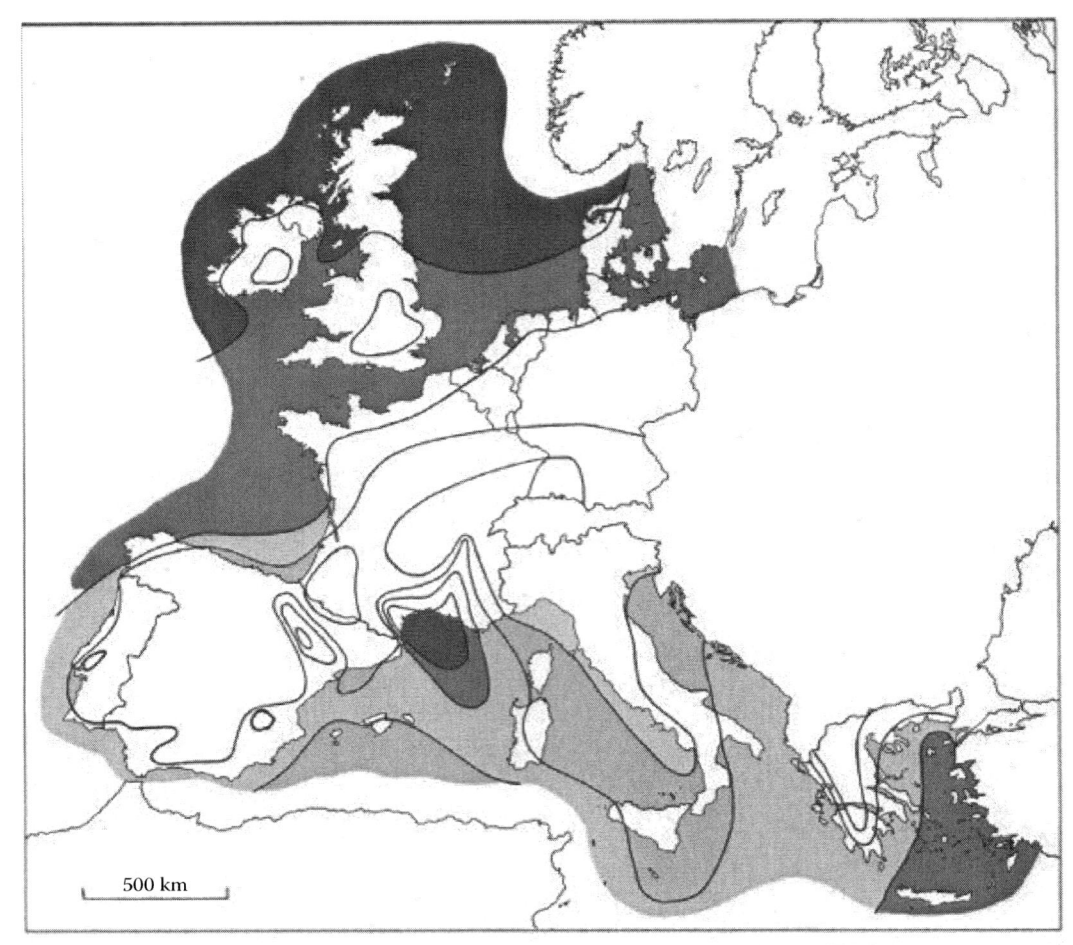

| Wind resources over open sea (more than 10 km offshore) for five standard heights |||||||||||
| 10 m || 25 m || 50 m || 100 m || 200 m ||
ms^{-1}	Wm^{-2}	ms^{-1}	Wm^{-2}	ms^{-1}	Wm^{-2}	ms^{-1}	Wm^{-2}	ms^{-1}	Wm^{-2}
> 8.0	> 600	> 8.5	> 700	> 9.0	> 800	> 1.0	> 1100	> 11.0	> 1500
7.0–8.0	350–600	7.5–8.5	450–700	8.0–9.0	600–800	8.5–10.0	650–1100	9.5–11.0	900–1500
6.0–7.0	250–300	6.5–7.5	300–450	7.0–8.0	400–600	7.5–8.5	450–650	8.0–9.5	600–900
4.5–6.0	100–250	5.0–6.5	150–300	5.5–7.0	200–400	6.0–7.5	250–450	6.5–8.0	300–600
<4.5	<100	<5.0	<150	<5.5	<200	<6.0	<250	<6.5	<300

FIGURE 4.7 European offshore wind resources at five heights in open sea. (*Source:* 1989 European Wind Atlas. (© Risø National Laboratory, Denmark.)

to changes in wind speed are slower. Wind turbines also have response times to changes in wind speed, so cup anemometers are adequate for determining wind energy potential. The advantages of sonic and hot wire anemometers are that they have no moving parts or response times in contrast to mechanical sensors. However, their higher cost has prevented much penetration into the wind resource assessment market.

An anemometer can measure the amount of wind that has passed (wind run). Based on the wind run, the average wind speed can be calculated for a specific period. An anemometer can also measure the fastest mile (maximum wind speed).

Previously, meters and strip charts that generated analog outputs were used. However, analyzing strip chart data is very tedious, and their time resolution is fairly coarse unless the paper feed rate is large. Today the major difference is the availability of microprocessors for sampling, storing, and even analyzing data in real time. Also, personal computers alleviate most of the problems in analyzing large amounts of data.

Digital instruments and digitized analog inputs typically have sample rates of 0.1 to 1 Hz (hertz = number/second). Values can be stored in a histogram of wind speeds or wind speed and other selected variables can be stored for selected averaging time periods along with standard deviations. Events such as maximums and times of occurrence can also be recorded and stored.

Microdata loggers were designed specifically for wind potential measurement and recording time sequence data (averaging time is selectable) on chips. The chips can store data from a number of channels, and the data loggers can even be queried by telephone (cell or direct), radio link, or satellite, so data are transmitted directly to a base computer. Now Internet connection is available. More detailed information on instrumentation and measurement can be found in Rohatgi and Nelson [18]. Also see the *Wind Resource Assessment Handbook* [25] for detailed information on wind measurement, instrumentation, and quality assurance.

The advantages of sonic detection and ranging (SODAR) and light detection and ranging (LIDAR) are that the instrumentation is at ground level and no tower is needed. Wind speeds can be measured to 500 m (SODAR) and even out to several kilometers (LIDAR). The disadvantage is the substantial cost for met towers over 60 m; the cost for a met tower of 150 m (the height to the top of a rotor on a large turbine) is very high. A short-term study [26] compared the relative accuracy of high-resolution pulsed Doppler LIDAR with a mid-range Doppler SODAR and direct measurements from a 116-m met tower that had four levels of sonic anemometers. The primary objective was to characterize the turbulent structures associated with the Great Plains low-level nocturnal jet. The actual measuring volumes associated with each of the three measurement systems varied by several orders of magnitude, and that contributed to the observed levels of uncertainty. The mean differences were around 0.14 m/sec.

The three general types of instrumentation for wind measurements are: (1) instruments used by national meteorological services, (2) instruments designed specifically for determining wind resources, and (3) instruments for high sampling rates to measure gusts, turbulence, and inflow winds for determining power curves, stress, fatigue, and other parameters of wind turbines.

The data collection by meteorological services is the most comprehensive and long term; however, in much of the world, the available data are almost worthless for determining wind power potential. The reasons are that (1) most stations are in cities and airports that are generally less windy; (2) sensors are mounted on buildings and control towers; (3) the quantity of data actually recorded is small (one data point per day or monthly averages); and (4) lack of calibration after installation. As an example of the problem of using meteorological data, the annual mean wind speed for Brownsville, Texas, is 5.4 m/sec, compared to 2.8 m/sec for Matamoros, Mexico, which is just across the Rio Grande River.

Costs for various types of instruments for measuring wind speed range from $400 hand-held anemometers and $1,200 data loggers with cell phones or Wi-Fi; sensors for multiple levels would add at least another $2,000. Companies sell instruments that sample at rates of 0.1 to 1 Hz and display their outputs on analog devices (meters and recorders) or digital devices (stored on chips). Instruments will record and analyze time sequence data. Wind speeds and directions can be stored for selected time intervals, power can be calculated, and options to measure selected events such as maximums, gusts, and times of occurrence are also available.

Companies that sell instrumentation specifically for wind measurements also sell digital readers and provide software for analyzing the data. Pole towers are available specifically for wind measurements from 10 m, $500; 60 m (with gin pole), $18,000; and to 80 m (with gin pole) $40,000. Guyed lattice towers can be obtained for higher heights. Pole towers of 50 and 60 m are normally used because they can be raised and lowered with gin poles, are tall enough to gather higher nighttime wind speeds, and are lower than the Federal Aviation Administration (FAA) height requirement of (61 m or 200 ft).

In many countries, mechanical anemometers were the normal measuring instruments but they require considerable maintenance and frequent calibration. The power from a cup anemometer drives a strip chart recorder or a counter. Because of the small number of data points, the Weibull distribution was widely used to estimate wind power potential. As an example of the problem, wind run data were collected three times a day from an anemometer at less than 2 m high at a national meteorological station in Jujuy, Argentina (Figure 4.8) to determine daily average wind speed. Due to height and blockage by trees and buildings, the wind power potential was vastly underestimated.

Data from Mexicali Airport, Mexico [27], reveal a trend in wind speed data over time (Figure 4.9). The number of 1-hr observations was fairly consistent from 1973 to 1999 when the

FIGURE 4.8 Meteorological station in Jujuy, Argentina.

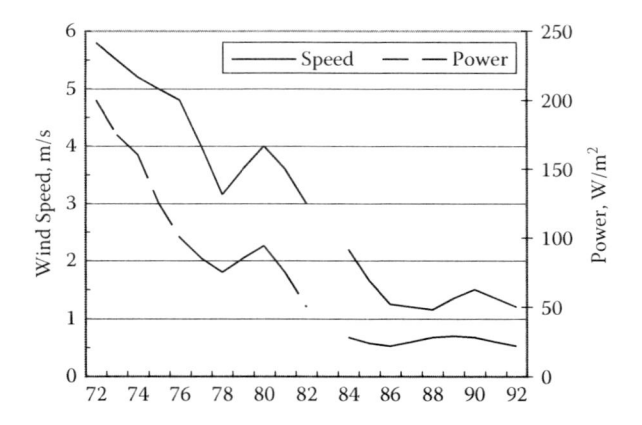

FIGURE 4.9 Wind speed data for Mexicali Airport, Mexico.

Wind Resource Assessment

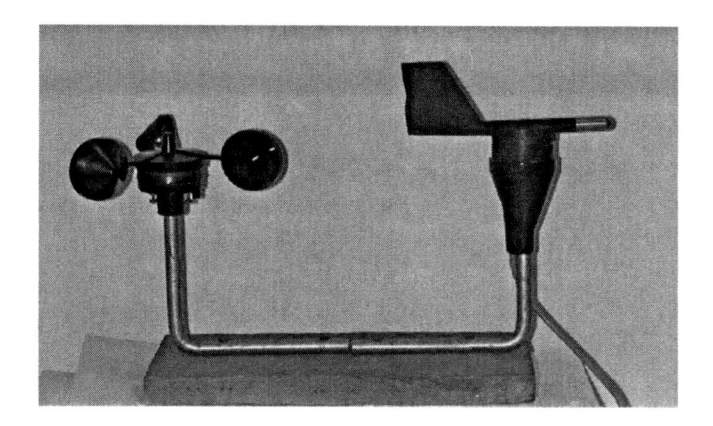

FIGURE 4.10 Maximum cup anemometer and wind vane. Anemometer is about 15 cm across.

airport operated. The downward trend indicates degradation of the anemometer from failure to maintain and recalibrate it or less exposure of the instrument because of increased vegetation or other obstructions. The wind power changed from 170 W/m^2 at the beginning to 25 W/m^2 at the end (a factor of 7).

4.5.1 Cup and Propeller Anemometers

A widely used anemometer for wind resource measurements has a circular magnet (four poles) in a cup housing and one or two coils for pick-up of the signal (Figure 4.10) that approximates a sine wave. The transducer counts zero crossings (sampling time is generally 1 Hz), so wind speed is related to number of counts. The advantage is that signals can be transmitted 150 m without loss of accuracy (none of the analog signal problems of attenuation and amplification). The accuracy of cup anemometers in wind tunnels is estimated at ±2% [28].

Another type of cup anemometer has a disk containing up to 120 slots and a photocell. The periodic passage of slots produces pulses in each revolution of the cup. This gives a better resolution and the sampling rate can be increased to 5 Hz.

The propeller anemometers (Figure 4.11) have faster responses and behave linearly in changing wind speeds. The wind speed is measured by measuring the voltage output of a DC generator. The propeller is kept facing the wind by a tail vane that also works as a direction indicator. The accuracy normally is about 2% for wind speed and direction.

Propellers are usually made of polystyrene foam or polypropylene. However, for turbulent winds, the values may be misleading in determining power curves for wind turbines. A propeller anemometer is better suited to measure the three components of wind velocity because it responds primarily to wind parallel to its axis. An array of three units in mutually perpendicular directions will measure the three components of wind.

4.5.2 Wind Direction

Wind direction is measured by a wind vane counterbalanced by a weight fixed on the end of a rod. However, the vanes of propeller anemometers form part of the propeller's axis and require minimum force to initiate movement. The threshold wind speed for this force is usually about 1 m/sec. Normally, the vane motion is damped to prevent rapid changes of direction. Wind vanes generally produce signals by contact closures or by potentiometers. The accuracy of potentiometers is higher than that achieved by contact closures, but the latter are less expensive.

FIGURE 4.11 Propeller anemometers for measuring in three directions. (Photo courtesy of R.M. Young.)

4.5.3 Instrument Characteristics

Sensors, transducers, and signal conditioners measure and transform signals for recording. Resolution is the smallest unit of a variable that is detectable by a sensor. Recorders may limit the resolution. Reliability is a measure of an instrument's ability to produce useful data over a period of time. The best indicator of reliability is the past performance of similar instruments.

Accuracy and precision are two separate measures of system performance that are often treated ambiguously. Accuracy refers to the mean difference between the output of a sensor and the true value of the measured variable. Precision refers to the dispersion about the mean. For example, an instrument may produce the same measured value every time but produce a value that is off by 50%. That system has high precision but low accuracy.

Accuracy, however, may be a function of time or depend on maintenance. Anemometers are calibrated in wind tunnels where the airflow is steady. Another method of calibrating anemometer performance for wind resource assessment (known as scale and offset) uses controlled velocity via a boom mounted on a truck. Generally, calibrated anemometers produce signals that are accurate to within 0.5 to 2.0% of the true wind speed. Under normal use in the atmosphere, good anemometers should be accurate to 2 to 4%.

The distance constant is the length of fluid flow past a sensor required to cause it to respond to 63.2% of a step change in speed. A step change is a change from one value to another value, similar to ascending a stairway one step at a time. The larger and heavier cup anemometers usually have distance constants of 3 to 5 m. For light-weight and smaller cup anemometers, such as those used to measure turbulence, the distance constant is typically about 1 m. The time constant is the period required for the sensor to respond to 63.2% of a step change in input signal.

The damping ratio is a constant that describes the performance of a wind vane in response to a step change in wind direction. The damping ratio is dimensionless and is generally between 0.3 and 0.7.

Wind Resource Assessment

The sample rate is the frequency (Hz) at which a signal is sampled and may include the time for recording data. Since a large amount of data requires large storage, wind speeds are averaged over a longer period and these values along with standard deviations are stored. Typical values for wind power analysis are sample rates of 1 Hz and averaging time of 10 min. Previously, 1-hr averaging times were used for many resource assessment projects.

4.5.4 Measurement

Anemometers mounted on towers should be mounted away from a lattice tower a distance of two to three tower diameters to reduce the effect of the tower on airflow. Solid towers should be mounted six tower diameters away. Met towers must be located away from obstacles such as trees and buildings.

The time and money spent for measuring wind resources depends on whether the plan is for a wind farm or a small wind turbine. The difference between class 3 and class 4 and above wind sites determines economic viability for wind farms. Individuals who install small wind turbines tend to overestimate wind resources before their turbines are installed and later bemoan the lack of wind.

Instrumentation for measuring turbulence and wind inflow for wind turbine response involves multiple anemometers and a higher sampling rate. A system for characterizing turbulence [29] developed and tested by Pacific Northwest Laboratory consisted of two towers and nine anemometers (Figure 4.12) and data sampling at 5 Hz. The propeller vane anemometers for horizontal

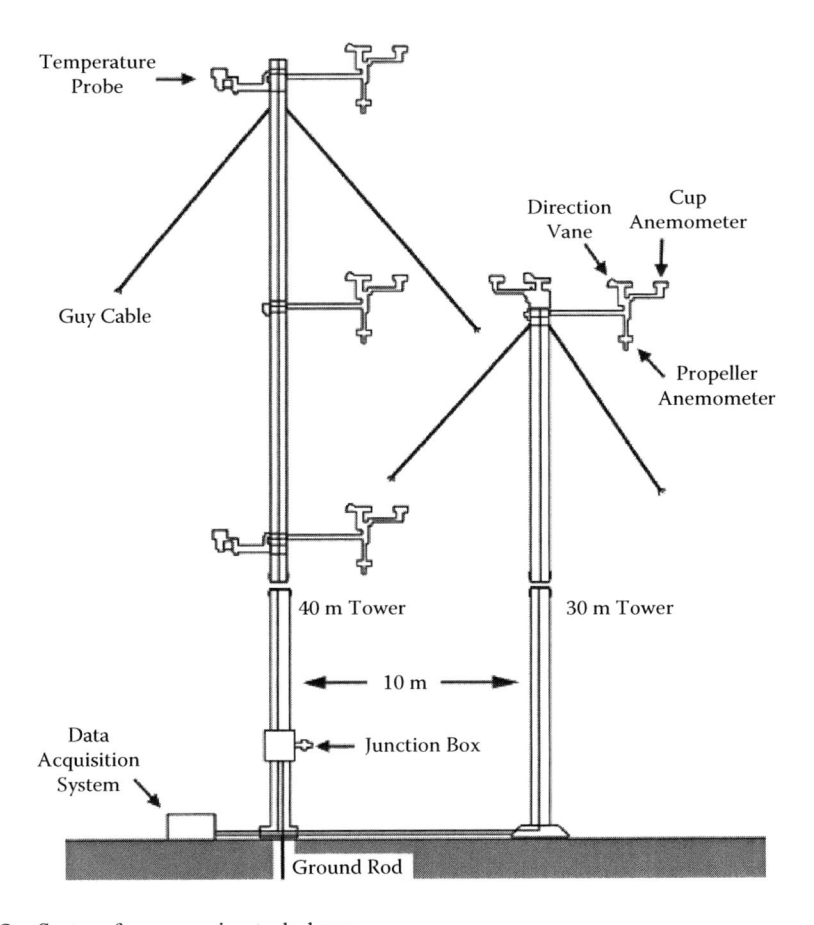

FIGURE 4.12 System for measuring turbulence.

4.5.5 Vegetation Indicators

Vegetation can indicate regions of high wind speed in areas where no measurements are available. Deformation or flagging of trees [18, p. 96] is the most common indicator (Figure 4.13). In some cases the flagging of trees is a more reliable indicator of wind resource than the data available.

For example, the Arenal region of Costa Rica has high winds that were measured (average for 12 stations) at 11 m/sec [30]. A meteorological station near Fortuna in the Arenal region was erected to collect hydrology data. The mechanical anemometer height is less than 2 m because the researchers were originally interested in determining evaporation, and the station was located near trees. Wind speed data indicated no wind power potential, but flagged trees in the area indicated high wind speeds. The Griggs and Putnam Index [31] for flagging coniferous trees (Figure 4.14) is related to annual mean wind speed [32] by

$$\bar{u} = 0.96\,G + 2.6$$

or a tree growing along the ground, the maximum deformation ratio is $D = 7$. The relation of deformation ratio to the mean annual wind speed, $\bar{u}$, was estimated for Douglas fir or Ponderosa pine trees [33]. From regression analysis of the data,

$$\bar{u} = 0.96\,D + 2.3$$

The relationship is used for both coniferous and hemispherical crowned trees. For coniferous trees, A is the angle formed by the crown edge and the trunk on the leeward side, B is the angle formed by the crown edge and the trunk, and C is the average angle of trunk deflection. For hemispherical crowned trees, A is the distance between the trunk and the crown perimeter on the leeward side, B is the distance between the trunk and the crown perimeter on the windward side, and C is the angle between the crown perimeter and the trunk on the leeward side.

The A/B ratio assumes that $1 \leq A/B \leq 5$. As a result, the minimum value of D is 1, which corresponds to no crown asymmetry, or trunk deflection C. Since the maximum deflection is 90 degrees

FIGURE 4.13 Flagging of trees. Left: tree on plains, Canyon, Texas (6 m/sec average wind speed at 10 m height). Right: tree at South Point, Hawaii (10 m/sec average wind speed).

Wind Resource Assessment

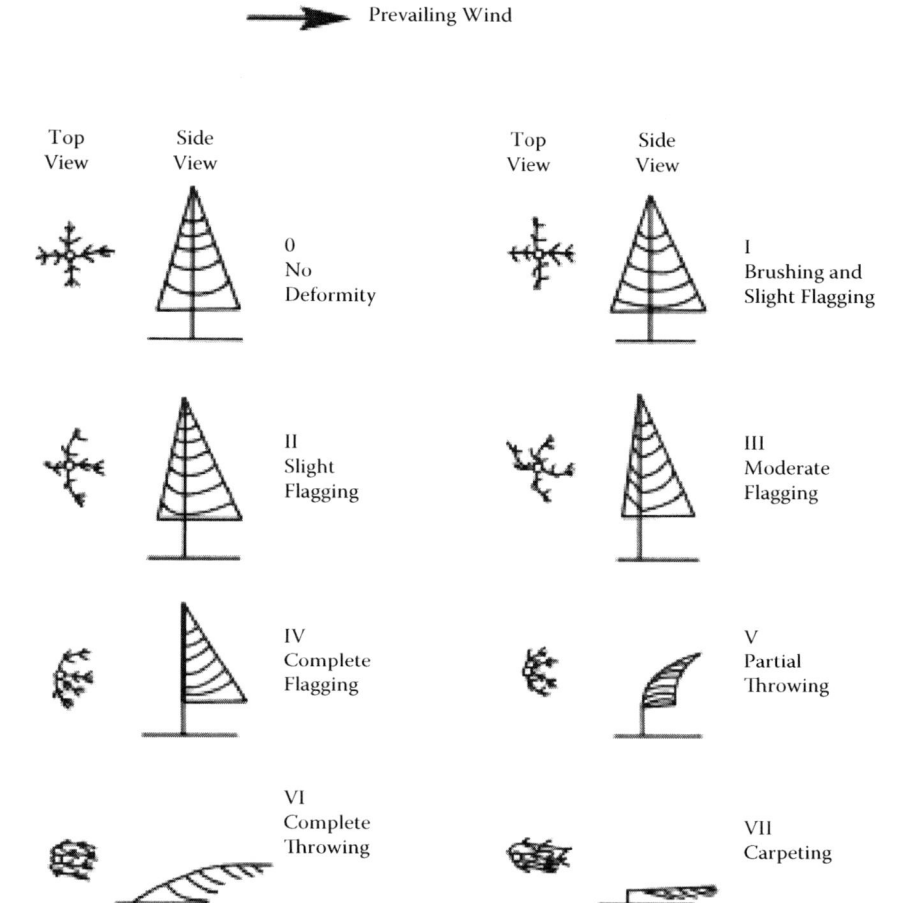

FIGURE 4.14 Griggs and Putnam Index of tree deformation.

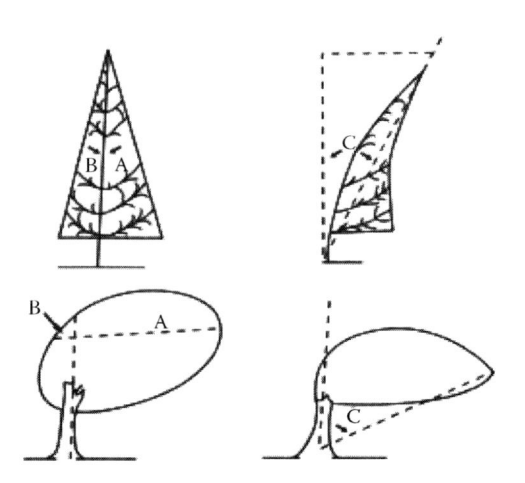

FIGURE 4.15 Estimation of wind speed by tree deformation.

for a tree growing along the ground, the maximum deformation ratio *is* D = 7. The relation of deformation ratio to the mean annual wind speed, <εθυ_0001.επσ>, was estimated for Douglas fir or Ponderosa pine trees. From regression analysis of the data,

$$<εθυ\_0001.επσ> = 0.95\ D + 2.3$$

Photographs can be used to determine the deformation ratio in lieu of direct examination. The deformation ratio and Griggs and Putnam Index give similar ranges of wind speeds.

The use of trees as indicators of wind speed is subject to a number of practical limitations. Of greatest concern is the exposure of a tree to the wind. The deformation should be viewed perpendicularly to the prevailing wind direction so that the full effects of flagging and throwing are considered. Trees selected as indicators must be well exposed to the prevailing winds. Seldom do trees in a forest extend far enough above the canopy to be in an airstream undisturbed by other trees. However, isolated trees or those in small, widely spaced groups should be favored as wind speed indicators. If several locations are to be compared, trees should be of nearly the same height and species. Near the seashore, flagging may be the result of sea spray (salt) and not totally due to the wind.

4.6 DATA LOGGERS

Data loggers for wind resource measurements are now the norm. Data are stored on data chips, and the chips are retrieved or the loggers send information to a base personal computer. The BASE program monitors telephone lines, answers calls, and determines the site calling and the status of the data card and call-in schedule (card unread, first call of six tries; card partially read, fourth call of six; etc.).

Time sequences generate large amounts of data. For example, suppose you want to measure wind speeds, wind directions, pressures, temperatures (1 Hz sampling rate), average values, and statistics stored every 10 min. That would occupy around 130 KB of data per month. Chips are available to store two and more years of data. You still need to retrieve the data at least monthly as a check on problems. Data generated by phones or satellite connections should be retrieved once per week.

The logistic problems must be resolved to ensure high data recovery and the quality of the data analyzed. Calibration and replacement of sensors must be part of a routine maintenance program. For example, anemometers should be replaced once every six months to two years, depending on the number of revolutions and the environment.

A quality assurance program for flagging suspect data is imperative. Data recovery should be around 95%. Sensor problems arise from equipment failure and inadequate data collection due to icing, lightning, and even vandalism. Data loggers and transmission problems can also lead to loss of data. Yearly failure rates are around 25% for sensors and 10% for data loggers. Rates may be higher for sites with very harsh conditions, for example, hail, lightning, dust and sand circulation, and extended periods of high winds.

Generally, two anemometers and one wind vane per level are installed in systems with two or more levels. If one anemometer is down, data can be collected by the second anemometer. If both are not operating, it is possible to estimate values based on data from another level and past wind shear values, so 95% data recovery is feasible.

Wind farm developers want an average wind speed (10 min or 1 h) so they can predict energy production. Available data analysis programs are fairly flexible. Monthly average, minimum, and maximum data for each sensor are available and selected graphs and tables can be generated.

Wind Resource Assessment

> **EXAMPLE OF SUMMARY REPORTS AVAILABLE BY MONTH FROM AN ANALYSIS PROGRAM**
>
> Comparison of hourly wind speeds (two anemometers at same height or at different heights)
> Frequency distributions (calculate wind power/area; see Figure 4.16)
> Frequency distribution graph
> Diurnal wind speed graph
> Average turbulence intensity (upper level, use prevailing wind anemometer)
> Wind rose graph
> Average wind shear table (between two heights)
> Average temperature graph

Data can be placed in spreadsheets for further analysis since most data loggers allow export. Another benefit is that data analysis is not tied to proprietary programs whose manufacturers may have trouble updating, especially if subcontractors developed the software programs.

4.7 WIND MEASUREMENT FOR SMALL WIND TURBINES

For a very small wind turbine (100 W to 3 kW), anemometers, data loggers, and analyses may cost more than the wind turbine. In one sense, a wind turbine is an anemometer and the energy produced is the measurement. It is wise to depend on historical and regional data to determine the feasibility of installing a small wind turbine.

Two other indicators of feasibility are the past historical use of farm windmills in the area and a check with other owners about the performance of other small wind turbine installations in the region. For a wind turbine of 10 to 50 kW, the investment is fairly large—$60,000 to $250,000. Inexpensive digital weather stations including data loggers are now available for $300 to $600 and the data loggers can be plugged into personal computers for analysis.

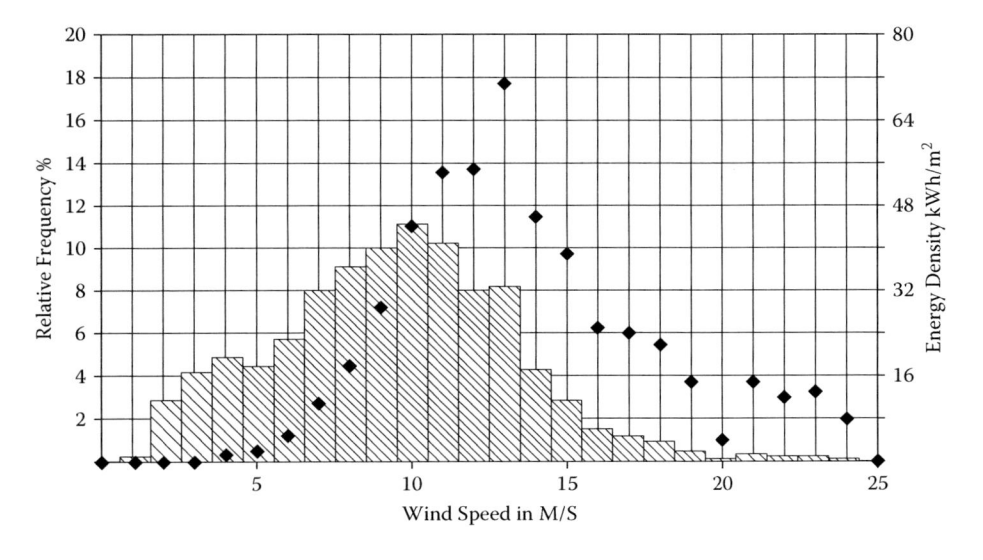

FIGURE 4.16 Example graph, frequency distribution plus energy, from analysis program, 50 m height, White Deer, Texas, April 1998.

These instruments are not suitable for collecting long-term data for wind resource assessment or for wind farms. If maps indicating sufficient winds are available and wind farms have already been installed in an area, there is no need to collect wind data before installing same size turbines. However, use caution when installing wind turbines within cities, even in windy areas, as the winds will be less than those indicated on wind maps.

LINKS

MAPS

Canada. www.windatlas.ca/en/index.php
Database of wind characteristics. www.winddata.com
NREL (International). www.nrel.gov/wind/international_wind_resources.html
NREL (U.S.). www.nrel.gov/wind/resource_assessment.html
Wind Atlases of the World. www.windatlas.dk/

OCEAN WIND DATA

College of Marine Studies, University of Delaware. Annual and monthly values. www.ocean.udel.edu/windpower/
European Wind Atlas. www.windatlas.dk/Europe/oceanmap.html
Galathea 3. www.galathea3.emu.dk
Galathea 3. Nine-month expedition. http://galathea3.emu.dk/satelliteeye/projekter/wind/back_uk.html
Ocean surface winds. http://manati.orbit.nesdis.noaa.gov/doc/oceanwinds1.html
DTU (Denmark, formerly Risoe). 2011 South Baltic Wind Atlas. http://130.226.56.153/rispubl/reports/ris-r-1775.pdf
Wind Resource Assessment Handbook (excellent source). www.nrel.gov/docs/legosti/fy97/22223.pdf

DATA LOGGER, SENSOR, AND TOWER INFORMATION AND PHOTOS[*]

www.campbellsci.com
www.ekopower.nl
www.nrgsystems.com
www.secondwind.com
www.wilmers.com
www.rohnproducts.com (towers)

REFERENCES

1. N.J. Cherry, D.L. Elliott, and C.I. Aspliden. 1981. World-wide wind resource assessment. In *Proceedings of Fifth Biennial Wind Energy Conference and Workshop*, Vol. II, p. 637.
2. NREL MapSearch. www.nrel.gov/gis/mapsearch
3. C.L. Archer and M.Z. Jacobson. 2005. Evaluation of global wind power. *Journal of Geophysical Research* 110, 12110. www.stanford.edu/group/efmh/winds/global_winds.html
4. Wind Atlases of the World. www.windatlas.dk/World/Index.htm
5. D.L. Elliot and W. R. Barchet. 1977. *Synthesis of national wind energy assessments*. BNWL-2220 WIND-5/UC-60. Battelle Pacific Northwest Laboratory.
6. D.L. Elliott et al. 1986. *Wind energy resource atlas of the United States*. DOE/CH 10093-4. http://rredc.nrel.gov/wind/pubs/atlas
7. National Wind Technology Center. www.nrel.gov/wind/resource_assessment.html
8. Wind Powering America. Wind maps. www.eere.energy.gov/windandhydro/windpoweringamerica/wind_maps.asp
9. Northwest Wind Power Mapping Project. http://www.windpowermaps.org/windmaps/windmaps.asp

---

[*] Listing does not imply endorsement.

10. Alternative Energy Institute, West Texas A&M University. www.windenergy.org/maps
11. Alternative Energy Institute, West Texas A&M University. www.windenergy.org/datasites
12. I. Troen and E.L. Petersen. 1989. *European wind atlas*. Denmark: Risoe National Laboratory. www.windatlas.dk.
13. National Wind Technology Center. www.nrel.gov/wind/international_wind_resources.html
14. Solar and Wind Energy Resource Assessment. http://en.openei.org/apps/SWERA/
15. Wind Atlases of the World. www.windatlas.dk/World/About.html
16. Database on wind characteristics. www.winddata.com
17. RETScreen International. Energy resource maps. http://www.retscreen.net/ang/energy_resource_maps.php
18. J. Rohatgi and V. Nelson. 1994. *Wind characteristics: An analysis for the generation of wind power*. Alternative Energy Institute, West Texas A&M University.
19. Remote Sensing Systems. Geophysical graphic images and binary data files on FTP server. ftp.ssmi.com/ssmia; climatological data sets (3-day, weekly, and monthly averages) for each geophysical parameter and satellite. http://www.remss.com
20. Jet Propulsion Laboratory, Physical Oceanography DAAC. Ocean winds. http://podaac.jpl.nasa.gov/index.html
21. D. Elliott and M. Schwartz. 2006. Wind resource mapping for United States offshore areas. In *Proceedings of Windpower Conference*. www.nrel.gov/wind/pdfs/40045.pdf
22. V. Nelson and K. Starcher. 2003. *Ocean winds for Texas*. AEI Report 2003-1. Alternative Energy Institute, West Texas A&M University.
23. State Energy Conservation Office of Texas. Texas coastal wind resource assessment. www.seco.cpa.state.tx.us/re_wind_maps.htm
24. Risoe National Laboratory of Denmark. www.eru.rl.ac.uk/POWER_project/POWER_project.htm
25. *Wind Resource Assessment Handbook*. 1997. NICH Report SR-440-22223. www.nrel.gov/docs/legosti/fy97/22223.pdf
26. N.D. Kelley, B.J. Jonkman, and G.N. Scott. 2007. Comparing pulsed Doppler LIDAR with SODAR and direct measurements for wind assessment. In *Proceedings of Windpower Conference*. www.nrel.gov/wind/pdfs/41792.pdf
27. M.N. Schwartz and D.L. Elliott. 1995. Mexico wind resource assessment project. In *Proceedings of Windpower Conference*, p. 57.
28. J.L. Obermier. 2000. Single variable comparison of calibrated results for NRG maximum 40 anemometers. In *Proceedings of Windpower Conference*.
29. L.L. Wendell et al. 1991. Turbulence characterization for wind energy development. In *Proceedings of Windpower Conference*, p. 254.
30. T. de la Torre. 1993. International developments: new wind power projects and wind electric power development in Costa Rica. In *Proceedings of Windpower Conference*, p. 429.
31. P.C. Putnam. 1948. *Power from the Wind*. New York: Van Nostrand Reinhold, p. 84.
32. E. W. Hewson and J. E. Wade. 1979. *A handbook on the use of trees as indicators of wind power potential*. Report RLO-2227. U.S. Department of Energy.

PROBLEMS

1. If there is a wind map for your nation, what is the wind speed or wind power potential near your location?
2. What is the average wind speed offshore in the ocean south of Cape Cod, Massachusetts? Use ocean wind data or Figure 4.1.
3. For South Dakota (Figure 4.2), make an educated guess of percent area with wind class 5 and above.
4. For the offshore wind map for Europe (Figure 4.7), what regions have the highest wind power/area?
5. What region of Nicaragua has the best wind power potential? Use NREL international wind maps.
6. What offshore location for Texas has the highest wind speeds at 10 m height? Use Figure 4.6 or map (Reference 22).
7. Go to www.windmap.org (2/27/2013). Stateline wind farm is located near the boundary of Oregon and Washington. The Columbia River forms the boundary as it comes out of

Washington. Just east of where it crosses the border, the boundary is straight. The wind farm is located there, primarily in Oregon. Use the Oregon wind map. What is the highest wind class for the project area? On a large map, zoom in once. Can you obtain a larger image by going to the interactive tool on the left navigation bar?

8. How far should an anemometer be placed away from a tall tower 13 cm in diameter?
9. You are installing anemometers on an existing guyed lattice tower (three sides, each side is 1.5 m wide) for radio communication. How far should the anemometer be placed away from the tower?
10. You are installing anemometers on a stand-alone, lattice tower for radio communication. The tower has three sides, and each side is 4 m wide at 10 m height. Compare the recommended length with a practical length of the boom (mounting pipe or bracket).
11. Why were the propeller anemometers for horizontal wind measurements on the turbulent characterization tower (Figure 4.12) replaced with cup anemometers?
12. Are there any examples of vegetation indicators of wind in your region? What wind speed do they indicate?
13. No tower is needed for a laser system that measures wind speed. What is the reason for not employing a laser system?
14. You want to measure wind speeds and directions at three levels (10, 25, and 50 m) at six sites (dispersed across your state) for two years. Estimate the cost for equipment and workers to handle installation, data collection, and data analysis. You may choose any type of data logger, tower, data retrieval, and analysis.
15. Compare the amount of storage needed for (a) 1 hr average and standard deviation (sample rate 1 Hz) for sixteen channels for one year and (b) 1 min average and standard deviation (sample rate 5 Hz) for sixteen channels for one year.
16. How many years of data are required to establish a database to which shorter term data for wind farms can be referenced? In other words, how many years of data are needed to generate a wind map for a large region or state?
17. Estimate the cost for installation of a 50-m guyed pole tower. Consider difficulty getting to the site when estimating travel costs.
18. Estimate the costs to install a 50-m guyed lattice tower (e.g., Rohn 25G or 45G); include travel cost based on difficulty getting to the site. Will you use a crane or attached gin pole for installation of the lattice tower?
19. Estimate the cost of installation of a 100-m guyed lattice tower. Include travel costs based on difficulty getting to the site. A crane will be used for installation. Also, FAA regulations require lights so the additional cost of transmitting power to a remote location must be considered.
20. Compare wind resource instruments (cost, sample rate, data storage, and data analysis) from two different manufacturers.
21. Are there any shareware programs for wind resource analysis?

5 Wind Turbines

Wind turbines are classified according to the interactions of the blades with the wind (aerodynamics), orientation of the rotor axis to the ground, and innovative or unusual types of machines. The aerodynamic interaction of the blades with the wind is caused by drag or lift or a combination of both.

5.1 DRAG DEVICES

In a drag device, the wind pushes against the blades or sails (Figure 5.1). Drag devices are inherently limited in efficiency since the speed of the device or blades cannot exceed the wind speed. The wind pushes on the blades of a drag turbine, forcing the rotor to turn on its axis.

Examples of drag devices are cup anemometers, vanes, and paddles that are shielded from the wind or change parallel to the wind on half the rotor cycle (Figure 5.2). Clam shells that open on the downwind side and close on the upwind side are also examples of drag devices. No drag wind turbines have been produced commercially because they are inefficient, and the blades require a lot of material. However drag devices are popular with inventors and homebuilders because they are easy to construct (Figure 5.3). Invariably, inventors become irate when they are told that the inefficient aerodynamics and amounts of material required for blades limit the commercialization of drag devices.

5.2 LIFT DEVICES

Most lift devices use airfoils similar to propellers or airplane wings for blades but other concepts have been used. Using lift, the blades can move faster than the wind and are more efficient in terms of aerodynamics and amount of material needed. The *tip speed ratio* is the speed of the tip of the blade divided by the wind speed. At the point of maximum efficiency for a rotor, the tip speed ratio is around 7 for a lift device and 0.3 for a drag device. The ratio of amount of power per material area for a lift device is around 75, again emphasizing why wind turbines using lift are used to produce electricity. The optimum tip speed ratio also depends on the solidity of the rotor. *Solidity* is the ratio of blade area to rotor swept area.

A single blade rotating very fast can essentially extract as much energy from the wind as many blades rotating slowly (Figure 5.4). A wind turbine with one blade will save on material but a counter weight is needed for balance. Most modern wind turbines have two or three blades because of other considerations, and most large turbines in the commercial market have three blades.

The MBB Monopteros and Flair designs were single-bladed wind turbines built in Germany. A one-blade (5 kW) unit was built by Riva Calzoni in Italy. The Monopteros had full-span pitch control and the rotor was upwind. MBB and Riva Calzoni collaborated on a 20-kW one-blade unit, and then Riva Calzoni built a 330-kW unit. Chalk [1] invented a rotor with a large number of blades based on the design of a bicycle wheel. Notice the large number of blades on the Honeywell unit in Figure 1.7. Some modern wind turbines have been built with four to six blades.

A Savonius rotor (Figure 5.5) is not strictly a drag device, but it has the same characteristic of large blade area to intercept area. This means more material for construction and problems arising

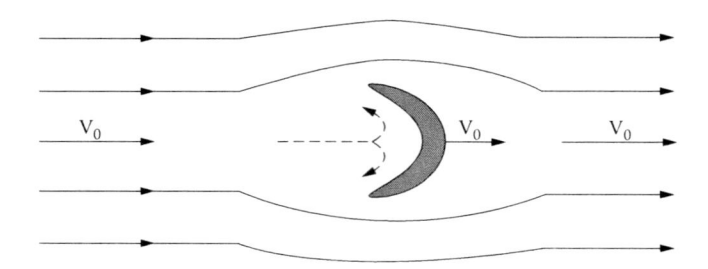

FIGURE 5.1 Drag device. An example is a sailboat moving downwind.

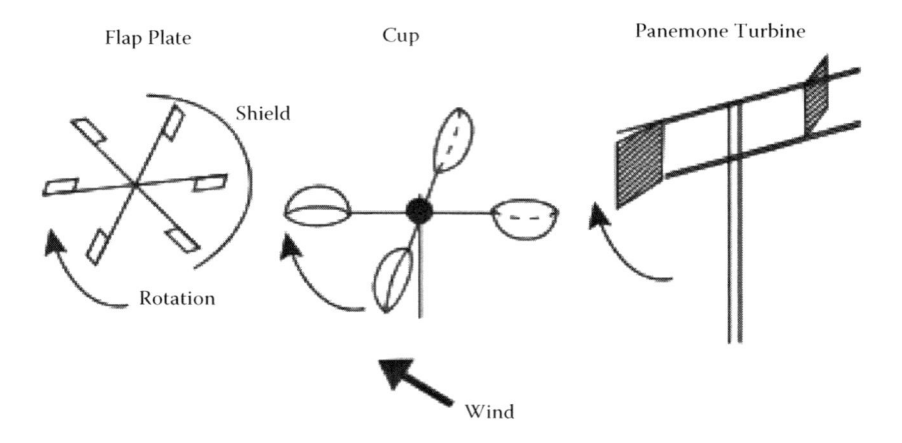

FIGURE 5.2 Drag wind turbines.

from force at high wind speeds even if the rotor is not turning. An advantage of the Savonius wind turbine is the ease of construction.

5.3 ORIENTATION OF ROTOR AXIS

Wind turbines are further classified by the orientation of the axis of the rotor to the ground: horizontal-axis wind turbine (HAWT) and vertical-axis wind turbine (VAWT; Figures 1.10 and 5.6). The rotors on HAWTs must be kept perpendicular to the flow of the wind to capture maximum energy. This rotation of the unit or rotor about the tower axis (*yaw*) is accomplished by tails on upwind units (small turbines up to 10 kW although some 50 kW units have tails), by *coning* on downwind units (Figure 5.7), or by electric motors or wind-propelled fan tail rotors to drive the unit around the yaw axis. *Coning* means the blades are at an angle from the plane of rotation.

VAWTs have the advantage of accepting wind from any direction. However, the Darrieus wind turbine is not reliably self-starting because the blades have to be moving faster than the wind to generate power. An induction or other type of motor or generator is used for start-up to get the blades moving fast enough to generate positive power. A giromill (Figure 5.8) may have articulated blades that can change angle on the rotational cycle and may be self-starting. Another advantage of VAWTs is that the speed increaser and generator can be at ground level. The two disadvantages are (1) the rotor is closer to the ground and (2) cyclic variation of power occurs on every revolution of the rotor.

FIGURE 5.3 Examples of drag devices. Top left (clockwise): (1) Around 10 m diameter, with flywheel supposed to store energy and reduce power variations. (2) Cups 1.2 m in diameter. Inventor predicted power output at 4 kW. (3) Panemone device. Blades move parallel to wind when moving upwind. (4) Shielded plywood sheets, 1.2 by 2.5 m. Notice the large wheel for speed increase to the generator. Inventor predicted output at 4 kW.

5.4 SYSTEM DESCRIPTION

A total system consists of the wind turbine and load. A typical wind turbine consists of the rotor (blades and hub), speed increaser (gearbox), conversion system, controls, and tower (Figure 5.9). The nacelle is the covering or enclosure. The output of the rotor (rotational kinetic energy) can be converted to electrical, mechanical, or thermal energy. Generally, the choice is electrical energy, so the conversion system is a generator.

Blade configuration may include a nonuniform platform (blade width and length), twist along the blade, and variable (blades can be rotated) or fixed pitch. The pitch is the angle of the chord at the tip of the blade to the plane of rotation. The *chord* is the line from the nose to the tail of the airfoil.

Components for a large unit mounted on a bedplate are shown in Figure 5.10. Most large wind turbines that are pitch regulated have full-span (blade) control, and in this case electric motors are used to rotate (change the pitch of the blades). All blades must have the same pitch for all operational conditions.

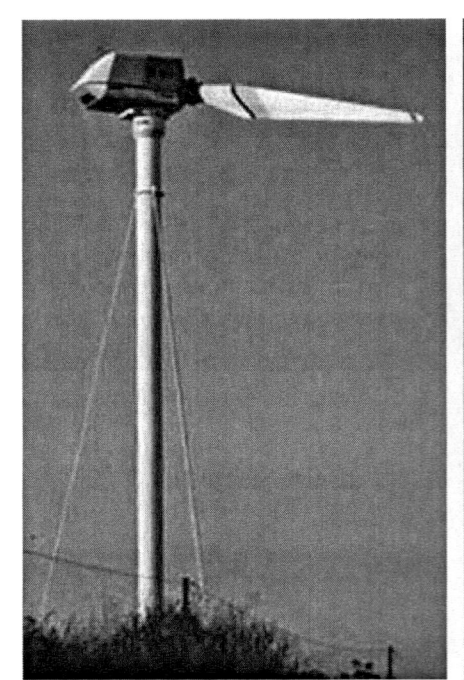

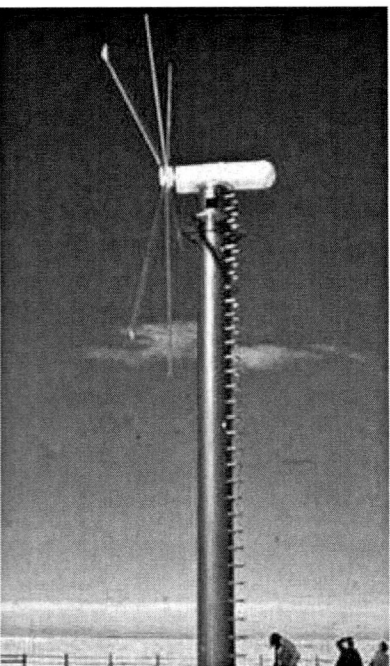

FIGURE 5.4 Left: One-blade Monopteros wind turbine, 475 kW, variable pitch blade, upwind, near Hamburg, Germany. Right: Six-blade Mehrkam wind turbine, Mehrkam, 40 kW, fixed pitch blades, downwind, U.S.

For units connected to a utility grid, 50 or 60 Hz, the generators can be synchronous or induction type connected directly to a grid, or a variable-frequency alternator or direct current generator connected indirectly to a grid through an inverter. Most direct current (DC) generators and permanent magnet alternators on small wind turbines do not have speed increasers. One type of large wind turbine has no gearbox; it has very large generators.

Some HAWTs use slip rings to transfer power and control signals from the top of the tower to ground level, while others have wire cords of extra length for absorbing twist. After a certain amount of twist, the excess must be removed by yawing the turbine or via a manual disconnect. For large wind turbines, the transformer or a winch may be located in the nacelle. A total system is called a wind energy conversion system (WECS).

5.5 AERODYNAMICS

The moving blades of a wind turbine convert part of the power in the wind to rotational power.

$$P = T \times \omega \tag{5.1}$$

where T is the torque (N–m) and ω (rad/sec) is the angular velocity. The same power can be transferred with a large T and small ω, or a small T and large ω. The torque–ω characteristics of the rotor should be matched to the torque–ω characteristics of the load.

Note: Θ is the angle expressed in units of degrees or radians. A radian is the angle at which the arc of the circle equals the radius, so circumference = 360 degrees = 2π radians, or 1 radian = 57.3 degrees. Angular velocity $\omega = \Delta\Theta/\Delta t$. Linear velocity of the tip of the blade is given by $v = \omega \times r$, where r = radius of the blade. For the same angular velocity, the larger the radius, the faster the tip

FIGURE 5.5 Savonius wind turbine (5 kW, each rotor was 3 m high by 1.75 m diameter) test at Kansas State University. (Photo courtesy of Gary Johnson.)

of the blade is moving. However, for the same tip speed ratio, an increased rotor size will result in fewer revolutions per minute (rpm) by the rotor. That is why small diameter rotors complete many revolutions per minute and large diameter rotors make few revolutions.

The torque–rpm relationship also explains why drag devices are not used to produce electricity. Drag devices have larger torques; the low rpm rate means the amount of power is low. Too many inventors of drag devices equate torque with power.

Based on conservation of energy and momentum, the maximum theoretical efficiency for the capture of wind power and wind energy is 59%. The highest experimental efficiencies for systems converting wind energy to electricity are around 50%.

Lift and drag forces of airfoils are measured experimentally in a wind tunnel as functions of the *attack angle* (angle of the relative wind to the chord of the airfoil; Figure 5.11). Lift is perpendicular and drag is parallel to the relative wind. The horizontal component of the lift on the blades that depends on the angle of attack makes the rotor turn about the axis (Figure 5.12). The relative wind experienced by the blade has two parts: the vector sum of the motion of the blade and the motion of the ground wind far away from the unit.

Maximum power output for any wind speed can be obtained by letting the revolutions per minute of the rotor for fixed-pitch operation increase as the wind speed increases or by changing the pitch of the blades to obtain the correct attack angle for constant rpm operation. A fixed-pitch blade or constant rpm rotor only reaches maximum power coefficient at a single wind speed. The *power*

FIGURE 5.6 Horizontal axis wind turbine, 10 m diameter, 25 kW rating and Darrieus vertical axis wind turbine, 17 m diameter, 100 kW, at U.S. Department of Agriculture's Agricultural Research Service installation in Bushland, Texas.

coefficient is the power output of the wind turbine divided by the power input (power in wind across the rotor area). Although rotor efficiency decreases above the point of maximum power coefficient for fixed-pitch blades, the power output of the wind turbine can remain high because available power increases as the cube of the wind speed.

Computer programs are available for estimating the aerodynamic performances of both HAWTs and VAWTs. Inputs include airfoil lift and drag versus attack angle, radius, twist and pitch of blades, and solidity. Wind speeds or tip speed ratios can be varied to determine power, forces, moments, and other parameters for each blade section and for total blades.

The theoretical values of torque versus rpm were calculated for a VAWT for constant values of wind speed (Figure 5.13). The design point was selected as a rated wind speed of 12.5 m/sec. The number of blades, airfoil, and other parameters were selected for a low-solidity rotor. Each point on the curves is an operating point (power) along lines of constant wind speed. Wind turbines can be operated at constant tip speed ratios (line B, maximum power coefficient), constant rpm (line A), or constant torque (line C). As noted, the rpm is variable along line B, which is the operation of maximum power coefficient. However, at some point the wind contains too much power, and the wind turbine is controlled to capture less power and in cases of very high winds to shut down.

Notice that the constant torque operation soon reaches very high values of rpm, so the wind speed range of operation is limited. For constant torque loads, high torque is necessary for start-up. Therefore, it is very difficult to connect a constant torque load to a wind turbine and obtain much efficiency. The other side of that is that high-solidity rotors like farm windmills have high starting torques at low winds and tip speed ratios around 1 and are thus too inefficient for generating electricity.

Wind Turbines

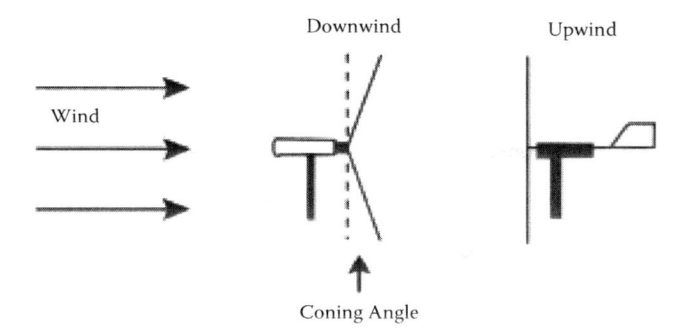

FIGURE 5.7 Diagrams of downwind unit with coning, passive yaw control, and upwind unit with tail for yaw control. Photos of Enertech downwind turbine (6.5 m diameter, 5 kW) and Hummingbird upwind turbine (6 m diameter, 5 kW) at Alternate Energy Institute Wind Test Center, Canyon, Texas.

5.6 CONTROL

Because wind power increases so rapidly, all wind turbines must have ways to dump power (not capture power) at high wind speeds. The methods of control are:

1. Change aerodynamic efficiency
   a. Variable pitch, feather, or stall
   b. Operate at constant rpm
   c. Spoilers

FIGURE 5.8 Giromill, 40 kW (rotor dimensions: 18 m diameter, 12.8 m height) at National Wind Technology Center of National Renewable Energy Laboratory.

    2. Change intercept area
        a. Yaw rotor out of wind
        b. Change rotor geometry
    3. Brake
        a. Mechanical or hydraulic
        b. Air brake
        c. Electrical (resistance, magnetic)

All these methods have been used alone or in combination for control in high wind speeds and for loss of load control. Rotor geometry was changed for two vertical-axis wind turbines. One rotor was a V shape that became flatter in high winds, and the other was a two-bladed giromill whose rotor geometry changed from an H shape to a ↔ shape. A blade was designed so that its length could change as the outer part moved into the rest of the blade.

For control in high winds, most small wind turbines and farm windmills have tails to yaw the wind turbine out of the wind (*furling* the rotor). For high speed wind control, the rotors of some wind turbines are rotated about the horizontal axis rather than yawed about the vertical axis. The results are the same; the intercept area is decreased.

A pitch control system is one method to control rpm, start-up (need high torque), and overspeed. Blades are in the *feather* position (chord parallel to wind) during shutdown. When the brake

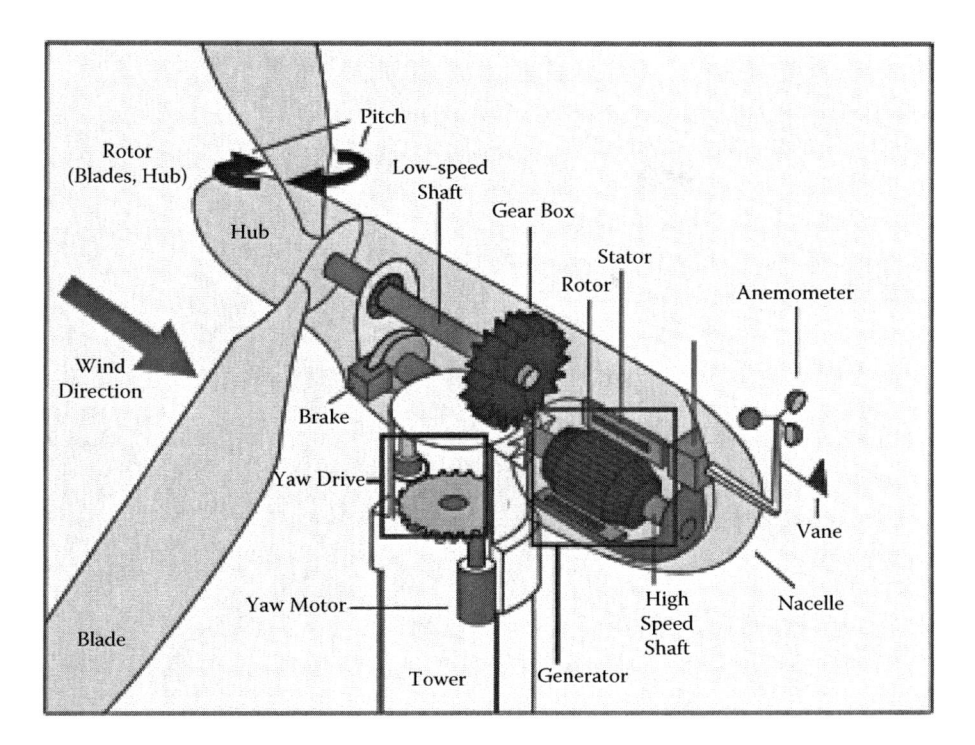

FIGURE 5.9 Major components of large wind turbine.

is released, the feather position provides starting torque, and the pitch is changed to the run position (pitch angle around 0 degrees) as revolutions increase. The blades are kept at the same pitch (run position) over a range of wind speeds. For high wind speeds and over-speed control, the blades are moved to the feather or *stall* position (blades perpendicular to wind, negative pitch) to shut the unit down. The pitch can be changed to maintain constant rpm for synchronous generators. For an induction generator, variable-speed generator, or alternator that operates over a range of rpm in the run position, the tip speed ratio over this range is constant and the unit operates at higher efficiency.

Fixed-pitch blades allow two possible operations: (1) constant tip speed ratio (variable rpm) providing maximum efficiency and (2) constant rpm. The blade must have enough twist to produce torque for start-up, or the induction motor or generator can start the rotor at the cut-in wind speed. Constant rpm operation with induction generators means that the maximum efficiency is reached only at the design wind speed. Above rated power, the output is controlled by the reduced aerodynamic efficiency (stall control).

Part of the control system can be electronic, generally a microprocessor or microcomputer (Figure 5.14). In constant rpm operation with an induction generator, the unit is connected to the utility line after the rpm exceeds the synchronous rpm of the generator. In reality, an induction generator does not operate strictly at constant rpm due to a small change in rpm (slip) with power output. Doubly fed induction generators have large rpm ranges, around 50%, and are used because of the increased aerodynamic efficiency of large wind turbines with blades in the run position.

5.6.1 NORMAL OPERATION

A *power curve* (*plot of* power versus wind speed) describes the normal operation of a wind turbine (Figure 5.15). Notice that the difference in power output at low wind speed is due

FIGURE 5.10 Suzlon, 64 m diameter, 1000 kW, induction generator, 4/6 pole. Winergy cutaway gearbox.

to differences in the electric efficiency of the generators. At the cut-in wind speed, the unit starts to rotate or produce power, then reaches rated power (based on size of generator) at the rated wind speed and continues to produce that power until the unit shuts down at the cut-out wind speed. Some wind turbines with fixed-pitch blades and induction generators continue to operate at any wind speed. Above the rated wind speed, the power output is constant or even decreases somewhat because of the decreasing aerodynamic efficiency with increasing wind speed.

The most important parameter in determining energy production is the rotor area because energy production will increase as the square of the radius. A larger generator does not necessarily mean more energy production because the efficiency at low wind speeds will change with generator size. Some large wind turbines have two generators and utilize the smaller generator at lower wind speeds to increase overall efficiency. Although a larger generator is probably desirable

Wind Turbines

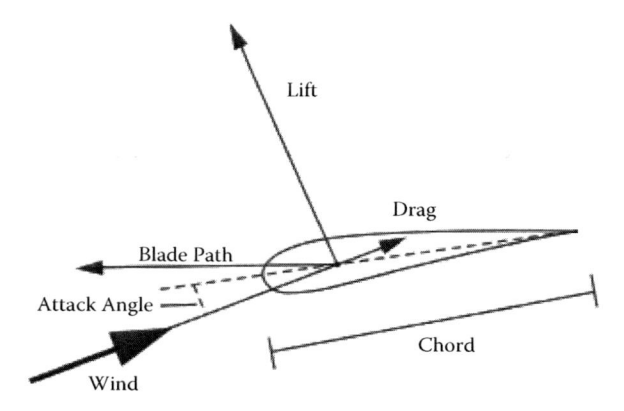

FIGURE 5.11 Forces on blade and lift and drag from airflow of relative wind.

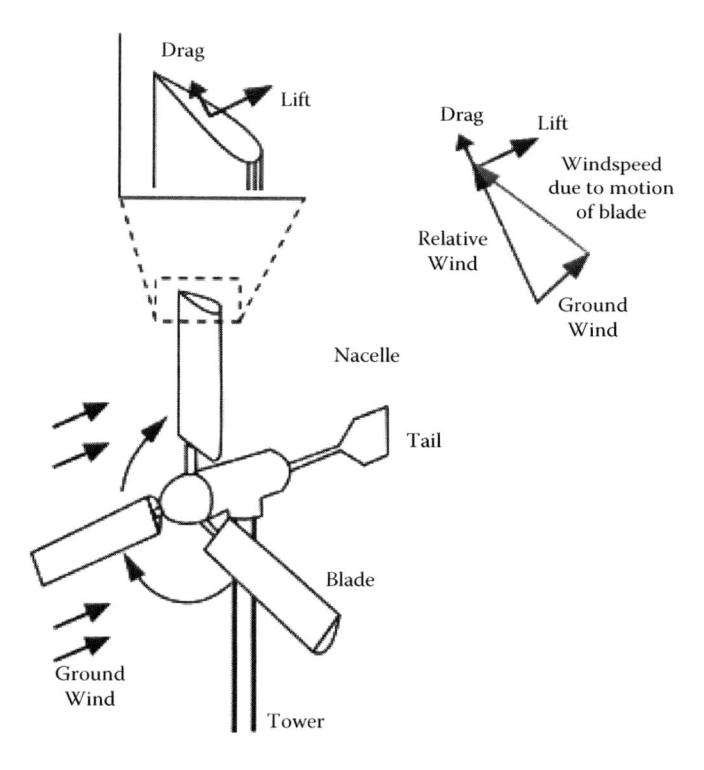

FIGURE 5.12 Wind produces forces on blade. Relative wind (experienced by blade) is the vector sum of blade speed plus ground wind. Rotor is perpendicular to ground wind.

in the best wind regimes, the optimum size for a rotor radius for a specific wind regime is still undetermined.

Manufacturers now offer different size generators (with different power ratings) for the same rotor diameter or the same size generator for different rotor diameters. Jay Carter, Sr. designed and built a wind turbine for both medium and good wind regimes. The adjustment is made simply by changing the sizes of the induction generators (30 kW, six poles and 50 kW, four poles).

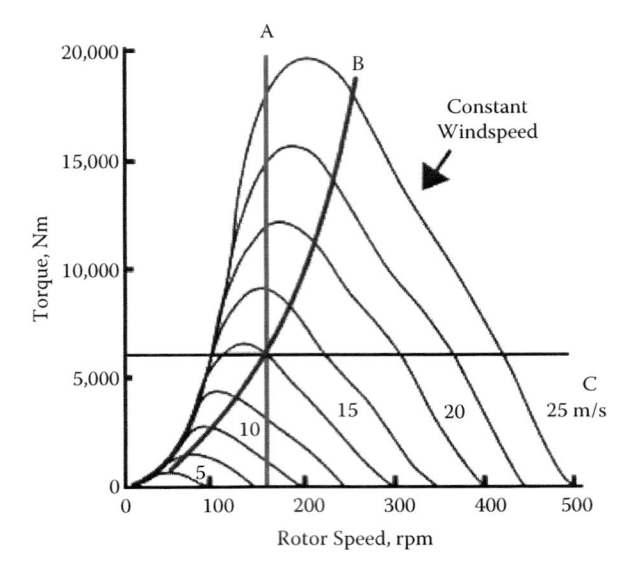

FIGURE 5.13 Theoretical curves of torque versus rpm for different wind speeds.

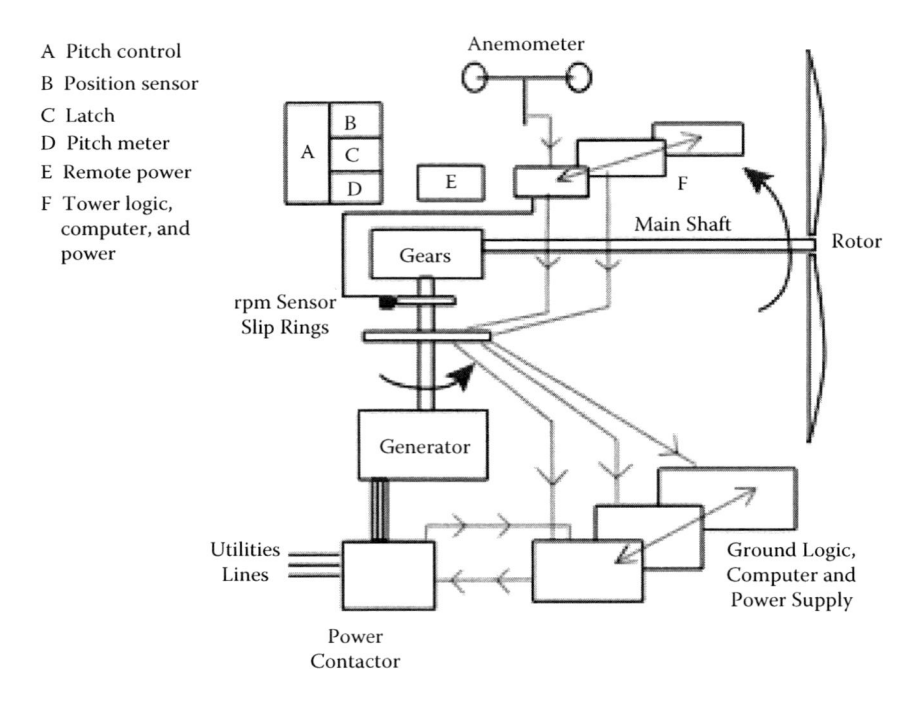

FIGURE 5.14 Block diagram of system with pitch control.

5.6.2 Faults

Wind turbines are shut down for faults such as loss of load, vibration, loss of phase, and current or voltage anomalies. Each of these safety features could save the unit, but the most important feature is a method of controlling the rotor during a loss of load (fault on the utility grid) during high winds (over-speed control). If the unit is not shut down within a few seconds, it will reach such high

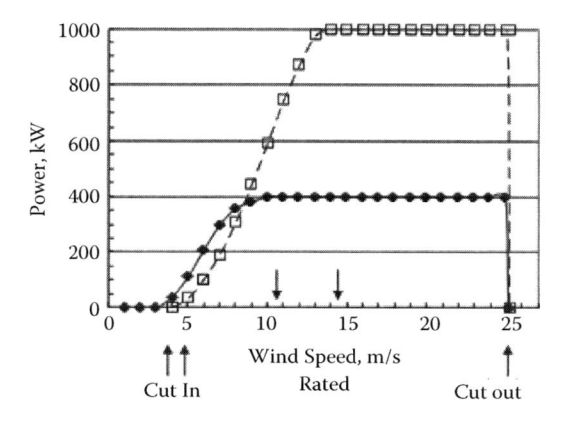

FIGURE 5.15 Power curves of rotor with two different generator sizes.

power levels that it cannot be shut down and will self-destruct. The large torque excursions and also the emergency application of mechanical brakes may damage the gearbox. Faults result in power spikes, large current, and voltage drops.

5.7 ENERGY PRODUCTION

Annual energy production is the most important factor for wind turbines. Of course, production is combined with economics to determine feasibility for installation of wind turbines and wind farms. Approximate annual energy can be estimated by the following methods:

1. Generator size (rated power)
2. Rotor area and wind map
3. Manufacturer's curve of energy versus annual wind speed

5.7.1 GENERATOR SIZE

This method gives a rough approximation because wind turbines with the same sized rotors may have different size generators:

$$AEP = CF \times GS \times 8{,}760 \tag{5.2}$$

where AEP = annual energy production, kWh/year; CF = capacity factor; and 8,760 = number of hours in a year.

The effect of the wind regime and the rated power for the rated wind speed can be estimated by changing the capacity factor. The *capacity factor* is the average power divided by the rated power (generator size). The capacity factor is estimated from energy production over a selected period, and in general, capacity factors are quoted on an annual basis, although some are calculated for a quarter of a year.

Capacity factors can also be calculated for wind farms, and they should be close to the capacity factors calculated for individual wind turbines. However, if a wind farm is composed of different wind turbines, the differences should be considered. For example, the Green Mountain Wind Farm at the Brazos near Fluvana, Texas, has 160 1-MW wind turbines; however, 100 have rotor diameters of 61.4 m and 60 have rotor diameters of 56 m. Therefore, the capacity factor will be larger for the units with the larger rotors.

Notice that capacity factor is like an average efficiency. In general, the generator size method gives reasonable estimates if the rated power of the wind turbine is around 11 to 13 m/sec. If the rated power is above that range or a wind regime is below class 3, the capacity factor should be reduced accordingly.

Example 5.1

Wind Turbine Specifications

Rated power = 25 kW at 10 m/sec
Rotor diameter = 10 m
Estimated capacity factor = 0.25

AEP = 0.25 × 25 kW × 8,760 hr/year = 55,000 kWh/year

For a poor wind regime, AEP would be closer to 30,000 kWh/year.

A capacity factor of 0.25 would suffice for a generator rated at a wind speed of 10 m/sec for a wind turbine in a medium wind regime. Wind farms are located in good to excellent wind regimes and capacity factors should be 32 to 45%. Capacity factors up to 50% were reported for a wind farm located in the Isthmus of Mexico and the Wildorado Wind Ranch near Amarillo, Texas has a capacity factor of 45%.

5.7.2 Rotor Area and Wind Map

The amount of energy produced by a wind turbine primarily depends on the rotor area, also referred to as cross-sectional area, swept area, or intercept area. The swept areas for different types of wind turbines can be calculated from the dimensions of the rotor (see Figure 1.10).

HAWT: πr^2, where r = radius.
VAWT: H = height and D = diameter of rotor:
   Giromill: $H \times D$
   Savonius: $H \times D$
   Darrieus: $0.65\, H \times D$

The annual average power/area can be obtained from a wind map, and the energy produced by the rotor can be calculated from

$$AEP = CF \times Ar \times WM \times 8.76 \qquad (5.3)$$

where Ar is the area of the rotor (m²); WM is the power/area from a wind map (W/m²); and 8.76 is the conversion factor that yields an answer in kWh/year. Again, the capacity factor reflects the annual average efficiency of a wind turbine, around 0.20 to 0.35.

Example 5.2

Use the wind turbine in Example 5.1, and from the wind map:

WM = 200 W/m²

Area = πr^2 = 3.14 × 25 m² = 78.5 m²

AEP = 0.25 × 78.5 m² × 200 W/m² × 8.76 kWh/year = 34,000 kWh/year

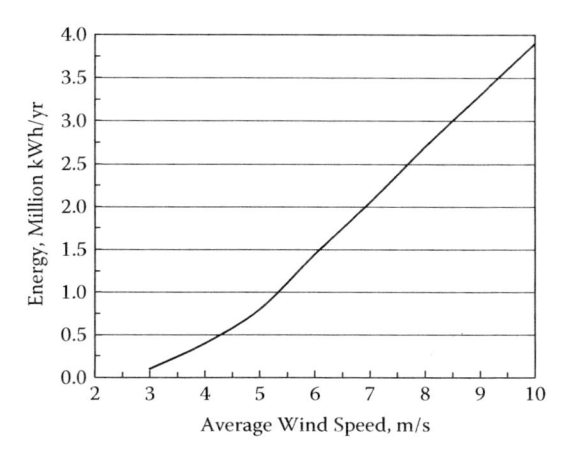

FIGURE 5.16 Estimated annual energy production based on annual average wind speed.

Notice the large difference in the answers for the two examples. The difference may arise from two factors: (1) generator size is too large for rotor size or (2) the wind regime is low, that is, the wind map value is low. With this estimate of energy production, the wind map value should be selected or estimated for the hub height of the wind turbine, especially when estimating energy production for large wind turbines.

5.7.3 Manufacturer's Curve

Manufacturers assume a Rayleigh distribution for a wind speed at 1 m/sec intervals and then calculate the annual energy production at standard density using the power curve for their wind turbines at a selected hub height. An example graph of the annual energy production versus average wind speed is given for a 1-MW wind turbine (Figure 5.16).

Notice the average wind speed at a location should be somewhat close to the hub height. At 10-m height, the average wind speed was around 6 m/sec for the High Plains of Texas (1,100 m elevation), and at 50 m height, the wind speed was 8.2 m/sec. Based on the graph, a wind speed of 8.2 m/sec means a turbine should produce around 2,800,000 kWh/year.

5.8 CALCULATED ANNUAL ENERGY

If the wind speed histogram or wind speed distribution is known from experimental data, a good estimation of energy production can be calculated from the histogram and the power curve for a wind turbine. Manufacturers supply power curves for their wind turbines, and most of the power curves are available online. For each interval (a bin width of 1 m/sec is adequate), the number of hours at that wind speed is multiplied by the corresponding power to find the energy. These values are added together to find the energy production for the total number of hours (Table 5.1).

This is the method that wind farm developers use to estimate the energy production. Wind speed histograms should reflect annual values, not the value for part of a year or even one year, which could be above or below the annual values. A one-year histogram could be adjusted to annual values if long-term regional data are available. Two to three years of wind speed data, averaged to an annual histogram, will suffice.

Wind speed histograms and power curves must be corrected for height and adjusted for air density due to location of the data compiled for the power curve. When a density correction is made from 1.2 to 1.1 kg/m^3 for the Texas Panhandle and an availability of 98% is assumed, that reduces production of 3,061,000 kWh/year to 2,750,000 kWh/year.

TABLE 5.1
Calculated Annual Energy Production for 1 MV Wind Turbine in the Panhandle of Texas

Wind Speed (m/sec)	Power (kW)	Bin (hr)	Energy (kWh)
1	0	119	0
2	0	378	0
3	0	594	0
4	0	760	171
5	34	868	29,538
6	103	914	94,060
7	193	904	174,281
8	308	847	260,760
9	446	756	337,167
10	595	647	384,658
11	748	531	396,855
12	874	419	366,502
13	976	319	311,379
14	1000	234	233,943
15	1000	166	165,690
16	1000	113	113,369
17	1000	75	74,983
18	1000	48	47,964
19	1000	30	29,684
≥20	1000	40	39,540
25	0		0
		8760	**3,060,545**

Availability is the time that a wind turbine is in operational mode, and it does not depend on whether the wind is blowing. Availability relates to the reliability of a wind turbine and reliability is affected by the quality of the turbine and operation and maintenance. Experimental values of availability of wind turbines in the field were poor for first production models. but availabilities of 98% have been reported for later units that have good programs of ongoing maintenance. Remember, a wind turbine does not have problems when the wind is not blowing. Therefore, preventive maintenance is imperative to maintain energy production.

Calculation of estimated energy production is simple using spreadsheets or by writing a program to perform the calculation from a histogram and power curve. The data will be in tabular form and can be graphed using spreadsheets or generic plot programs.

5.9 INNOVATIVE WIND POWER SYSTEMS

Innovative or unusual wind systems (Figure 5.17) must be evaluated in the same way as other wind turbines. The important parameters are system performance, structural requirements, and quantities and characteristics of materials. Innovative ideas include tornado types, tethered units to reach the high winds of the jet stream, tall towers that use rising air, tall towers for humid air, torsion flutter, electrofluid, diffuser augmented, Magnus effect, and other systems. Many of these have been reported in *Popular Science* [2–4]. Most innovative concepts remain at the experimental or feasibility stage and not all are recent inventions. For example, sail wings, wings on railroad cars and the Magnus effect (Madaras' concept of rotating cylinders on railroad cars) have been around a long time.

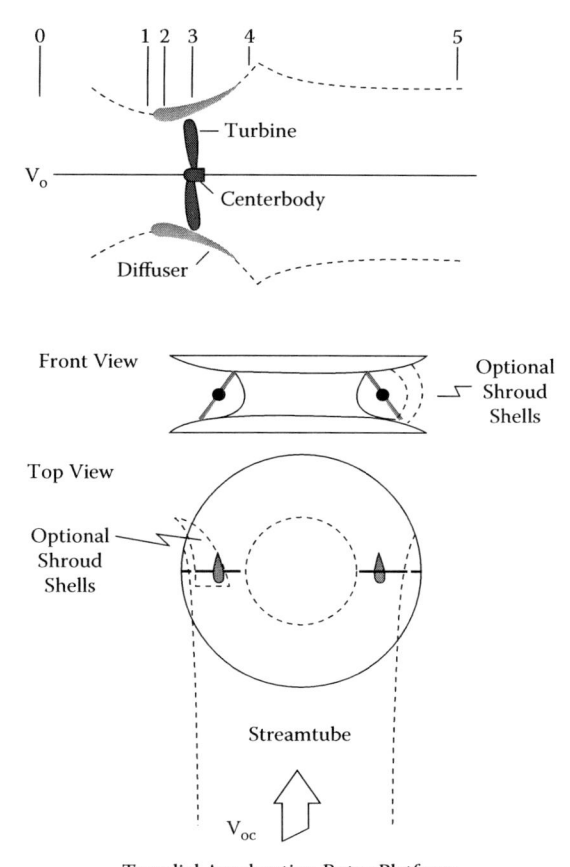

Wind Turbines

FIGURE 5.17 Examples of innovative wind turbines. Tower with greenhouse at bottom.

The West German government funded the construction of a 200-m tall tower in Spain in the 1980s [5]. A 240-m diameter greenhouse at the bottom provided the hot air to drive the air turbine rated at 75 kW and located inside the tower. A private entrepreneur in California [6] constructed a Magnus type wind turbine 17 m in diameter with a purported rated capacity of 110 kW (Figure 5.18). The unit was later moved to the wind test site of Southern California Edison located in San Gorgonio Pass. A small wind turbine with spirals on the cylinders was built. (Figure 5.19). A built-in motor spins the cylinders when wind makes the rotor rotate due to the Magnus force on the cylinders. The unit is 11.5 m in diameter and rated power is 12 kW.

The most different concept is the electrofluid unit that has no moving mechanical parts. The wind carries the moving charge to generate electricity for a load. A somewhat similar device consists of a balloon covered with a thin conductive layer. Static electricity generated by wind friction would be conducted through a cable to the surface [7]. Oscillations of piezoelectric polymers driven by the wind would also make a unique type of wind turbine. One idea was to place such devices along highways to use the turbulent wind generated by passing trucks and cars.

The Solar Energy Research Institute (SERI), later renamed the National Renewable Energy Laboratory (NREL), was the lead agency in innovative concepts (Table 5.2), and reports on projects funded by SERI are available in conference proceedings [8–10]. The U.S. Department of Energy (DOE) discontinued funding for this program after a few years.

FIGURE 5.18 Magnus effect wind turbine at Southern California Edison test site.

FIGURE 5.19 Spiral Magnus wind turbine (11.5 m diameter, 12 kW). Model shows spiral (helix fins) on cylinders. (Photos courtesy of Mecaro Japan and Charlie Dou.)

Winglets or tips (dynamic inducers) on the ends of the blades [11] that reduce the drag from the tip vortex were tested by Aerovironment and the University of Delft, The Netherlands. The results were inconclusive due to the variability of the wind speeds. In some cases, energy production could be improved, but the cost of the winglets could be offset by increasing the radius of the blades. Where wind speed variability is not a major factor, winglets can reduce drag and increase lift, as they do on some airplanes.

TABLE 5.2
Solar Energy Research Institute, Innovative Wind Program

Project	Contract
Innovative wind turbines (VAWT)	West Virginia University
Tornado type wind energy system	Grumman Aerospace
Diffuser-augmented wind turbine	Grumman Aerospace
Wind/electric power-charged aerosol	Marks Polarized
Electrofluid dynamic wind generator	University of Dayton
Energy from humid air	South Dakota School, M&T
Madras rotor power plant, phase I	University of Dayton
Vortex augmenters	Polytechnic Institute, New York
Yawing wind turbine, blade cyclic pitch	Washington University, St. Louis
Oscillating vane	United Technologies
Dynamic inducer	AeroViroment

A simple sail wing consisting of a pipe spar and a trailing cable was designed and built by Sweeney [12]. The advantages are light weight and ease of repair. The patent rights were purchased by Grumman. The company built a couple of prototypes but never put the unit into production. WECS Tech installed a number of sail wing units on a wind farm in Texas and others on wind farms in California. The operating history was very poor, as high winds destroyed the sails within a short time. The same sail wing design was used on a prototype project by the Instituto de Investigación Electricas in Mexico.

The idea of a confined vortex (tornado) was invented by T.J. Yen. The U.S. Department of Energy (DOE) funded theoretical and model studies of the concept. Another concept was using unconfined vortices produced along the edges of a delta wing and placing two rotors at those locations. Again, DOE funded model studies. Existing structures could be modified or new buildings could incorporate features to increase wind speed that would be captured by a wind energy conservation system (WECS). Since wind speed increases with height, a large amount of energy could be obtained by placing rotors in low-altitude jets by use of tethered balloons or airfoils.

Another idea is to use lift translators with horizontal or vertical axes. This concept is similar to the idea of railroad cars with wings, except that cables hold the sails or airfoils and the wind turbine resembles a moving clothes line. Both concepts need wind from a predominant direction because the large units cannot be oriented. A number of foundations were constructed, and a few lift translators were built during the early 1980s in California, but they were never really operational.

An idea for reducing weight was to use cables to provide tension to support long cage-containing blades—similar to the use of cables on suspension bridges. An oscillating vane or airfoil could extract energy from the wind, but the intercept area is fairly small for the amount of material required.

Numerous designs and several wind turbines utilizing different combinations and unusual blade shapes were built. A few examples are Darrieus or giromill wind turbines with Savonius rotors on the inner shafts for start-up torque (Figure 5.20), wind turbines with double rotors (some rotors close together, some farther apart), multiple vertical or horizontal rotors on a single shaft, double-bladed giromills, and blades with nontraditional shapes (e.g., helical curves) on horizontal or vertical axes. A wind system with three stacked Darrieus units (4 kW each) was built at a newspaper office in Florida. Other units utilized enclosures to increase wind speed or were designed to be incorporated into tall buildings.

Because of the stronger and more consistent winds at higher altitudes and thus better energy production, prototypes of tethered systems were tested in recent years. One system used a helium filled balloon (Figure 5.21) while others used kites, propellers, and wings and propellers (Figure 5.22).

FIGURE 5.20 Wind–photovoltaic street light. Darrieus rotor (1 m diameter) and Savonius device for starting (wind = 300 W; photovoltaic = 120 W).

The Noah wind turbine had two five-blade rotors (Figure 5.23) placed close to one another. The wind rotors counter-rotated, with one connected to the stator and the other to the rotor of a generator so no gearbox was needed. The wind turbine had a unique over-speed control consisting of a counterweight that tilted the rotor assembly to the horizontal position and then had to be reset manually.

Another system utilized multiple rotors on a coaxial shaft [13]. The line of the rotors was kept at an angle to the wind to improve influx of the wind to the downwind rotors. Units with two to seven rotors have been built (two and three blades) with rated power from 2 kW (2.4 m diameter, two rotors, 3.7 m apart) to 4 kW. One unit contained thirteen two-bladed rotors, each with a diameter of 0.5 m, and rated at 400 W. For a number of rotors on a single shaft, the almost ultimate wind turbine is the Sky Serpent (Figure 5.24).

Lagerway built a unit with two conventional wind turbines (25 kW each) mounted on a horizontal cross-beam at the top of a tower. The company also built another that resembled a tree. It had two more levels accommodating a total of six wind turbines.

A shroud, diffuser, duct, or even a building can increase wind flow and thereby increase the output of a wind turbine [14]. Prototypes have been built and tested and a few models are available on the market (Figure 5.25). It is generally cheaper to use longer blades than to construct a shroud or modify a building to enhance wind speed. Also, should wind turbines on a building be fixed in place or able to rotate around the structure to the highest wind speed locations?

A helical structure instead of a toroidal acceleration rotor platform (Figure 5.17) was developed at Cleveland State University [15]. The structure (Figure 5.26) has four fixed wind turbines and the whole structure is rotated in yaw to keep the turbines perpendicular to the wind. An aluminum frame (1.4 metric tons) is covered with white plastic pieces to form a helix, and the wind system rises around 12 m above the upper concourse of Progressive Field, the Cleveland Indians' ball park. The unit is expected to generate around 40,000 kWh per year.

FIGURE 5.21 Prototype 105 m diameter tethered balloon filled with helium. Skystream wind turbine in center (3.7 m diameter, 2.1 kW. (Photo courtesy of Altaeros Energies.)

5.10 APPLICATIONS

The kinetic energy of the wind can be transformed into mechanical, electrical, and thermal energy. Historically, the transformation was mechanical, and the end use was grinding grain, powering ships, and pumping water [16,17].

The applications can be classified as wind-assist and stand-alone systems. A wind-assist system, the wind turbine works in parallel with another source of energy to provide power. The advantages of such systems are that power is available on demand, no storage is required, and power source and load are better matched. Stand-alone systems provide power only when the wind is blowing and the output is variable unless a storage system is connected. In a wind–diesel application is a wind-assist system. The wind turbine is primarily a fuel saver. Another emerging application is a hybrid system for powering villages and telecommunications equipment.

5.10.1 ELECTRICAL ENERGY

Most wind turbines are designed to provide electrical energy. In a wind-assist system, wind turbines are connected to the utility line directly through induction generators and synchronous generators or indirectly by variable-frequency alternators and DC generators connected through inverters. The utility line and generating capacity of a power station act as a storage system. For stand-alone systems, battery storage is the most common option.

FIGURE 5.22 Prototype tethered wing with four wind turbines (0.75 m diameter, 30 kW, 8 m wind span). (Photo courtesy of Makani Power.)

FIGURE 5.23 Wind turbine with double rotor.

The U.S. Department of Agriculture (USDA) in Bushland, Texas, and the Alternative Energy Institute (AEI) at West Texas A&M University evaluated stand-alone electric-to-electric systems for pumping water [18]. The wind turbine generator was connected directly to an induction motor or submersible pump run at variable rpm. The advantages of such a system are higher efficiency and higher volumes of water—enough for a village water supply and low-volume irrigation. Such systems are now commercially available.

Wind Turbines

FIGURE 5.24 Doug Selsam with his Sky Serpent (25 rotors on single shaft with 3 kW generator). (Photo courtesy of Doug Selsam.)

FIGURE 5.25 Small shrouded wind turbines, rotor diameter 0.68 m, rated 200 W. Can be mounted in clusters, each rotates independently. Company has larger shrouded turbine rotor diameter 9.8 m, rated 50 kW.

5.10.2 Mechanical Energy

The major use for windmills has been the pumping of water. Farm windmills are well designed to pump small volumes of water at low wind speeds. Because a farm windmill has a large number of blades (vanes), it will start under a load because it has a large torque. However, many blades require a lot of material and the units are inefficient at high wind speeds. Power ratings are around 0.5 kW for a 5-m diameter rotor.

The Brace Research Institute combined a modern three-bladed wind turbine, a transmission from a truck, and a conventional centrifugal pump on a prototype project to pump irrigation water

FIGURE 5.26 Wind amplification turbine system at Progressive Field, Cleveland, Ohio. Height = 6.4 m, diameter = 6.7 m, inner spiral diameter = 4.6 m. Each wind turbine is 2 m in diameter, rated for 1.6 kW. (Photo courtesy of Majid Rashidi.)

on the Island of Barbados [19, 20]. The rotor was not self-starting, and the blades of fiberglass were expensive. A person had to manually shift the transmission to match the load of the pump to the output of the wind turbine at different wind speeds.

In 1976, AEI and USDA studied the feasibility of using wind turbines for pumping irrigation water with positive displacement pumps and airlift pumps. It is difficult to match the power output of the wind turbine with the power needed by the irrigation pump. Calculated maximum efficiencies were very low, on the order of 10%, for both types of pumps.

The airlift pump has the advantages of no moving parts in the well, and the wind turbine does not have to be located at the well. Airlift pumps were in use at the turn of the century for pumping water from mines, but were replaced by other types. Koenders Windmills makes an airlift pump (www.koenderswindmill.com) and two U.S. companies manufactured a wind-powered airlift pump to compete with the farm windmill; however, only Airlift Technologies has units for sale today.

For maximum efficiency, the submergence (depth of pump below water level) should be equal to the lift. Wells with little water at large depths present problems for airlift pumps. Also, there is the problem of load matching between the wind turbine and the air compressor, a constant torque device, and the inherent inefficiencies.

A wind turbine can be connected mechanically to another power source to serve as a wind-assist system for pumping water. The other power source could be an electrical motor or internal combustion engine. Both systems have been tested.

Wind Turbines

5.10.3 Thermal Energy

Thermal energy can be obtained directly by churning water or another fluid with viscosity. The load matching between the wind turbine and the churn is very good. A prototype system for providing heat to a dairy was tested by a research group at Cornell University [21–23]. Conversion of electrical to thermal energy by resistance heating has been tested a few times [24], and one company marketed such a wind system.

5.10.4 Wind Hybrid Systems

A large market exists for wind-assist to diesel-generated electricity systems for isolated communities, businesses, farms, and ranches [25]. About 2 billion people live without electricity, and hybrid systems consisting of wind, photovoltaic, hydro or diesel, battery storage, and an inverter are now part of the planning process to provide alternating current (AC) electricity for villages needing energy of 20 to 200 kWh per day [26, 27]. Hybrid systems have also been installed in very remote locations such as military facilities and telecommunication systems. For telecommunications, the emphasis is on continuous power, so redundancy is important to achieve high reliability.

NREL established a site dedicated to hybrid systems for village power. The Renewables for Sustainable Village Power (RSVP) project database covered about 150 projects (50 involved wind) in over thirty countries. Project information included basic concepts, technological requirements, economic and financial data, host country descriptions, lessons learned, pictures and graphics, and contact information. The database is now archived and is not available online.

A large number of hybrid projects have been installed since 2004. For example, China now has over 700 village installations (16-MW capacity) powered by mini hydro, photovoltaic, or wind–photovoltaic hybrid systems [28]. Alaska has 63 MW of wind–diesel systems [29] and China has also installed a few wind–photovoltaic–diesel systems [30].

5.11 SUMMARY

Applications will be considered in more detail after we learn more about design and construction of wind turbines. Electricity generation is the most used application of wind power. The problem of load matching when pumping water for irrigation must be a design consideration.

LINKS

Altaeros Energies (tethered wind). www.altaerosenergies.com
Bergey Windpower (wind–CAD performance models; spreadsheet). http://bergey.com/technical
Global Village Energy Partnership International. www.gvepinternational.org
Kite Gen (tethered wind). www.kitegen.com
Makani Power (tethered wind). www.makanipower.com
Sky Windpower (tethered wind). www.skywindpower.com
Twind (tethered wind). www.twind.eu/wp2/

REFERENCES

1. E. F. Lindsey. 1974. Windpower. *Popular Science*, July, 54.
2. B. Kocivar. 1977. Tornado turbine. *Popular Science*, January, 78.
3. V. Chase. 1978. Thirteen wind machines. *Popular Science*, September, 70.
4. J. Schefter. 1983. Five wild windmills. *Popular Science*, June, 76.
5. B. Juchau. 1983. A 650-foot power tower. *Popular Science*, July, 68.
6. J. Schefter. 1983. Barrel-bladed windmill: power from the Magnus effect. *Popular Science*, August, 70.

7. G. Lorente. 1982. Nuevo concepto de generador elctro-eólico. *Metalurgia y Elctricidad*, 532, 51.
8. SERI (Solar Energy Research Institute). 1980. In *Proceedings of Second Wind Energy Innovative Systems Conference, Vols. I and II*, pp. 635–638 and 938–1051.
9. SERI (Solar Energy Research Institute). 1981. In *Proceedings of Fifth Biennial Wind Energy Conference and Workshop, Vol. I*, p. 415.
10. American Solar Energy Society. 1983. In *Proceedings of Sixth Biennial Wind Energy Conference and Workshop*.
11. D. Scott. 1983. Tip-vane windmill doubles output efficiency. *Popular Science*, September, 78.
12. S. Kidd and D. Garr. 1972. Electric power from windmills? *Popular Science*, November, 70.
13. D. Selsam. www.selsam.com.
14. Y. Ohya, T. Karasudani, T. Nagai et al. 2011. Development of shrouded wind turbines with wind-lens technology. Poster presented to European Wind Energy Association. http://www.riam.kyushu-u.ac.jp/windeng/img/aboutus_detail_image/EWEA2011_poster.pdf
15. Cleveland State University. Wind Amplification Turbine System. www.csuohio.edu/engineering/wind/research/
16. V. Nelson, N. Clark, and R. Foster. 2004. *Wind Water Pumping*. CD, Alternative Energy Institute, West Texas A&M University. www.windenergy.org
17. V. Nelson, N. Clark, R. Foster et al. 2005. *Bombeo de agua con energía eólica*. CD, Alternative Energy Institute, West Texas A&M University. www.windenergy.org
18. Agricultural Research Service, U.S. Department of Agriculture. www.cprl.ars.usda.gov; Alternative Energy Institute. www.windenergy.org
19. R.E. Chilcott and E.B. Lake. 1966. Proposal for the establishment of a 10-hp windmill water pumping pilot plant in Nevis, West Indies. Brace Research Institute Publication I.45.
20. T. A. Lawand. 1968. The evaluation of a windmill water pumping irrigation system. Brace Research Institute Publication I.58.
21. W.W. Gunkel et al. 1981. *Wind energy for direct water heating*. Report DOE/SEA-3408-20691/81/2 available from NTIS.
22. D.H. Lacey and W.W. Gunkel. 1980. Operating performance and observed performance of a SEWCS for direct water heating. In *Proceedings of National Conference*, V. Nelson, Ed., p. 96.
23. M. Rolland and D. Cromack. 1980. Wind-driven fluid devices for water heating. In *Proceedings of National Conference*, V. Nelson, Ed., p. 93.
24. M. Edds. 1980. UMASS wind furnace performance and analysis. In *Proceedings of National Conference*, V. Nelson, Ed., p. 142.
25. R. Hunter and G. Elliot, Eds. 1994. *Wind–Diesel Systems: A Guide to the Technology and Its Implementation*. Cambridge, UK: Cambridge University Press.
26. L. Flowers et al. 1993. Decentralized wind electric applications for developing countries. In *Proceedings, of Windpower Conference*, p. 421.
27. L. Flowers et al. 2000. Renewables for sustainable village power. In *Proceedings of Windpower Conference*. CD; also available at RSVP website.
28. C. Dou and J. Graham, Eds. 2005. *China Village Power Project Development Guidebook*. CD, in Chinese and English.
29. Alaska Center for Energy and Power. www.uaf.edu/acep/
30. C. Dou, Ed. 2008. Capacity building for rapid commercialization of renewable energy in China. Beijing: Chemical–Industrial Press.

PROBLEMS

1. Estimate the difference in the amount of material in the rotor for a giromill and a Savonius rotor with H = 10 m, D = 10 m.
2. A wind turbine is rated at 300 kW. Estimate annual energy production using the generator size method.
3. For a 1.5-MW wind turbine, estimate the annual energy production for a good site using the generator size method.
4. For a conventional HAWT, radius of 50 m, estimate annual energy output for a good wind region (use class 4, 5, or 6) from the U.S. wind power map.

Wind Turbines

5. For a Darrieus unit, 34 m diameter by 42.5 m height, estimate annual energy output for two different regions from the European wind map.
6. For a giromill, H = 10 m, D = 12 m, estimate annual energy output for two different regions from the U.S. wind power map.
7. From manufacturer's curve (use Figure 5.16) for annual energy, estimate the annual energy production for a region where the average wind speed is 9 m/sec.
8. Calculate the power from Figure 5.13 at 20 m/sec for the VAWT for the following conditions (remember to convert rpm to rad/sec).
    a. Wind turbine operating at 160 rpm (line A).
    b. Wind turbine operating at maximum power coefficient (line B).
    c. Wind turbine operating at constant torque (line C) of 6,000 Nm.
9. From Figure 5.13, the design wind speed is 12.5 m/sec (where lines A, B, and C cross). What is the torque? What is the rpm measure? What is the power?
10. Calculate the wind speed frequency distribution for the data in the table below.

Wind Speed (m/sec)	Power (kW)	Bin (hr)	Energy (kWh)	Distribution	Frequency
1	0	119	0		0.014
2	0	378	0		0.043
3	0	594	0		0.068
4	0	760	171		0.087
5	34	868	29,538		0.099
6	103	914	94,060		0.104
7	193	904	174,281		0.103
8	308	847	260,760		0.097
9	446	756	337,167		0.086
10	595	647	384,658		0.074
11	748	531	396,855		0.061
12	874	419	366,502		0.048
13	976	319	311,379		0.036
14	1000	234	233,943		0.027
15	1000	166	165,690		0.019
16	1000	113	113,369		0.013
17	1000	75	74,983		0.009
18	1000	48	47,964		0.005
19	1000	30	29,684		0.003
≥20	1000	40	39,540		0.005
25	0		0		
		8760	**3,060,545**		1.00

11. Calculate the annual energy production for a mean wind speed of 8.2 m/sec and average air density of 1.1 kg/m^3. Use the Rayleigh distribution to obtain a wind speed histogram. Use the power curve from Table 5.1.
12. Refer to Figure 5.15. What are the cut-in and rated wind speeds for the 1,000-kW unit?
13. Refer to Figure 5.15. What are the cut-in and rated wind speeds for the 400-kW unit?
14. For large wind turbines, what is the primary method of control for power output?
15. For large wind turbines, what is the primary method of control for shutdown for high winds?
16. For loss of load caused by fault on the utility line, how much time is available for shutdown of the wind turbine?

17. Are there any wind hybrid systems for village power in your country? If yes, select one project; briefly describe its location, power rating, main components (wind, PV, diesel, batteries), and approximate output in kilowatt hours per day.
18. For innovative wind systems, what are two or three major problems with tethered wind turbines?
19. A wind amplification system (Figure 5.26), is expected to produce 40,000 kWh per year. Use a wind map and area of the wind energy conversion system to estimate annual energy production. Is the result larger or smaller than the estimate given? In your opinion, is the cost for the extra structure economical? Justify your answer.
20. What is the estimated increase in energy per year for tethered systems?

6 Design of Wind Turbines

6.1 INTRODUCTION

The design of wind turbines developed from a background of work on propellers, airplanes, and helicopters. Computer codes developed for analyzing aerodynamics, forces, and vibration have been modified to examine wind turbines. Theory and experimental procedures are well developed, and no scientific breakthroughs are needed. However, there are problems of predicting loads from unsteady aerodynamics. These loads lead to material fatigue and less life than predicted by the design codes. Part of the time, wind turbine blades operate in regions of large attack angles that are very different from the stresses imposed on airplane wings.

Someone made the comment that you could use brooms for turbine blades and the rotor would turn. Of course, the efficiency would be low, control would be a problem, and the strength would not be adequate. A large number of airfoils developed for wings on planes and sailplanes were later used for wind turbine blades.

In wind turbine history, aerospace engineers thought that the design and construction of wind turbines would simply involve the transfer of technical knowledge from airplanes and helicopters. However, this was an erroneous conclusion. One big difference is that airplanes and helicopters move in response to large loads from wind gusts, whereas a wind turbine is tied to the ground. Because power in the wind increases as the cube of the wind speed, the blades must have the strength and flexibility to withstand highly variable loads and must include control mechanisms for shedding power in high winds.

A lot of research and development, primarily by universities and national laboratories, and later by the manufacturers of wind turbines, created today's wind industry. The design of wind turbines requires a broad cross section of knowledge of aerodynamics, mechanical engineering, electrical engineering, electronics, materials and industrial engineering, civil engineering, and meteorology. The design process is iterative from first concept to final design, and it is easier to fix problems at the design stage than deal with retrofits in the field.

6.2 AERODYNAMICS

The analysis of aerodynamic performance begins with a disk or area in a stream flow of air. Conservation of energy and momentum are used to determine the limit on the amount of extractable energy.

Forces of lift and drag on airfoils are measured experimentally in wind tunnels. Since early measurements were made for use with airplanes, a lot of airfoil data [1] are available from national laboratories. Almost any shape can serve as an airfoil, even a flat plate, and the design of airfoils is almost an art. As wind turbine blades operate in different wind speeds than airplane wings, airfoil data with low Reynolds numbers [2] became available.

Most of the lift and drag data were limited to attack angles up to stall and a few degrees past stall (beyond a stall point, an airplane loses lift, stalls, and falls). Lift and drag data for attack angles up to 180 degrees were available only for a few airfoils. Airfoils exhibiting large ratios of lift to drag were developed for sailplanes.

Choices of airfoils for wind turbines depend on the ratio of lift to a drag and a number of other factors. Because the requirements are different for wind turbines, airfoils were designed specifically for wind turbines starting in the late 1980s. A major change was designing airfoils that were less sensitive to surface roughness.

Different theories (strip theory, circulation, vortex shedding) and experimental data on airfoils are used to predict the rotor performances of wind turbines. This theoretical performance can be checked against the measured outputs of models in wind tunnels, truck testing for small-diameter units, or field testing (atmospheric) of wind turbines. At one time, larger turbines mounted on railroad flat cars were used for controlled speed testing.

Overall efficiencies of the rotor, drive train, and energy converter (generator, etc.) must be tested. Complete analyses covering design of wind turbines, primarily rotors and structures, can be found in more advanced texts [3–13]; however, a knowledge of basic physics is useful for a qualitative understanding of rotor performance.

6.3 MATHEMATICAL TERMS

The momentum of a particle equals the mass times the velocity. Boldface indicates a vector that has both magnitude and direction. In two dimensions, two components are required to define a vector; in three dimensions, three components are required. In an analytical representation, a vector can be represented by its components along two axes (perpendicular and orthogonal for this presentation).

$$p = mv \qquad (6.1)$$

Any particle can be treated as a single particle with the mass M concentrated at a point (center of mass R). Position vector is indicated by r.

$$MR = m_1 r_1 + m_2 r_2 + \ldots + m_i r_i \qquad (6.2)$$

Forces on particles make them accelerate. Newton's second law describes the dynamics of motion; force is the change in momentum over the change in time. In other words, force is required to change the momentum of a particle. That could mean a change in speed or a change in direction of the motion of the particle. There is also a force if a change in mass occurs. Mass will be considered constant for this discussion.

$$F = \frac{\Delta p}{\Delta t}, \text{ newton (N)} \qquad (6.3)$$

Torque makes a particle turn around some point and can be thought of as the lever arm times the force. A larger torque can be obtained by increasing the length of the lever arm or increasing the force.

$$T = r \times F \qquad (6.4)$$

where the cross product means that two vectors produce a single vector whose direction is perpendicular to the plane of the two vectors.

If a mass is attached to a rod that is free to rotate about its end (Figure 6.1) and a force is applied, the torque will make the mass rotate, and power will be available. The amount of power is the product of the torque and angular velocity [Equation 95.1)] and the power is available at the shaft. Most operations of transferring shaft power utilize large ω because of structural considerations.

$$P = T\omega, \text{ W} \qquad (6.5)$$

Also, a rotating object will have rotational kinetic energy.

$$KE_{rot} = 0.5 \, m \, v^2 = 0.5 \, m \, r^2 \, \omega^2, \text{ J} \qquad (6.6)$$

Design of Wind Turbines

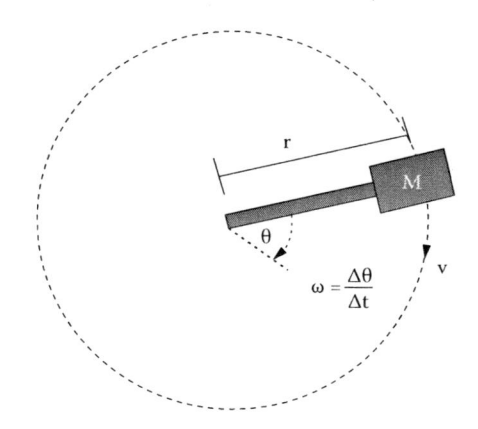

FIGURE 6.1 Mass rotating about point.

where the speed of the rotating mass depends on the radius $v = \omega \times r$. The power coefficient is the power delivered by the device divided by the power available in the wind. Since the area cancels out, the power coefficient C_P is

$$Cp = \frac{power\ out}{power\ in} = \frac{power\ out}{0.5\rho v^3} \quad (6.7)$$

The work or energy to move an object is the force times the distance through which it moves. Remember, work is a scalar (it has only value, no direction). Also note that no work (gain or loss of energy) is done if the force is perpendicular to the motion. An example is the motion of the moon around the earth.

$$W = F \cdot \Delta r = F \cdot (r_f - r_i) \quad (6.8)$$

The dot between the vectors means only the parallel component of the **F** is used ($W = F \cos\theta\ \Delta r$), where Δr = final position – initial position, and θ is the angle between **F** and **r**. We then divide both sides of Equation 6.8 by time:

$$\frac{W}{t} = \frac{F \cdot \Delta r}{t}$$

Thus, the power is

$$P = F \cdot v \quad (6.9)$$

6.4 DRAG DEVICE

The power from a drag device (Figure 5.1) can be calculated from the force on the device and the velocity of the device **u**. From Equation (6.9), $P = F \times u$ since force and speed are in the same direction. The force/area of the air on a stationary object at wind speed v is

$$\frac{F}{A} = 0.5\rho v^2 C_D \quad (6.10)$$

where C_D is the drag coefficient. Drag coefficients for different shapes are given in *Marks' Handbook* [14], but the simplest procedure is to use $C_D = 1$ for round pipes and wires and flat plates perpendicular to the wind. Flat plates at an angle to the wind will experience some lift and drag like airfoils, and C_D data are available.

The force/area is also the pressure, so the wind blowing against an object creates pressure. If the winds are high enough, as in hurricanes and tornadoes, the pressure will destroy buildings and topple trees and power poles.

Based on Equations (6.9) and (6.10), the power loss due to drag from struts can be calculated. Notice that the power loss is proportional to velocity cubed:

$$P = 0.5\rho v^2 C_D A \quad v = 0.5\rho v^3 C_D A \tag{6.11}$$

The power loss from struts for a 4-kW giromill was so large that the struts were redesigned to an airfoil shape to reduce drag. Notice that fuel efficiency for vehicles is improved by reducing the drag coefficient and decreasing speed. Automobile manufacturers are doing everything possible to decrease drag coefficients, even by small increments.

The power coefficient for a drag device can be calculated from the relative wind speed experienced by the device and the device speed. The relative velocity of the wind as measured by a sensor mounted on the drag device is $v_r = v_0 - u$ where v_0 is the wind speed and u is the speed of the device. The power per unit area from Equation (6.9) is

$$\frac{P}{A} = 0.5\rho v_r^2 C_D u = 0.5\rho(v_0 - u)^2 C_D u \tag{6.12}$$

Notice that at $u = 0$ and $u = v_0$, the power is zero. In other words, no power is output if a drag device is not moving, and a drag device cannot move faster than the wind. From Equations (6.7) and (6.12), the maximum power coefficient for a drag device can be calculated. The maximum power coefficient $C_{P(max)} = 4/27 = 0.15$ occurs when the drag device is moving at $u = 1/3$ the wind speed. This maximum power coefficient is for a drag coefficient around 1.

Some drag devices can have coefficients greater than 1 and the maximum power coefficient could be as high as 20%. The maximum power coefficient can be found using calculus or can be estimated from a spreadsheet or graph of P/A versus wind speed [Equation (6.12)] for various values of u from 0 to v_0. Low efficiency is another reason no commercial drag devices for generating electricity are available.

6.5 LIFT DEVICE

A lift device can produce about 100 times more power per unit surface area of blade than a drag device. See Rohatgi and Nelson [15, chap. 6] for more details.

Example 6.1

Suppose we have a two-blade wind turbine. Each blade is 5 m long and 0.1 m wide. As a drag device, the capture cross section is 1 m². As a lift device in a HAWT, its capture cross-sectional area is 78.5 m². If the difference in efficiencies is included, the ratio of the power out per blade area for the lift device over the drag device is over 300.

An example of a lift device is a sailboat—a lift translator (Figure 6.2) whose sails form an airfoil. Notice that a sailboat moving downwind (drag device) moves much slower than a sailboat moving perpendicular to the wind (lift device). Using lift translators to generate power has been proposed. The problems are the large speeds of the devices because lift devices can move faster than the wind,

Design of Wind Turbines   119

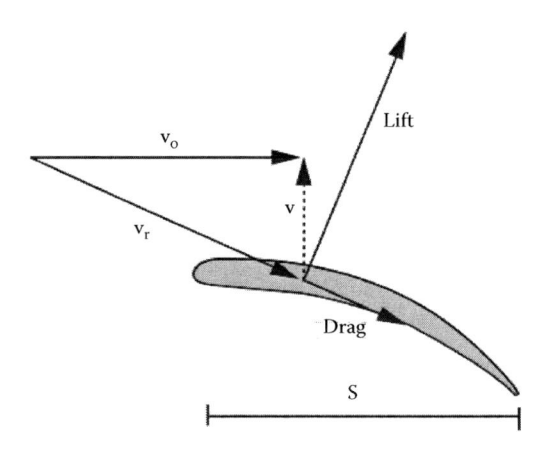

FIGURE 6.2 Lift translator. Direction of motion v is perpendicular to ground wind v_o. S is length of cross-sectional area of blade or sail.

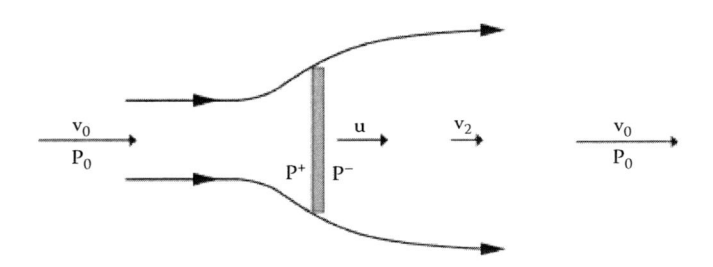

FIGURE 6.3 Wind speeds and pressures at infinity, at disk, and behind disk.

proximity to the ground, and the need for a predominant wind direction. Some lift translators were built, but never operated successfully.

The simple analysis for a lift device assumes streamline flow (irrotational, incompressible fluid) and conservation of energy and momentum. The wind speed interacts with the disk (propeller, rotor, or screw), and a pressure drop occurs across the disk (Figure 6.3). The thrust (force) loading T is uniform across the disk and no friction or drag force is present. At large distances behind and in front of the disk, the wind speed and pressure will have the same values. As stated earlier, the pressure p is the force/area.

The concept of conservation of momentum is momentum in = momentum out. The mass flow $\Delta m/\Delta t$ across any area is constant. Across the area of the disk, the mass flow is the product of air density (ρ), area (A), and wind speed; so for the three regions

$$\frac{\Delta m}{\Delta t} = \rho A_0 v_0 = \rho A u = \rho A_2 v_2$$

Using Equation (6.3):

$$T = \frac{\Delta P}{\Delta t} = \frac{\Delta m}{\Delta T}(v_0 - v_2) = \rho A u (v_0 - v_2) \qquad (6.13)$$

Also, the thrust loading on the disk due to the pressure difference across the disk is

$$T = A(p^+ - p^-) \qquad (6.14)$$

Bernoulli's theorem relates the velocity and pressures in streamline flow (kinetic energy and pressure are constants for horizontal flow). If the velocity increases, the pressure decreases; the two are related through conservation of energy and momentum. The wind speed and pressure upstream and downstream of the disk are related by:

Upstream	Disk	Downstream
$0.5\rho v_0^2 + p_0 = 0.5\rho u^2 + p^+$		$0.5\rho u^2 + p^- = 0.5\rho v_2^2 + p_0$

From the two equations, take the pressure difference $(p^+ - p^-)$ and substitute into Equation (6.14):

$$T = 0.5\rho A(v_0^2 - v_2^2) \qquad (6.15)$$

The thrusts are equal, so we set Equation (6.13) equal to Equation (6.15):

$$\rho A u(v_0 - v_2) = 0.5\rho A(v_0^2 - v_2^2) = 0.5\rho A(v_0 + v_2)(v_0 - v_2) \qquad (6.16)$$

From Equation (6.14), the wind speed at the disk is the average of the wind speeds before and after the disk (wake).

$$u = 0.5\,(v_0 + v_2) \qquad (6.17)$$

The axial interference factor is defined by the ratio to which the wind speed is reduced by the disk.

$$\alpha = \frac{v_0 - u}{v_0} = 1 - \frac{u}{v_0} \text{ or } u = v_0(1 - \alpha) \qquad (6.18)$$

Substitute into Equation (6.17) and the wake wind speed is

$$v_2 = v_0(1 - 2\alpha) \text{ or } \alpha = \frac{v_0 - v_2}{2v_0} \qquad (6.19)$$

If the disk or rotor absorbs all the energy, $v_2 = 0$ and $\alpha = 0.5$. That is physical nonsense because all the mass would pile up at the rotor. The power is equal to the change in kinetic energy from upstream to downstream: $P = \dfrac{\Delta KE}{t} = \dfrac{KE_{us} - KE_{ds}}{t} = 0.5\dfrac{m}{t}v_0^2 - 0.5\dfrac{m}{t}v_2^2 = 0.5\rho A u(v_0^2 - v_2^2)$ and the value of the axial interference factor is substituted into the equation to obtain the power/area for a lift device:

$$\frac{P}{A} = 0.5\rho v_0^3 \, 4\alpha(1 - \alpha)^2 \qquad (6.20)$$

A lift device can produce much more power per area of blade than a drag device (Figure 6.4). Notice the small black area is for the drag device that reaches a maximum around 0.22 at a speed ratio of 0.3. The maximum for the lift device is around 15 at a speed ratio 2/3 of the ratio of lift to drag coefficients. For this example, the power per area of blade was calculated for the drag device with a coefficient of 1.5. For the lift device, the ratio of lift coefficient to drag coefficient was 10. Thus, the lift device can easily produce fifty times the power/area—another reason drag devices are not used to produce electricity although a company in South Africa has a farm windmill that has an option for an electric generator.

Design of Wind Turbines

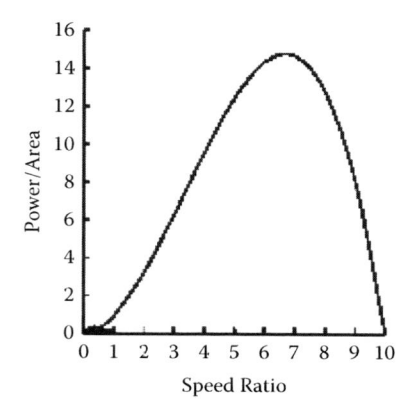

FIGURE 6.4 Comparison of power and area for a translating drag device (small solid curve) and translating lift device versus speed ratio of device to wind.

6.5.1 Maximum Theoretical Power

The maximum power/area can be found by plotting the curve P/A versus α from Equation (6.20) or by using calculus. The answer is $\alpha = 1/3$ or 1. Of course, $\alpha = 1$ means no reduction of wind speed and so the disk would not take out any power. For $\alpha = 1/3$, the maximum power is

$$\frac{P}{A} = 0.5 \rho v_0^3 \frac{16}{27} \tag{6.21}$$

The maximum power coefficient from Equation (6.6) is $C_P = 16/27 = 0.59$. Real rotors will have smaller power coefficients due to drag, tip and hub losses, losses from rotation of the wake, and frictional losses; however, measured values can reach 50% (which includes drive train and generator). This is another reason lift devices are used to generate electricity instead of drag devices; the maximum theoretical power coefficients are 50% versus 20%, respectively. However, a farm windmill that has some of the same characteristics as a drag device (large solidity, low tip speed ratio) is well designed to pump low volumes of water.

6.5.2 Rotation

Angular momentum is

$$L = r \times p \tag{6.22}$$

Angular momentum, like momentum, is always conserved. Based on the concept of conservation of angular momentum, a rotating disk will impart rotation to the wake in the opposite direction of the disk (Figure 6.5). From the conservation of energy concept,

$$KE_{up} = \text{energy extracted (by rotor)} + KE_{wake} + KE \text{ (rotation of wake)}$$

The torque acting on the rotor makes it rotate and power can be extracted. To obtain maximum power, a high angular velocity Ω and a low torque T are desirable because a large torque will result in a large wake rotational energy (angular velocity of wake = ω).

$$\text{Power (rotor)} = T\Omega$$

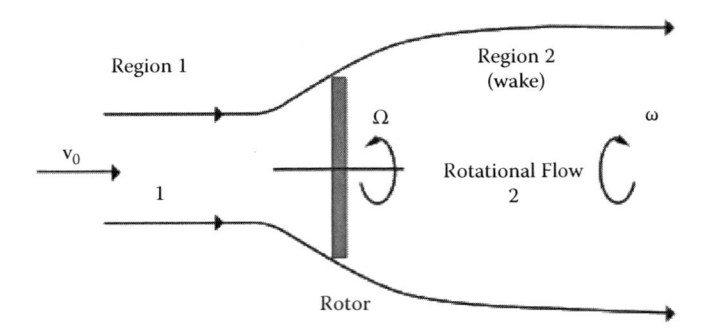

FIGURE 6.5 Rotor imparts rotation to wake.

A similar analysis, as previously described, can determine the power extracted when conservation of angular momentum is included. An annular ring is considered and an angular (tangential) induction factor α' is used. The main difference is that rotor velocity is a function of the radius, so the values must be calculated for the annular ring.

6.6 AERODYNAMIC PERFORMANCE PREDICTION

The ratio of lift to drag for airfoils is around 100, so the two forces that act at the quarter chord of the airfoil are represented by a (tangential) force that makes the blade rotate and a (perpendicular) force trying to push the rotor over. If these lift and drag forces are calculated for a blade, the tangential and perpendicular forces are calculated and the rotor performance can be predicted. If the angle between the blade path and the wind at the blade is Φ (Figure 5.9), the tangential and perpendicular forces are

$$F(\tan) = L \sin \Phi - D \cos \Phi \qquad (6.23)$$

$$F(\text{per}) = L \cos \Phi + D \sin \Phi$$

Notice that the perpendicular force will be larger than the tangential force and at 90 degrees there is only drag.

A number of computer programs can predict aerodynamic performance of wind turbines [16]. These are based on the momentum (or strip) theory that assumes that each element of a blade (Figure 6.6) can be analyzed independently from the others, and the two-dimensional data for lift and drag coefficients can be used at the center of the section. Performance predictions of power, torque, force, and power coefficient can be obtained for a blade (rotor) using a numerical technique. Values are calculated for sections of the blade and then summed to calculate total performance.

Drag and lift coefficients versus angle of attack and Reynolds number are available for many airfoils. In general, the coefficients are given for attack angles near zero to a few degrees past stall. Stall occurs when lift decreases and drag increases steeply. The problem in calculating performance predictions is using the correct inflow angle to the blade because the angle depends on the wind speed at the blade.

The relative wind speed must be corrected for the actual speed at the blade using the axial interference factor α and the rotational interference factor α'. At each section of the blade, an iterative procedure is used to calculate the angle of the inflow to the airfoil. Because sections of the blade may operate at high angles of attack, lift and drag data for the angles from a flap plate or other actual measured data from some airfoil are added to the tabular values. Tip losses and hub losses can be included along with wind shear and yaw (off-axis components). The main limitations of the programs are the treatment of unsteady aerodynamics in the region of dynamic stall and the use of two-dimensional data for lift and drag.

Design of Wind Turbines

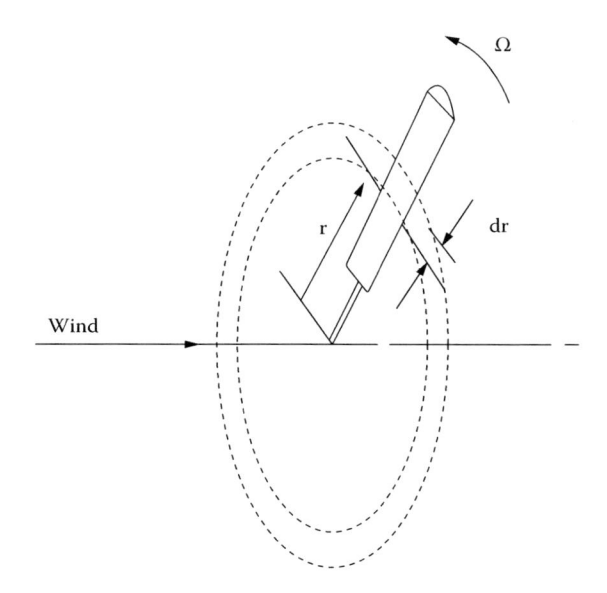

FIGURE 6.6 Blade element.

Rotors for vertical axis wind turbines present another problem since the blades go through attack angles of 360 degrees and the blades are curved for Darrieus wind turbines. A number of performance models for the Darrieus rotor have been formulated [3,17–19]. In general, symmetrical airfoils are used, so lift and drag data are needed from 0 to 180 degrees. The operation of a vertical axis wind turbine at an attack angle of 90 degrees also means no lift, so the torque and power are negative and a cyclic variation occurs on every revolution [20].

From observations of the flow field of a Savonius rotor, an analytical model was developed for the analysis of performance [3]. Two major discernible features of the flow field are: (1) counter-rotating vortices are shed from the vane tips when a vane is approximately at right angles to the flow, and (2) the vortices move rearward at approximately the free stream speed. The model was adequate in that it predicted a power coefficient around 0.30 at a tip speed ratio around 1, which is in line with field data and wind tunnel tests for Savonius rotors.

Dynamic stall produces higher loads on blades and larger power output than the predictions from the performance codes using steady-state data for lift and drag. Dynamic stall may occur during operation in high winds due to a gusts on constant pitch blades or on variable pitch blades in the run position. During this increasing angle of attack, a vortex forms near the leading edge and moves to the trailing edge of the blade, resulting in higher lift, hence the name. After the vortex is shed off the trailing edge, deep stall occurs.

The other condition for occurrence of dynamic stall in high winds is during shutdown as variable pitch blades are moved to the feather position. Westinghouse wind turbines installed in Hawaii and rated at 600 kW had this problem as power spikes to 800 kW occurred during shutdown. The solution was to change the blade pitch in the run position to lower the rated power, so when a spike occurred during high wind shutdown, the loads and power were not too high. Now lift and drag data for some airfoils are available that show dynamic stall for changing attack angles. These data can be used for performance prediction.

The dynamic stall vortex has been visualized and also noted by the analysis of time-varying surface pressure data from field tests and wind tunnel experiments [21]. Blades with pressure taps were used for an unsteady aerodynamics experiment [22] that included a test of an extensively instrumented wind turbine in the giant (24.4 × 36.6 m) NASA Ames wind tunnel. Results from computer models at high wind speeds under stall were significantly different; power predictions ranged from

TABLE 6.1
Sample Output from PROP93

Propprint 3

Blade element data for delta beta = 0.00, X = 6.11, yaw = 0.00

Element	1	2	3	4	5	6	7	8	9	10
Theta	180	180	180	180	180	180	180	180	180	180
Vel	1.00	1.00	1.00	1.00	1.00	1.00	1.00	1.00	1.00	1.00
A	0.296	0.140	0.188	0.204	0.230	0.213	0.195	0.206	0.231	0.308
AP	0.073	0.021	0.016	0.012	0.010	0.008	0.006	0.006	0.006	0.007
CL	0.813	1.005	1.160	1.206	1.334	1.311	1.168	1.037	0.918	0.772
CD	0.014	0.098	0.053	0.043	0.020	0.019	x0.016	0.014	0.013	0.011
PHI	49.92	42.48	27.54	20.14	15.45	13.03	11.35	9.74	8.34	6.72
ANG	7.92	19.18	15.74	14.84	13.35	12.93	11.35	9.74	8.34	6.72
TC	0.384	0.526	0.622	0.656	0.707	0.665	0.609	0.610	0.609	0.572
QC	0.040	0.059	0.073	0.075	0.083	0.079	0.074	0.073	0.069	0.056
PC	0.243	0.363	0.443	0.459	0.508	0.485	0.453	0.443	0.421	0.344
TD, lb/ft	2.64	6.03	11.90	17.57	24.37	28.01	30.31	35.04	329.6	41.60
QD, ft-lb/ft	4.38	10.92	22.21	32.26	45.86	53.47	59.02	66.73	71.85	65.54
PD kW	0.024	0.298	0.606	0.880	1.251	1.458	1.610	1.820	1.959	1.788
Rey, *10^6	0.920	0.862	0.922	0.931	0.910	0.868	0.890	1.004	1.132	1.132
Rotor 2 blades	Pitch	X	TC	QC	PC	V$_0$ m/s	TD lb	MD ft-lb	QD ft-lb	PD kW
	0.0	6.1	0.614	0.070	0.427	10.0	752	3,984	1,372	23.3

Note: Output for one blade (Carter 25, 10 m diameter, pitch = 0°), divided into ten stations, and then the total is summarized at the bottom. Wind speed is 10 m/s and tip speed ratio, X = 6.11.

30 to 275% of the measured values. Hence aerodynamic performance prediction programs are used as design tools, not final answers.

Aerodynamic performance prediction programs [3] are now available for personal computers, and include menu-driven interactive editing and graphical displays to facilitate use as design tools. The inputs to PROP93 program [23] include blade characteristics (number, length and hub cut-out, platform, twist, and pitch), lift and drag coefficients of airfoils for different angles of attack, and operating characteristics, such as tip speed ratio, rpm, and wind speed. The tabular output of PROP93 in metric or English units may be directed to a screen, printer, or data file. For the selected input in the example (Table 6.1), the rotor is predicted to produce 23.3 kW at 10 m/sec.

Graphs of the standard output parameters can be displayed as functions of blade station, pitch, wind speed, or tip speed ratio. Calculated values can then be compared with experimental values. These steady-state programs do not predict the high loads seen in the field from gusts or from changing the pitch to feather in high winds (dynamic stall).

Graphs of the planform (Figure 6.7) lift and drag data can be produced. Sample output graphs (Figure 6.8 through Figure 6.10) are for a Carter 25 wind turbine, NACA 2300 series airfoil. Smoother graphs would be obtained by using twenty data stations. These blades had large twist and larger chord toward the root and the same chord and twist from the midpoint, which produced an aerodynamic efficiency close to the theoretical limit. Notice the twist is to obtain the correct angle of attack due to the different inflow wind due to the contribution of the blade speed, which is slowest at the root.

Also, twist on the inward part of the blade increases the torque for starting rotation. Note that for constant pitch blades with little twist, starting torque is insufficient and the rotor needs to be motored for start-up. Variable pitch blades are in the feather position, which produces enough torque

Design of Wind Turbines 125

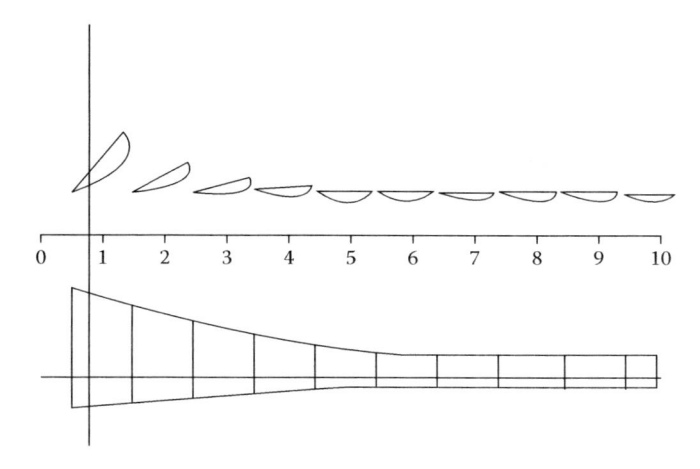

FIGURE 6.7 Twist and planform for Carter 25 wind turbine blade. Blade is divided into ten sections for analysis. The station is at midpoint of the section.

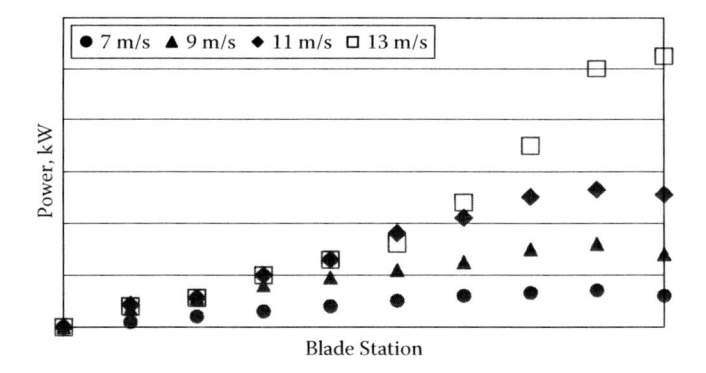

FIGURE 6.8 PROP93: Prediction of power output for one blade by blade station for four wind speeds. Tip speed ratio = 6.1.

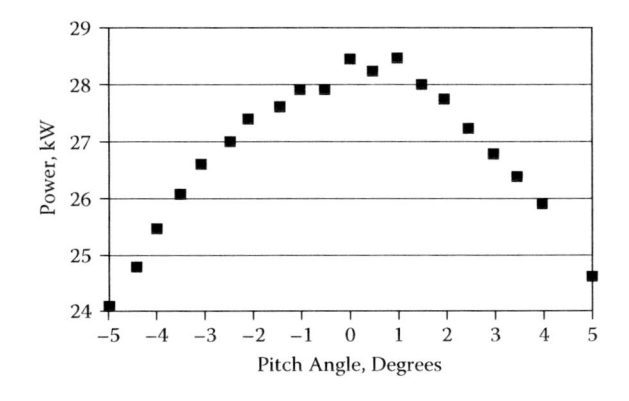

FIGURE 6.9 PROP93: Prediction of rotor power output for different pitch angles at 10 m/sec. Carter 25 wind turbine is a fixed-pitch constant rpm machine.

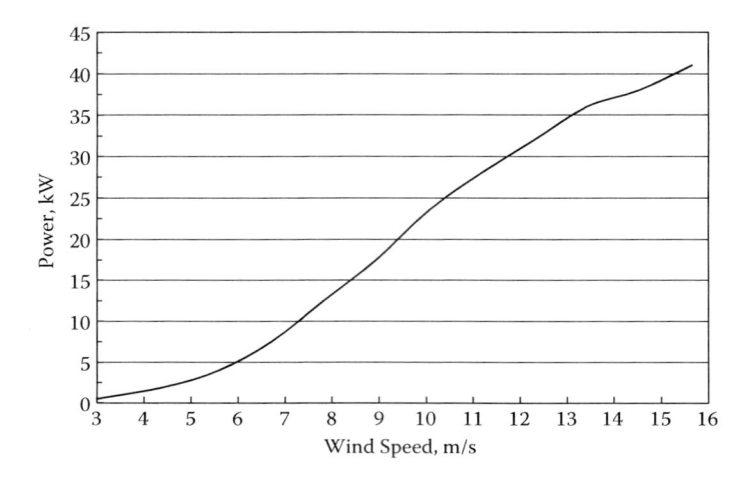

FIGURE 6.10 PROP93: Theoretical power curve for Carter 25 rotor. Tip speed ratio = 6.1.

for start-up. Notice that for constant tip speed ratio, the power continues to increase with wind speed (Figure 6.10).

Tangler and Kocurek [24] provided guidelines for input of post-stall airfoil data for the prediction of peak and post-peak rotor power for performance programs using blade element momentum theory. A steady-state data set from the rotor test in the unsteady aerodynamics experiment was used for the global post-stall method for predicting post-stall 3-D airfoil characteristics to be used with 2-D airfoil data.

PROPID [25] is a personal computer program for the rotor design and analysis of horizontal axis wind turbines, and the executable program is available online [26]. The strength of the method is its inverse design capability. PROPID is based on the PROPSH blade element/momentum code, and includes a 3-D post-stall airfoil performance synthesization for better prediction of peak power at high wind speeds.

Most wind turbine blades use the same airfoil for the entire blade; however, twist and chord length change from the root to the tip of the blade. The surface of the blade should have a smooth transition along the length. The Alternative Energy Institute (AEI) also fabricated test blades for the Carter 25, which used new airfoils designed specifically for wind turbines by the National Renewable Energy Laboratory (NREL) [27].

The criteria for the design of thin airfoils were high lift to drag for the inboard blade portion, restraining maximum lift coefficient of the outer part of the blade to limit peak power, and provide insensitivity to surface roughness. Because three different airfoils were used, a computer program was developed to calculate blade fairness (no waves) along the blade. The program used cubic splines under tension [28,29] and is available from the Alternative Energy Institute.

The basic input to the program consists of specified airfoils, blade radius, root cut-out, and wind distribution. Additional input can be specific: spanwise airfoil stations, specified twist, and taper distributions. Different tension parameters result in a different continuous spanwise airfoil distribution. Optimization is achieved by iteration through computer codes to determine the surface based on annual energy output and predicted blade load history for a specified wind distribution. Computer design of blades is of little value if a blade cannot be constructed practically. Therefore, the program permits various input constraints on twist, taper, and sharpness of edges and corners.

The blade fairness program determined the airfoils at ten sections from the three input airfoils (Figure 6.11). The templates were cut out on a numerical control milling machine and assembled with the proper twist (Figure 6.12). The blade templates were then used to construct a plug, from which top and bottom molds were constructed. After fabrication of the skins, the blades were attached to a Carter 25 spar and hub and tested in the field in a side-by-side comparison with

Design of Wind Turbines

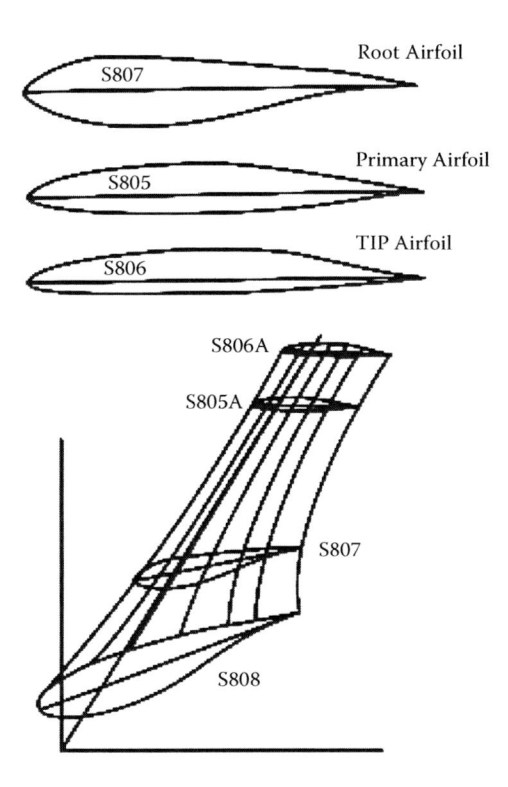

FIGURE 6.11 Thin airfoil series for wind turbine blade and input placement for 5-m blade.

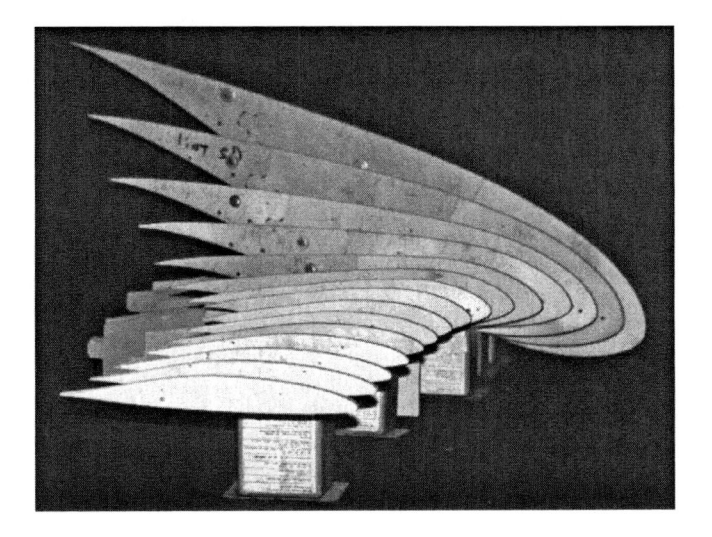

FIGURE 6.12 AEI blade templates using new thin airfoils for Carter 25 wind turbine.

a production unit [30,31]. Data were collected at low, medium, and high wind speeds for clean, medium, and heavy surface roughness conditions.

The roughness conditions were simulated with the application of grit on 2.5-cm wide tape on the upper (0.02 chord) and the lower (0.05 chord) leading edges. Results of the tests showed little power difference at low wind speed. The reduced power from the outer part of the blade could not be tested since the teetering hub reduced high flap loads. However, the new airfoils were much less sensitive to surface roughness at medium and high wind speeds.

Essentially the same amount of power can be obtained from one blade rotating fast, more blades rotating more slowly, or the same number of blades with different chord lengths. Performance prediction programs show that as solidity increases for a given rotor area, the tip speed ratio that gives the maximum power coefficient becomes smaller. For a given size rotor operating at fixed rpm, different size generators (rated power) can be placed on a unit by increasing the rated wind speed.

In the past, many wind turbines were built with the same 10-m diameter and their rated powers were 8, 12, 15, 25, 40, and 90 kW. Today, most wind turbines have rated powers at wind speeds from 10 to 13 m/sec. The new standard (Small Wind Certification Council) for small wind turbines in the U.S. is rated power at 11 m/sec.

The design engineers of wind turbines follow a number of rotor parameters to select: airfoil, planform, solidity, number of blades, radius, tip speed ratio (variable or fixed), and others. The most efficient blade from an aerodynamic view is generally more difficult to construct from a practical and manufacturing standpoint. Early blades (and propellers) were made from wood, and a commonly used airfoil was the NACA 4400 series because its bottom side was flat. Other airfoils with better lift to drag such as the NACA 23000 series and the LS1 airfoil were used. These airfoils had cambers curved on the bottom sides that made them somewhat more difficult to construct.

An aerodynamically efficient blade will have the largest twist and chord at the root that then decreases toward the tip; however, because of other considerations, the inner part of the blade is designed generally to achieve some efficiency and starting torque because the outer third of the blade generates most of the power. Therefore, that part of the blade must be aerodynamically efficient. Finally, the design of the tip of the blade is important for noise considerations and to reduce tip losses if possible. The outer portion of the General Electric blade is now swept back and the Skystream has sweep blades, which means the outer portion is curved like a scimitar (sword).

Other parameters, for example, are the design point, wind speed for the rated power (which primarily determines rotor area), and tip speed ratio determined by the solidity of the rotor. In general, the blade tip speed is limited to roughly 70 m/sec since the blade tips cause excessive acoustical noise at higher tip speeds. For an offshore wind turbine, noise is not an important issue. Besides rotors, an offshore system also includes a hub that may include components for adjusting blade pitch; a drive train and gearbox in most cases; a generator; yaw control; a tower; and a control system.

6.7 MEASURED POWER AND POWER COEFFICIENT

A common specification is the power output of the wind turbine versus wind speed depicted by a power curve. The curve generally includes all efficiencies from wind to electrical output in addition to rotor efficiency. Since all wind turbines must control power output at high wind speeds, efficiency decreases at some point. Control can be implemented by changing blade pitch or by operating fixed pitch blades at constant angular speed. Operating at fixed pitch is also called stall control. Power curves are obtained by the method of bins, so in reality, a power curve is a band of values, not a line.

The experimental power and power coefficient curves (Figure 6.13) apply to a wind turbine that has an induction generator, operation at constant angular speed, and fixed pitch, which means it is stall controlled. Therefore, it reaches maximum power coefficient at only one point, and the decreased aerodynamic efficiency at wind speeds above this point make the power coefficient also decrease. The increased power in the wind and the decreased aerodynamic efficiency combine to give a constant power output above 12 m/sec. The high efficiency that includes drive train and generator is possible because this unit has an almost optimal blade (taper, twist, and thickness).

Besides the tip and hub losses of the blades, a further reduction of the power coefficient will occur due to the inefficiencies of the mechanical system (drive train, coupling) and the generator. Under optimum design conditions, the modern two- or three-bladed rotors at tip speed ratios in the

Design of Wind Turbines

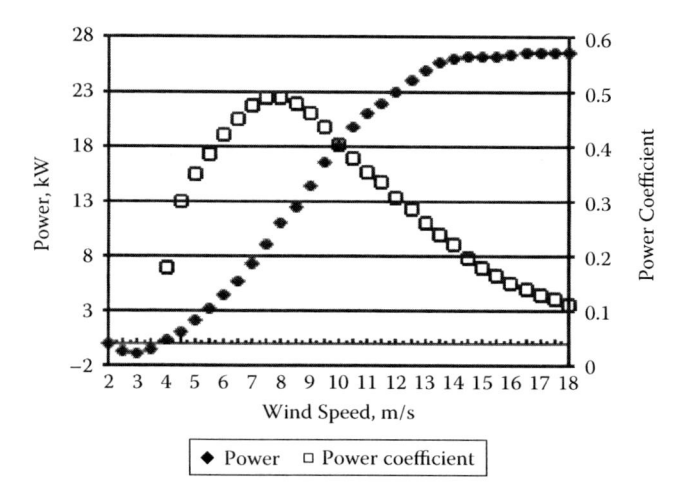

FIGURE 6.13 Experimental power and power coefficient for a Carter 25 rated 25 kW, 10 m in diameter. At 3 m/sec, the turbine uses power (energy for field coils of induction generator).

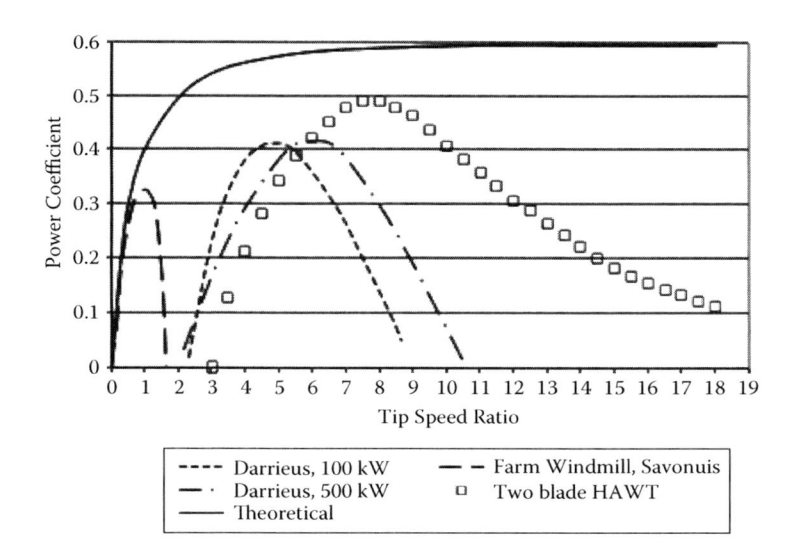

FIGURE 6.14 Experimental power coefficients for farm windmill, Savonius, 100-kW Darrieus, 500-kW Darrieus, horizontal axis wind turbine, and Carter 25 compared to theoretical values. (*Sources:* J. van Meel and P. Smulders. 1989. World Bank Technical Paper 101; M. Khan. 1978. *Wind Engineering,* 2, 75, with permission; T.D. Ashwill. 1992. Sandia National Laboratory Report 91-2228; J.H. Strickland. 1975. Sandia National Laboratory Report 75-0431.)

range of about 4 to 10 will have power coefficients of about 0.4 to 0.5 (Figure 6.14). The power coefficients for farm windmills and Savonius rotors are essentially the same, with a maximum just over 0.3. The maximum power coefficients for the vertical axis wind turbines are just over 0.4—lower than those for the horizontal axis wind turbines. This is one reason vertical axis wind turbines are not commercially available for wind farms.

The three methods of regulating output are passive stall in which the wind turbine operates at fixed rotational speed with fixed pitch blades; active stall in which the wind turbine operates at fixed rotational speed with adjustable pitch; and variable pitch in which the wind turbine operates at variable rotation speed with adjustable pitch blades. The last method is the most efficient aerodynamically, but the method of control chosen is always a trade-off between energy production and cost.

Control of rotor rpm using adjustable pitch includes full-span control (pitch motors located in the hub; variable pitch tips; and ailerons (flaps on airplane wings) to control aerodynamics even though pitch is not adjustable. The variable pitch tips and ailerons have pitch motors in the blades. The most common method for large wind turbines is full-span control, although wind turbines have been built with the other two methods. The MOD-2 and MOD-5, large prototype turbines funded by U.S. government, had tip controls.

Ailerons are moved to the low-pressure side of a blade to reduce lift, in contrast to flaps on airplanes that are moved in the opposite direction to increase lift. A NASA–DOE program investigated ailerons both theoretically and experimentally for application to medium and large wind turbines [35]. Zond built and installed 12 500-kW units with aileron control near Fort Davis, Texas, as part of the Utility Wind Turbine Verification Program [36]. However, after four years of operation they were dismantled; one reason was maintenance problems with the ailerons. Finally, the control system for the Italian Gamma 60 1.5-MW wind turbine with fixed pitch blades was yawing the rotor and that presented the problem of differences in lift on the blades on each cycle.

Efforts have been made to develop passive pitch control techniques that adjust the blade pitch angle without need for actuators [37]. One concept is the self-twisting blade in which the blade spar at the hub is flexible, and the thrust and centrifugal forces on the blade cause it to twist to the feathered position.

United Technologies Research Center built a 10-m diameter unit with the two blades (constant chord, no twist) attached to a flexbeam (Figure 6.15) attached in the middle to the drive shaft. The twist was sufficient to provide torque for start-up, and pendulum weights outside the plane of rotation moved toward the plane of rotation and provided proper pitch angle for the run position. Also, the weights provided control at high winds by twisting the blades toward stall. One problem was that the flexbeam moved toward a different set twist over time and that reduced the starting torque.

The Proven wind turbine had a flexible hinge (delta-3) near the root of the blade [38]. As rotor rpm increased, the blades were forced outward. The blade pitch changed toward stall and the blades

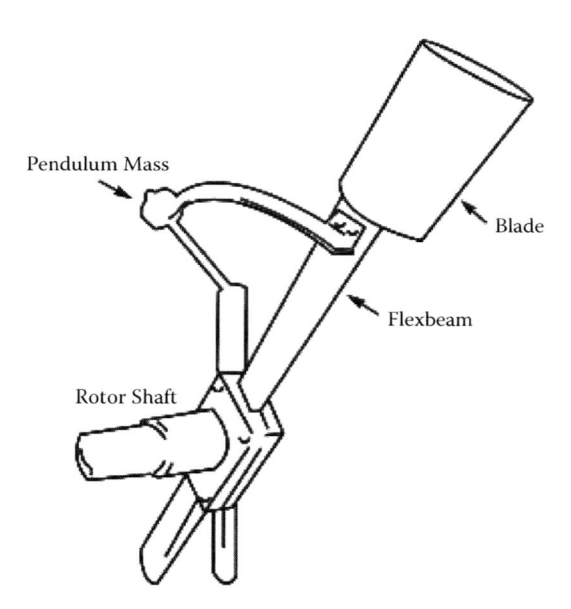

FIGURE 6.15 Passive control with flex beam and pendulum weights for units at constant rpm operation.

coned, thus reducing the rotor swept area. Even in high winds, the rotor rpm is limited, and it can continue to produce power.

6.8 CONSTRUCTION

6.8.1 BLADES

For years, small wind turbines blades were carved from single pieces of wood or from wood blocks glued together from several pieces. The material properties of wood are good: strength, flexibility, and resistance to fatigue. Machines could carve up to four blades from a master blade. However, for large blades, solid wood was not acceptable, as the weight became excessive.

One type of larger blade was constructed like an airplane wing with a spar, ribs, and a covering. The spar is the load-bearing element and the ribs form the airfoil shape. As noted earlier, fabrication of blades depends on design, materials, and the construction processes, all of which are related. Wind turbine blades have been made from a number of materials, for example, aluminum, fabric, or metal covers on ribs and spars (like an airplane wing). Other examples are a sail wing made of fabric attached to a leading edge spar; a laminated wood composite shell; fiberglass-reinforced plastics (FRPs) and carbon fibers; pultruded FRPs; extruded aluminum (blades for vertical axis wind turbines); and small turbine blades made from injection molds.

Pultruded blades are made by pulling the fiberglass and other parts through a die and applying epoxy at the same time; then the blades are cut to length. Extruded blades are made by pushing the material through a die. Blades for Darrieus wind turbines are bent to curvature after extrusion. Cross sections of some blades illustrating different manufacturing processes are presented in Figure 6.16.

A blade on a wind turbine goes through more fatigue cycles in one year than airplane wings undergo in their lifetimes. Therefore, fatigue is the major concern of wind turbine manufacturers and operators because wind loads are large and variable. The fatigue properties of metal blades built for wind turbines are not satisfactory and many failures resulted. Carbon filaments are used in blades because they are stronger, even though they are more costly than glass filaments. The limitation on pultruded FRP blades is the constant chord with no twist. FRP blades incur additional costs for master molds and there is a trade-off between automated winding of filaments and hand lay-up. Molds are expensive and dies for the extruded aluminum blades are even more expensive.

The material for blades is predominantly FRPs, for both spar and blade skin, which also supports the load. Large wind turbine blades are constructed very differently from airplane wings. One basic concept is two glass fiber shells attached to two rigid beams, or a glass fiber shell with one beam (Figure 6.17). The technology, from design to process, is discussed by LM Wind Power, the world's leading supplier of blades [39].

Small wind turbine manufacturers have also switched from wood to FRP blades. Composite wood laminate blades (6.5 m long) have been successful on 50-kW units.

The Sandia National Laboratory concentrates research on the aerodynamic and structural designs of wind turbine blades [40]. Topics include adaptive structures, thick airfoils, material and fatigue, manufacturing issues, design tools and applications, and sensors and nondestructive inspection. Reports in all of these areas are listed online. Another aspect of its program is a joint project with NREL to collect experimental inflow and turbine response data for a long-term inflow and structural test. One of the instrumented wind turbines is a GE 1.5-MW unit.

Vortex generators have been used to improve aerodynamic efficiency. Novel ideas for improving efficiency of blades include dimples (like golf balls), scales (like sharks), and bumps on leading edges (like tubercles on whales). There is always a trade-off between increased efficiency and construction costs for novel airfoils. The simple solution is to increase the blade length a little on conventional blades to collect the same energy.

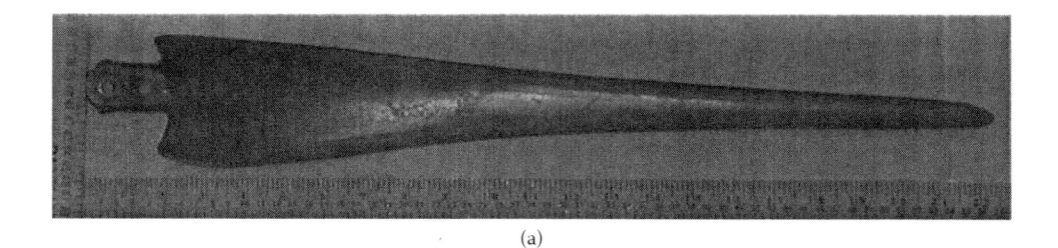

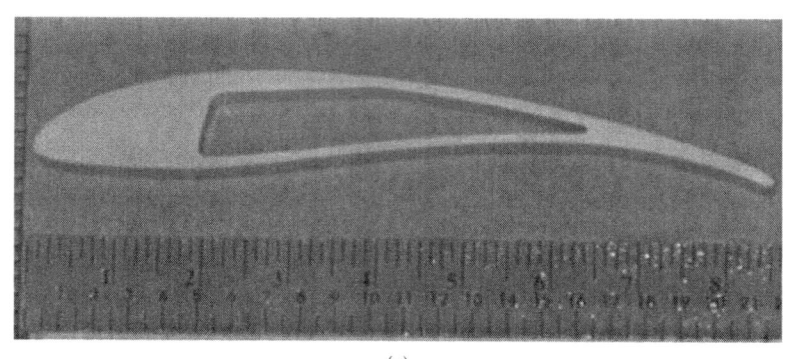

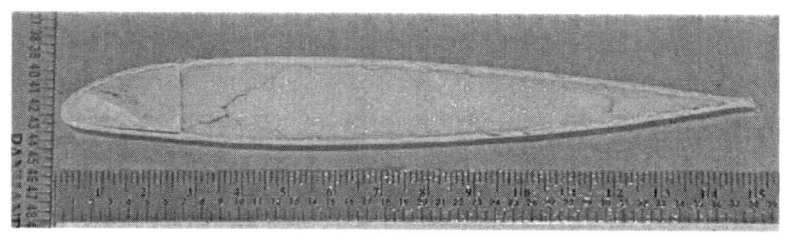

FIGURE 6.16 Blade cross sections from different manufacturing processes. (a) Injection mold blade for 300-W unit has carbon filaments. (b) Solid wood 4400 series airfoil for 4-kW unit. (c) Pultruded FRP special airfoil for 10-kW unit. (d) Pultruded FRP airfoil 23012 for 25-kW unit (weight in nose and foam keep skins from flexing). (e) Laminated wood composite blade for 50-kW unit.

Design of Wind Turbines

(e)

(f)

(g)

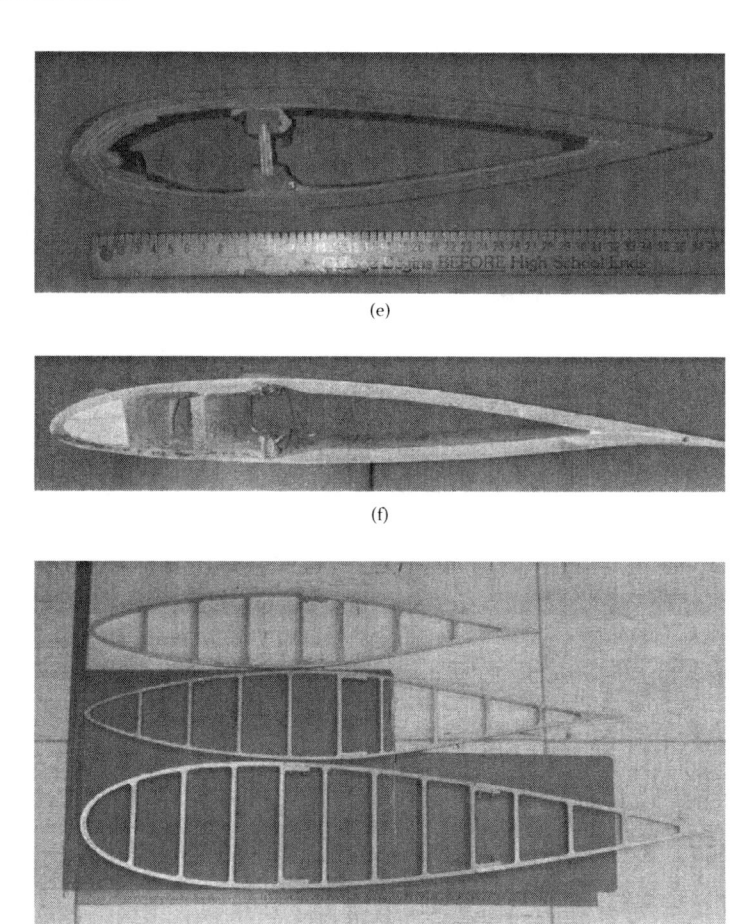

FIGURE 6.16 (*Continued*) Blade cross sections from different manufacturing processes. (f) FRP LS1 airfoil LS1 for 300-kW unit, hand lay up in three molds, top and bottom skins attached to nose, D-spar with lead weight. (g) Extruded aluminum for 500-kW VAWT with 34-m test bed, bottom chord 1.2 m from three pieces, others from two pieces (top chord 0.9 m).

Presently Vestas makes a 164-m diameter (80-m blade) 8-MW unit and Siemens produces a 75-m blade [41]. Wind turbines of 10 to 20 MW require blades of 85 to 130 m length. Carbon fiber, maybe even carbon nanotubes, will eventually replace FRPs.

One of the main concerns is developing a procedure and mechanism for shutdown caused by over-speed. If a load loss arises when a utility transmission line goes down due to an ice storm or during high winds while a wind turbine is operating at rated power, the power of the rotor has to be controlled within 5 to 10 sec. If the condition produces too much power to control even with the application of a mechanical brake, the unit will self-destruct or a few high wind speed shutdowns will stress the drive train to the point where it must be replaced. For light-weight blades on wind turbines operating at constant rpm, rotor control period is 4 to 5 sec.

AEI and the U.S. Department of Agriculture's Agricultural Research Service installed and tested over 80 prototype and first-production wind turbines, from 50 W to 500 kW. Most units had failures within one year, and some of the failures resulted in loss of the rotor or even the destruction of the unit. When a rotor is in a runaway condition, the only remedy is to move upwind and wait a while.

FIGURE 6.17 Blade cross sections for megawatt wind turbine.

The USDA program for wind energy was terminated in 2012. AEI moved its test center, which was adjacent to the campus, to WTAMU's Nance Ranch and now the Regional Wind Test Center has space for testing twenty small wind turbines.

If a mechanical brake is part of a system for over-speed control, it should be installed on the low-speed shaft. If the brake is on the high-speed shaft and the drive train fails, the brake is useless. The failure does not have to be mechanical. For example, a 500-kW VAWT was lost because of a sequence of events that the software control program did not anticipate. The procedure for shutdown was to cut off the load and apply the mechanical brake. A high wind gust called for shutdown but the gust was short. The software ordered the brake to release, but the load was not reconnected because the time delay had not been reached. The turbine went into high rpm and the brake was applied again. Due to the high power, the brake soon burned up and the rotor was in the runaway condition. Within a short period, one blade broke loose and cut the guy wires and the unit fell.

All the blades must have the same pitch setting or a cyclic forcing function will affect the drive train and other components. One extreme case was a 40-kW wind turbine whose three blades had dihedral spars. The system was to change its position to feather for shutdown and change rapidly to feather for over-speed. An attachment mechanism connecting rods to the middle of each blade had some play in it so the pitch of each blade changed on every rotation and it was different from one side to the other. In moderate winds, the stable rotor position was yawed 45 degrees to the wind. Along with the wear problem caused by the yaw, the unit did not produce much power.

Another concern is yaw rate, especially for flexible blades. A rotor has angular momentum, and when the brake is applied, a wind turbine will tend to rotate about the yaw axis. The rate of yaw, which is motor driven on large turbines, is limited. On some smaller turbines the rate of yaw is limited by a yaw damper. The rate is limited because a change in angular momentum produces torque.

$$T = \Delta L/\Delta t \quad (6.24)$$

where the torque is in the direction perpendicular to the plane of rotation of the rotor. Therefore, a large change in angular momentum of the rotor caused by a large change in wind direction or a change in yaw due to shutdown for over-speed, results in a force perpendicular to the plane of

Design of Wind Turbines

rotation. For flexible blades, this force may be large enough to cause the blades to strike the tower. In the worst case, the blades break off at the roots.

Another example of fast yaw rate is for small wind turbines with flexible blades, downwind, with coning. Suppose a wind turbine is not operating due to no or little wind at night. The next day the winds are from the opposite direction and the unit starts with the rotor in the upwind orientation, which is possible, and the rotor will even track the wind. However, the condition is unstable and eventually the wind direction will change or the wind speed will increase enough for the rotor to suddenly change from the upwind to a stable downwind condition. This very fast yaw rate exerts large flat forces on the blades that will bend the blades. The solution is using a yaw damper, move the rotor farther from the tower or Install stiffer blades.

The guided tour of the Danish Wind Industry Association is excellent. The association also maintains data on testing wind turbine blades [42]. One problem with fatigue testing of large blades by vibration is the long time required to reach enough cycles so that fatigue becomes noticeable.

Nondestructive testing by acoustic emission is one way to monitor the progression of fatigue and may even predict where failures will occur [43]. Two fiberglass blades were tested by dynamic loading on a full-scale blade testing facility. The acoustic emission signatures focused on counting, amplitude distribution, and location and provided assessments of damage status, failure modes, and failure locations. The damage development in composite laminates under fatigue progresses from matrix cracking, crack coupling with interfacial debonding, delamination, fiber breaking, and fracture.

A general observation is that low acoustic emission amplitudes are associated with matrix damage, while high acoustic emission amplitudes are related to fiber failure. Mechanical properties such as natural frequency, elastic modulus, and tip deflection were measured during the fatigue tests, and changes in those properties indicate degradation.

Blades are loaded to static failure in flapwise bending, and some blades are tested to failure for edgewise bending. The National Wind Technology Center has a facility for static and dynamic load testing of blades that includes nondestructive techniques such as photoelastic stress visualization, thermographic stress visualization, and acoustic emission. A new facility for testing blades up to 90 m long is now available in the U.S. [44]. A facility in the U.K. can test blades up to 100 m In length and blade test facilities are available in other countries. New blade designs (12 m long for a constant rpm 100-kW unit) using carbon filaments for more strength at less weight were fabricated and tested [43]. All the blades survived the specified test loads and two designs exceeded them significantly.

In the final analysis, the important factor is energy produced by a wind turbine at the most economical cost per kilowatt hour. A rotor design study considered four basic configurations: upwind three blades, upwind two blades, downwind three blades, and downwind two blades [45]. The cost of energy was estimated with improvements and compared to baseline turbines of 750 kW, 1.5 MW, and 3.0 MW. Two conclusions were drawn: (1) the cost of energy would be reduced by up to 13%—a small saving relative to the magnitude of the load reduction, and (2) more than 50% of the cost of energy was unaffected by rotor design and system loads.

6.8.2 Other Components of System

For large wind turbines, the most common configuration is three blades made from FRPs, upwind, drive train, asynchronous generator, and tubular steel tower. The driver is the rotor and the dynamic loads are transferred to the rest of the system (drive train, generator, and tower). The difference between variable and constant rpm operation is that part of a wind load can be absorbed by inertia of the rotor in variable rpm operation. This reduces the severity of the loads for the drive train and generator.

Computer codes are available for the prediction of the wind turbine loads and responses. The NWTC has a tool kit for creating wind turbine models [46] for input into a multibody dynamics code (commercial). FAST (fatigue, aerodynamics, structures, and turbulence) factors can be used to model two- and three-bladed horizontal axis wind turbines. The code models the wind turbine as a combination of rigid and flexible bodies. For example, two-bladed teetering hub turbines are modeled as four rigid bodies and four flexible ones. The rigid bodies are the earth, nacelle, hub, and optional tip brakes (point masses). The flexible bodies are blades, tower, and drive shaft. The model connects these bodies with several degrees of freedom: tower bending, blade bending, nacelle yaw, rotor teeter, rotor speed, and drive shaft torsional flexibility.

The flexible tower has two modes of vibration, and the blades have two flapwise modes and one edgewise mode. Flutter is the coupling between blade flap and edge modes of vibration, and was actually used as a method of over-speed control for a 300-W wind turbine. The blades were constructed of carbon filaments formed in an injection mold, so the high strength allowed flutter. In all other cases, a blade that enters flutter will generally fall in a short time.

All wind turbines and blades have natural frequencies (modes) of vibration. The models predict the modes or the modes can be found experimentally. During operation, especially constant rpm operation, major modes must be avoided, for example, the natural frequency of guy wires for vertical axis wind turbines. For constant rpm operation, the drive train may incorporate a torque damper. Monitoring of acoustic emissions can be used to determine future drive train problems, thereby reducing costs by preventive maintenance.

Wind turbines are set on various types of towers—pole, guyed pole, pole or guyed pole with gin pole. Operation and maintenance functions can be at ground level (Figures 6.18 and 6.19) for guyed lattice, lattice, and tube towers. Most towers are made of steel, although concrete has been used and fiberglass is being considered. Lattice towers were common in the early wind farms in California; however, the later large wind turbines with hub heights from 50 to 100 m used tubular steel towers. To date, the record hub height is a 160-m lattice tower for a 2.5-MW wind turbine constructed by Fuhrländer.

Towers must be strong enough to support the weight at the top and resist the movement of the wind forces trying to push the tower over. Movement during operation at rated power in high winds can be quite large. Among the several types of foundations, a lot of rebar and concrete are required for large wind turbines. Examples of foundations are piers and bells (at the bottom) for each leg of lattice towers and various types of anchors, primarily piers for guyed pole towers. Foundation types for large turbines on tubular towers are piers, piers with concentric cylinders, or pads on the ground (Figures 6.20 and 6.21). Piers can be drilled in suitable ground (not rock) with augers. A hole for

FIGURE 6.18 Small (1.8 kW) wind turbine mounted on-10 m pole tower, no guy wires, with gin pole. Note sweep blades.

Design of Wind Turbines

FIGURE 6.19 International Wind Systems 300-kW system on 49-m pole tower with guy wires and gin pole.

FIGURE 6.20 Placing rebar for pad foundation for 2-MW wind turbine. Finished pad required 32 metric tons of rebar and 270 cubic meters of concrete.

a pier 5.5 m in diameter and about 9.5 m deep (Figure 6.22) for a 3-MW wind turbine was drilled with a giant auger. One advantage of concentric cylinders is less concrete; the inner can is backfilled with dirt. Also, the rods can be stressed after the concrete is poured to obtain a stronger foundation.

Wind turbines with downwind rotors will experience a reduction of wind speed due to tower shadow and a cyclic driving force. One result is the generation of repetitive sounds as the blades pass behind the tower. Repetitive sounds are more annoying than the normal chaotic noises generated

FIGURE 6.21 Finished pad foundation for 2.3-MW wind turbine. Note copper wire for grounding.

FIGURE 6.22 Pier foundation, concentric cans, for 3-MW wind turbine. Right: bolt rods are in place and starting to attach rebar.

by passing wind. For most wind turbines, noise attenuates to an acceptable level not too far from the turbine. However, the MOD-1, downwind unit emitted low-frequency sound waves, and under certain atmospheric conditions, the noise was at unacceptable levels at considerable distances from the turbine. It was strong enough to shake the dishes on the shelves of some homes. The solution was to reduce the rotor speed (less power) by replacing the generator.

6.9 EVOLUTION

Since 1970, the design of modern wind turbines evolved from the extremes of large utility scale wind turbines and small wind turbines. Large wind turbine development was funded primarily by governments, and only prototypes were built and tested. Small wind turbines were built in large numbers by private manufacturers for an emerging commercial market.

In the U.S., NASA–Lewis began with the MOD-O design, a two-blade, downwind 100- to 200-kW turbine that progressed to the design of the two-blade, 7-MW MOD 5. This design was reduced to 3.2 MW, and one prototype that had steel blades, teetered hub, upwind, and tip pitch control was built. The tip pitch control was driven by motors in the blades that created maintenance

Design of Wind Turbines   139

problems. The Hamilton-Standard WTS-4 was a 4-MW wind turbine with two blades, downwind, pitch control, and teetered hub.

The Schachle-Bendix wind turbine had an interesting concept: a variable speed hydraulic drive in the power train connected to hydraulic drives on the ground to drive the generator. The losses in the hydraulic drive were high, and the unit reached a power output of only 1.1 MW rather than the designed 3 MW. The unit was mounted on a tripod truss tower that rotated on a track; the tower was yawed for control.

Several large prototype systems were built in Europe. In Denmark, the wind turbines were the Nibe A and Nibe B, three blades, upwind, 630 kW, fixed pitch blades for A and variable pitch blades for B; the Tvind, three blades, variable pitch, upwind, 2 MW; and the Tjaereborg, three blades, upwind, 2 MW.

In Sweden, four 2- to 3-MW wind turbines were built. One had an angle gear drive to the generator in the top of the tower, so slip rings were not needed to transfer power. A second unique feature was a carriage assembly on rails on the side of the tower to raise and lower the entire assembly, nacelle, and rotor. In Germany, the largest wind turbine was the Growian I, two blades, variable pitch, downwind, 3 MW. Other megawatt prototypes were built in Italy, The Netherlands, Spain, and the U.K. The largest VAWT was a 4-MW Darrieus unit built in Canada.

A table of wind energy systems exceeding 500 kW through 1993 describes thirty-five units [7, Table 3-2]. Divone described the evolution of modern wind turbines from 1970 through 1990, with emphasis on description and operational notes of large units [7, chap. 3].

At the other extreme were wind turbines rated from 20 to 100 kW and designed for the commercial market. All sorts of designs were built and sold. Common types in the U.S. were two blades, fixed pitch, teetered hubs, downwind; three blades, fixed pitch, downwind; and three blades, variable pitch, downwind. U.S. Windpower installed over 4,000 units in California. In Europe the three-bladed, upwind, constant rpm, stall control units predominated. The different designs and their evolution through the mid-1980s are clearly shown by data sheets of wind turbines in Europe [48].

The technical data on wind turbines in commercial operation in the United States [7, Appendix C] through the mid-1980s covered the total number of installed units of a given configuration. As a result of the growth of the wind farm market in California and later in the 1990s in Europe, evolution continued toward larger wind turbines. By 2008, the predominant wind turbines were megawatt size, three blades, rigid hub, variable pitch (full-span control), variable and constant rpm operation, upwind.

As a result of the evolution, the manufacturers of small wind turbines became the winners over the large prototypes. Surprisingly, U.S. Windpower, the early leader in number of units installed, also built over 300 larger (300 kW) units but was unable to continue in business.

A Vestas, 90-m diameter, 3-MW wind turbine was installed near Gruver, Texas (Figure 6.23). For installation, an 800-metric ton crane was needed. Twenty trucks were needed to haul the crane to the site and another ten were required to transport the turbine and tower. The nacelle weighed 70 metric tons; the rotor, 41 metric tons; and tower, 160 metric tons. The 80-m tall tower consisted of four sections on a foundation that required 460 m^3 of concrete including a small pad for the transformer.

The sizes of commercial wind turbines (2013) increased from 4 to 8 MW (164 m diameter) and are primarily designed for offshore use. Designs for 10- to 20-MW (200 m diameter) systems are being developed, but transportation will probably require delivery of the turbine in sections and assembly on site.

6.10 SMALL WIND TURBINES

Generally small horizontal axis wind turbines are kept facing into the wind by tails. The control mechanism to reduce power in high winds is to offset the rotor axis from the pivot point axis of connection to the tower (Figure 6.24). The result is more force on one side of the rotor that tries to move the rotor parallel to the wind. However, the wind force on the tail keeps the rotor

FIGURE 6.23 Photos showing stages of erection of 3-MW wind turbine.

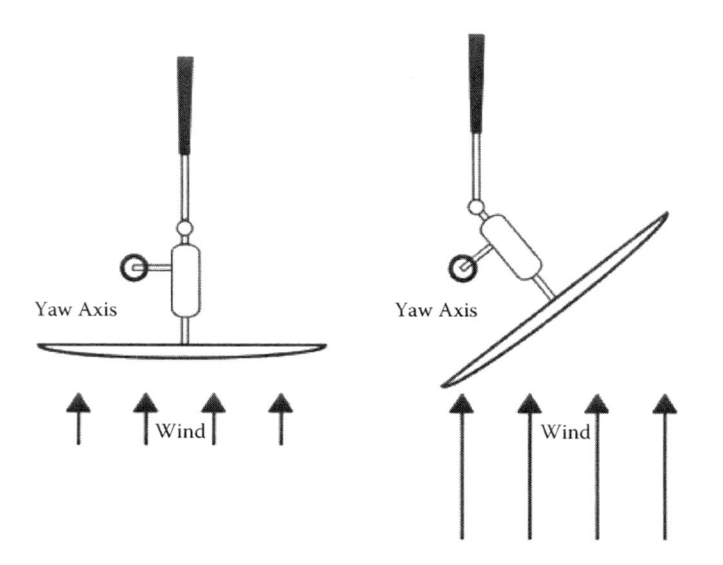

FIGURE 6.24 Furling diagram of rotor axis offset from tower (yaw axis) with hinged tail.

Design of Wind Turbines

perpendicular to the wind. In high winds the unequal force on the rotor is greater than the force of the tail; therefore, the rotor moves parallel to the wind. The tails of very small rotors may be fixed. During medium to high winds with rapid changes in direction, a small turbine may make a complete 360-degree revolution around the yaw axis.

Most wind turbines have hinges for their tails. During high winds, the rotor moves to a position closer to parallel to the tail (furling) [49–51]. When the winds decrease, the tail returns to a position perpendicular to the rotor by a force exerted by springs or gravity. Dampers, like shock absorbers, can keep the furling and restoration to normal operation from happening too rapidly. A farm windmill uses springs with adjustable lengths to restore force. One mechanism that uses gravity is installing the tail hinge at a slight inclined angle to the vertical plane.

Performance was measured for a 2-kW wind turbine for water pumping. Changes in the offset of the rotor axis to the yaw axis, length of tail boom, area of tail, and pitch angle were analyzed [52]. Four different tails and two different yaw axis offsets were tested because furling behavior was critical to the performance [53]. Overall, nine configurations including two sets of blades with different pitch angles were tested to try to improve performance at low wind speeds.

The pivot point does not have to be around the vertical axis (yaw); it can also be about a horizontal axis and would produce vertical furling. The rotor and generator on the North Wind high-reliability turbine had a horizontal pivot for the rotor and generator, a coil spring damper, and a gravity restoring mechanism. Another horizontal pivot was unique in that the rotor was downwind and the tail with flat plate and fins hung down (moderate winds to 13 m/sec). In high winds, the force on the rotor and also on the flat part of the tail moved the tail and rotor, the alternator to the horizontal position, and caused vertical furling (Figure 6.25). There is no hinge on the tail and the restoring force is gravity.

Small wind turbines are usually mounted on pole and lattice towers. Very small wind turbines are mounted on almost any structure—buildings and even on short poles (stub masts) mounted on sailboats.

In the 1980s, small wind turbines were made by a large number manufacturers. The viable numbers in the U.S. and Europe were reduced to a few. New subsidies after the turn of the millennium increased the number of manufacturers in the U.S. and Europe who now produce a large number of vertical axis designs. However as in the past, only a few will be long-term commercial operations.

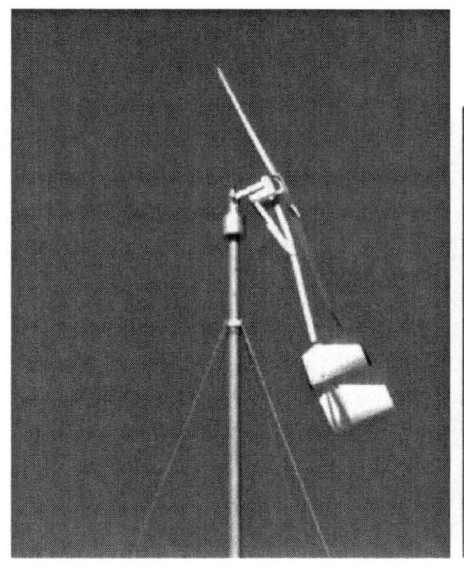

FIGURE 6.25 Rotor axis offset from horizontal pivot point for control. Left: run position at wind speeds below 13 m/sec. Right: furled position in high winds.

REFERENCES

1. I.A. Abbott and A.E. Von Dolnhoff. 1959. *Theory of Wing Sections Including a Summary of Airfoil Data.* New York: Dover.
2. S.J. Miley. 1982. *A catalog of low Reynolds number airfoil data for wind turbine applications,* RFP-3387. Department of Aerospace Engineering, Texas A&M University.
3. R.E. Wilson, P.B.S. Lissaman, and S.N. Walker. 1976. *Aerodynamic performance of wind turbines.* ERDA/NSF/04014-76-1, UC-60. Available from NTIS.
4. D.M. Eggleston and F.S. Stoddard. 1987. *Wind Turbine Engineering Design.* New York: Van Nostrand Reinhold.
5. D. Le Gouriérès. 1982. *Wind power plants, theory and design.* Oxford, UK: Pergamon Press.
6. L. L. Freris, Ed. 1990. *Wind Energy Conversion Systems.* Englewood Cliffs, NJ: Prentice Hall.
7. D. A. Spera, Ed. 1994. *Wind Turbine Technology.* New York: ASME Press.
8. NTIS. Wind Energy Conversion Reports. C00-4131-Ti: Methods for Design Analysis of Horizontal Axis Wind Turbines; Aerodynamics of Horizontal Axis Wind Turbines; Dynamics of Horizontal Axis Wind Turbines; Drive System Dynamics; Experimental Investigation of a Horizontal Axis Wind Turbine; Nonlinear Response of Wind Turbine Rotor; Effects of Tower Motion of the Dynamic Response of Windmill Rotor; Free Wake Analysis of Wind Turbine Aerodynamics; Aerodynamics of Wind Turbine with Tower Disturbances.
9. E. H. Lysen. 1983. *Introduction to Wind Energy.* Amersfoort, Netherlands: Consultancy Services: Wind Energy for Developing Countries.
10. T. Burton, D. Sharpe, N. Jenkins et al. 2001. *Wind Energy Handbook.* New York: John Wiley & Sons.
11. J.F. Manwell, J.G. McGowan, and A.L. Rodgers. 2002. *Wind Energy Explained.* New York: John Wiley & Sons.
12. M.O.L. Hansen. 2008. *Aerodynamics of Wind Turbines.* 2nd ed. London: Earthscan.
13. J. Pramod. 2011. *Wind Energy Engineering,* New York: McGraw Hill.
14. E.A. Avollone, T. Baumeister III, and A. Sadegh, Eds. 2002. *Marks' Standard Handbook for Mechanical Engineers,* 11th ed. New York: McGraw Hill.
15. J. S. Rohatgi and V. Nelson. 1994. *Wind characteristics: an analysis for the generation of wind power.* Alternative Energy Institute, West Texas A&M University.
16. J.L. Tangler. 1983. *Horizontal axis wind system rotor performance model comparison: a compendium.* RFP-3508, UC-60, Rockwell International and Wind Energy Research Center (now National Wind Technology Center).
17. R.J. Templin. 1974. *Aerodynamic performance theory for the NTC vertical axis wind turbine.* LTR-LA-160, National Research Council of Canada.
18. R.J. Maraca et al. 1975. *Theoretical performance of vertical axis windmills.* NASA TM TMS-72662, NASA Langley Research Center.
19. J.H. Strickland. 1975. *The Darrieus turbine: A performance prediction model using multiple stream tubes.* SAND 75-0431. Albuquerque, NM: Sandia National Laboratory.
20. R.E. Akins, D.E. Berg, and W.T. Cyrus. 1987. *Measurements and calculations of aerodynamic torques for a vertical axis wind turbine.* SAND 86-2164. Albuquerque, NM: Sandia National Laboratory.
21. S. Schreck and M. Robinson. 2007. *Wind turbine blade flow fields and prospects for active aerodynamic control.* NREL/CP-500-41606. www.nrel.gov/wind/pdfs/41606.pdf
22. D. Simms et al. 2001. *Unsteady aerodynamics experiment in the NASA Ames wind tunnel: a comparison of predictions to measurements.* NREL/TP-500-29494.
23. J. McCarty. 1993. PROP93: Interactive editor and graphical display. In *Proceedings of Windpower Conference,* p. 495.
24. J. Tangler and J.D. Kocurek. 2005. *Wind turbine post-stall airfoil performance characteristics guidelines for blade-element momentum methods.* NREL/CP-500-36900. Presented at 43rd AIAA Aerospace Sciences Meeting. www.nrel.gov/docs/fy05osti/36900.pdf
25. M.S. Selig and J.L. Tangler. 1994. A multipoint inverse design method for horizontal axis wind turbines. In *Proceedings of Windpower Conference,* p. 605.
26. M.S. Selig. PROPID for horizontal axis wind turbine design. www.ae.illinois.edu/m-selig/propid.html
27. J.L. Tangler and D.M. Sommers. 1993. NREL airfoils families for HAWTS. In *Proceedings of Windpower Conference,* p. 117.
28. B.C. Andrews. 1993. Optimal design of complex wind turbine blades. Master's thesis, West Texas A&M University (includes user's manual).

29. B. Andrews and K. Van Doren. 1988. A method for geometric modeling of wind turbine blades. In *Proceedings of Seventh ASME Wind Energy Symposium*, p. 115.
30. K.L. Starcher. 1995. Atmospheric test of special purpose thin airfoil family. Master's thesis, West Texas A&M University.
31. K.L. Starcher, V.C. Nelson, and J. Wei. 1996. Test results of NREL 10m, special-purpose family of thin airfoils. In *Proceedings of Windpower Conference*, p. 261.
32. J. van Meel and P. Smulders. 1989. *Wind Pumping: a Handbook*. World Bank Technical Paper 101, p. 30.
33. M.H. Khan. 1978. Model and prototype performance characteristics of a Savonius rotor windmill. *Wind Engineering*, 2, 75.
34. T. D. Ashwill. 1992. *Measured data for the Sandia 34-meter vertical axis wind turbine*. SAND 91-2228. Albuquerque, NM: Sandia National Laboratory.
35. D.R. Miller and P.J. Sirocky. 1985. Summary of NASA/DOE Aileron Control Development Program for Wind Turbines. In *Proceedings of Windpower Conference*, p. 537.
36. DOE-EPRI Wind Turbine Verification Program. www.p2pays.org/ref/11/10097.pdf
37. P. Veers, G. Bir, and D. Lobitz. 1998. Aeroelastic tailoring in wind-turbine blade applications. In *Proceedings of Windpower Conference*, p. 191. www.sandia.gov/wind/other/AWEA4-98.pdf
38. Proven Wind Turbines (now Kingspan). http://wind.kingspan.com/index.aspx
39. LM Wind Power. www.lmwindpower.com
40. Sandia National Laboratory. Wind concepts, analysis, and design tools. http://energy.sandia.gov/?page_id=352
41. Wind turbine with record breaking rotors. www.siemens.com/innovation/en/news/2012/e_inno_1223_2.htm
42. Danish Wind Industry Association. Blade testing. www.windpower.org/en/tour/manu/bladtest.htm
43. J. Wei and J. McCarty. 1993. Acoustic emission evaluation of composite wind turbine blades during fatigue testing. *Wind Engineering*, 6, 266.
44. Massachusetts Clean Energy Center. www.masscec.com/index.cfm/pid/10463 and J. Cotrell, W. Musial, and S. Hughes. 2006. *Necessity and requirements of a collaborative effort to develop a large wind turbine blade test facility in North America*. Technical Report NREL-TP-500-38044. www.nrel.gov/wind/pdfs/38044.pdf
45. J. Paquette, J. van Dam, and S. Hughes. 2007. Structural testing of 9-m carbon fiber wind turbine research blades. Paper presented at Wind Energy Symposium. www.nrel.gov/wind/pdfs/40985.pdf
46. D.J. Malcolm and A.C. Hansen. 2006. *WindPACT turbine rotor design study, June 2000–June 2002*. Subcontract Report NREL/SR-500-32495. www.nrel.gov/wind/pdfs/32495.pdf
47. J. Jonkman. FAST: an aeroelastic design code for horizontal axis wind turbines. http://wind.nrel.gov/designcodes/simulators/fast
48. J. Schmid and W. Palz. 1986. *European wind energy technology: state of the art of wind energy converters in the European community. Solar Energy R&D in the European Community, Series G, Wind Energy*, Vol. 3. Boston: D. Reidel.
49. National Wind Technology Center. Lancaster Tokyo Furling Workshop presentations. http://wind.nrel.gov/furling/presentations.html
50. J.M. Jonkman and A.C. Hansen. 2004. *Development and validation of an aeroelastic model of a small furling wind turbine*. NREL/CP-500-30589. www.nrel.gov/docs/fy05osti/39589.pdf
51. M. Bikdash. 2000. Modeling and control of a Bergey-type furling wind turbine. http://wind.nrel.gov/furling/bikdash.pdf
52. B. Vick et al. 2000. Development and testing of a 2-kilowatt wind turbine for water pumping. AIAA-2000-0071. In *Proceedings of 19th ASME Wind Energy Symposium*, p. 328.
53. B. Vick and R.N. Clark. 2000. Testing of a 2-kilowatt wind-electric system for water pumping. In *Proceedings of Windpower Conference*. CD.

PROBLEMS

1. A blade is 12 m long and weighs 500 kg. The center of mass is at 5 m. What is the torque if the force is 320 Nm?
2. Find the power loss for three struts on a HAWT. The struts are 4 m long and 2.5 cm in diameter. Rotor speed is 180 rpm. Use numerical approximation by dividing struts into 1-m sections and calculate at midpoint of section. Then add the values for each section. $C_D = 1$.

3. Calculate the power loss for the struts on a VAWT. Center tube torque tube diameters = 0.5 m. Struts are at the top and bottom, 2 m long from torque tube to blades, diameter = 5 cm, rotor speed = 80 rpm. C_D = 1. Calculate numerically (see problem 2) or use calculus.
4. (a) For those who know calculus, find the value of u (speed of drag device) that produces the maximum C_P for a drag device. Use Equation (6.10), where v_0 is the wind speed at infinity. (b) For those who do not know calculus, find the value of u that produces the maximum C_P for a drag device by plotting the curve [Equation (6.12)] for different values of u (between 0 and 1).
5. Aerodynamic efficiency can be maintained for different solidities of a rotor. If solidity increases, will you increase or decrease the tip speed ratio?
6. Explain the difference in performance of a wind turbine if it (a) operates at a constant tip speed ratio; and (b) operates at constant rpm.
7. What is the maximum theoretical efficiency for a wind turbine? What general principles were used to calculate this number?
8. If the solidity of the rotor is very small, for example, a one-bladed rotor, what is the rpm for maximum C_P compared to the same size rotor with higher solidity?
9. (a) For those who know calculus, calculate the value of axial interference factor for which C_P is a maximum for a lift device and show that this gives a maximum C_P = 59%. (b) For those who do not know calculus, find the value of α that produces the maximum C_P by plotting the curve [Equation (6.20)] for different values of α.
10. A rotor reaches maximum C_P at a tip speed ratio of 7. Calculate rotor rpm for four wind turbines (diameters of 5, 10, 50, and 100 m) at wind speeds of 10, 20, and 30 m/sec.
11. A wind turbine that operates at constant rpm will reach maximum efficiency at only one wind speed. What wind speed should be chosen?

For Problems 12 through 18, the specifications for a wind turbine are induction generator (rpm = 65), fixed pitch, rated power = 300 kW, hub height = 50 m, rated wind speed = 18 m/sec, tower head weight = 3,091 kg, two-blade rotor; mass of one blade = 500 kg, hub radius = 1.5 m, and rotor radius = 12 m.

12. How fast is the tip of the blade moving?
13. How fast is the blade root (at hub radius) moving?
14. Put the mass at the midpoint and calculate the kinetic energy for one blade. Assume the mass of the blade is distributed evenly over ten sections. What is the kinetic energy for one blade?
15. At rated wind speed, calculate the torque since you know power and rpm (remember angular velocity—rad/sec).
16. At 10 m/sec, what is the thrust (force) on the rotor trying to tip the unit over? Calculate for that wind speed over whole swept area.
17. If the unit produced 800,000 kWh per year, calculate output per rotor swept area.
18. Calculate the annual output per weight on top of tower (kWh/kg).

For Problems 19 through 25, specifications for a wind turbine are induction generator (rpm = 21), variable pitch, rated power = 1,000 kW, hub height = 60 m, rated wind speed = 13 m/sec, tower head weight = 20,000 kg, three-blade rotor, mass of one blade = 3,000 kg, hub radius = 1.5 m, and rotor radius = 30 m.

19. How fast is the tip of the blade moving?
20. How fast is the blade root (at hub radius) moving?
21. Place the mass at the midpoint of the blade and calculate the kinetic energy for one blade. Assume the mass of the blade is distributed evenly over ten sections. Now what is the kinetic energy for one blade?

Design of Wind Turbines

22. Calculate the torque at the rated wind speed. You know the power and rpm (remember angular velocity—rad/sec).
23. At 15 m/sec, what is the thrust (force) on the rotor trying to tip the unit over? Calculate for that wind speed over the whole swept area.
24. If the unit produces 2,800,000 kWh per year, calculate the specific output (annual kWh per rotor area).
25. Calculate the annual output per weight on top of tower (kWh/kg).
26. For a 12-m blade, center of mass at 5 m, and weight = 500 kg, calculate the angular momentum if the rotor operates at 60 rpm.
27. For the blade in Problem 26, the angular momentum is around 8×10^4 kg m²/sec. Calculate the torque on the blade at that point if the angular moment of the rotor is stopped in 5 sec. Use Equation (6.24). Then estimate the force trying to bend the blade.
28. Why are the blades for large wind turbines made from fiberglass-reinforced plastics?
29. Why are yaw rates limited on large wind turbines and yaw dampers installed on small wind turbines?
30. How does furling work on small wind turbines?
31. For loss of load on small wind turbines connected to a utility grid, how long can over-speed shutdown take?
32. For megawatt-size wind turbines, what is the most common configuration?
33. Go to a website of any manufacturer of small wind turbines. Note the type of blade construction and material.
34. List two methods of nondestructive testing and briefly describe them.

7 Electrical Issues

7.1 FUNDAMENTALS

Electricity and magnetism are concerned with charges and their movements. The fundamental ideas of electricity and magnetism are discussed in introductory physics texts. The following terms are given as a background about generators and controls.

Current: The current is the flow of charge q (electrons in most cases) past some point. Charge is measured in coulombs. Direct current (DC) results from flow in one direction, and alternating current (AC) results when the flow changes direction. The frequency (number of cycles per second) is measured in hertz (Hz).

$$I = \frac{\Delta q}{\Delta t}, \text{ampere (A)} \tag{7.1}$$

For electric utilities in the U.S., the voltage and current change sixty times per second (60 Hz). Other countries use 50 Hz for their utility systems. If the utility voltage or current is plotted versus time, it looks a sine curve (Figure 7.1).

Voltage: It takes energy to move charges around, and the potential energy (PE) to move charge divided by the charge is called the potential difference and is measured in volts (V). For AC, the voltage also changes with time, just like the current.

$$V = \frac{PE}{q}, \text{volts (V)} \tag{7.2}$$

Resistance: There is a resistance to the flow of a charge across different elements in a circuit. A circuit consists of a source (voltage), current through the wires, and a load or resistance.

$$R = \frac{V}{I}, \text{ohm } (\Omega) \tag{7.3}$$

In metals the amount of current is linearly proportional to the voltage—a relationship known as Ohm's law.

$$V = IR \tag{7.4}$$

Also, in metals the resistance increases with temperature, which means more energy is lost as the temperature increases because of the current.

Power: The power in a circuit is the voltage times the current:

$$P = VI \tag{7.5}$$

The power lost due to heating of the conductor (metal) depends on the square of the current:

$$P = VI = I^2 R \tag{7.6}$$

The implication is that electric power needs to be transmitted at high voltages. In the summer, as air temperature increases, the transmission lines are further limited in the amount of power they

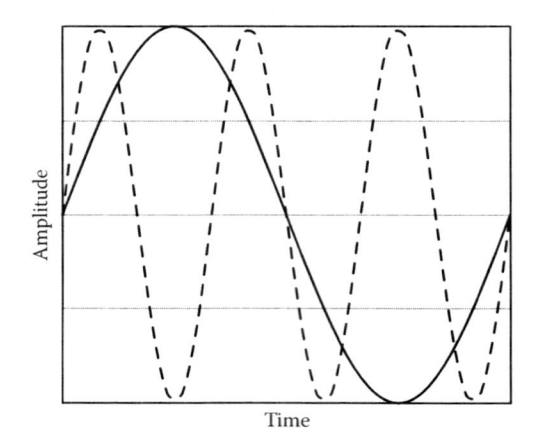

FIGURE 7.1 Two sine waves with different frequencies.

can carry. High current and high temperatures also lead to more sag in transmission lines. Wind turbines with generators at 240 or 480 V must be fairly close to the load or utility line. At higher voltages, smaller diameter wire can be used. Transformers change the voltage, so every large turbine in a wind farm will have a transformer to increase the voltage for transmission. The transformer may be at the top in the nacelle, which means the power wires down the tower can be smaller.

Capacitance: Capacitors are devices for storing charge. An example of a capacitor is two metal plates separated by a small distance. Capacitors are not used for long-term storage because the charge leaks away.

Inductance: Inductors are devices for storing magnetic fields. An example of an inductor is a coil of wire.

Electric field: An electric field E originates or terminates on charged particles. If a charged particle feels a force, it is in an electric field.

$$E = \frac{F}{q} \tag{7.7}$$

Magnetic field: A magnetic field B arises from moving charges or intrinsic spin (a property of particles just like charge is a property of particles). Some materials have magnet fields and are called permanent magnets. Permanent magnet alternators use rare earth atoms that cost more than iron, nickel, and cobalt. China is the source of a lot of the rare earths—a strategic asset that goes beyond wind turbines because rare earths are also needed for hybrid and electric vehicles.

If a moving charge feels a force at right angles to its motion, it is in a magnetic field. Also, changing electric fields create changing magnetic fields, and changing magnetic fields create changing electric fields. Maxwell formulated the theory of electromagnetism in all of its elegance of four equations, appropriately called Maxwell's equations. The theory serves the entire electric power industry and led to communication via electromagnetic waves that we accept as common today.

If charged particles are placed in external electric fields and moving charged particles are placed in external magnetic fields, a force is exerted on the charged particles. The amount of force depends on the strength of the electric and magnetic fields, the amount of charge, and the velocity of the charge.

$$F = q E + q (v \times B) \tag{7.8}$$

Electrical Issues

This equation is the basis for understanding the conversion of electric energy to mechanical energy (motor) and the conversion of mechanical energy to electric energy (generator).

Motor: A loop of wire contains moving charges (current) due to a connection to an electric plug. The loop is in an external magnetic field; therefore, a force on the charges and a torque on the wire constitute a motor (Figure 7.2). The torque on the loop is given by the current in the wire I, the area of the loop A, and the strength of the magnetic field B. The motor contains a coil of wire with many loops. Check the links listed at the end of this chapter for information on how an electric motor works.

$$T = I\,(A \times B) \tag{7.9}$$

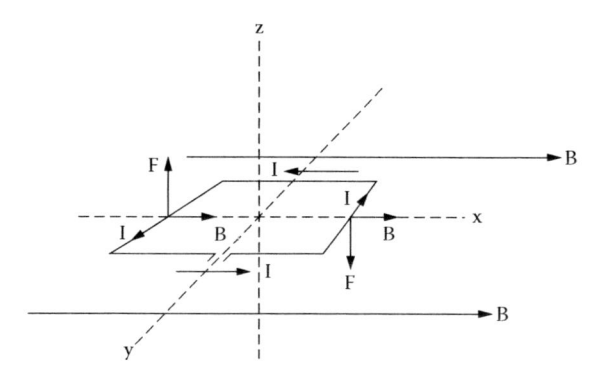

FIGURE 7.2 Forces on sides of a current-carrying loop in external magnetic field. The result of the set of forces produces torque T that makes the loop rotate (motor).

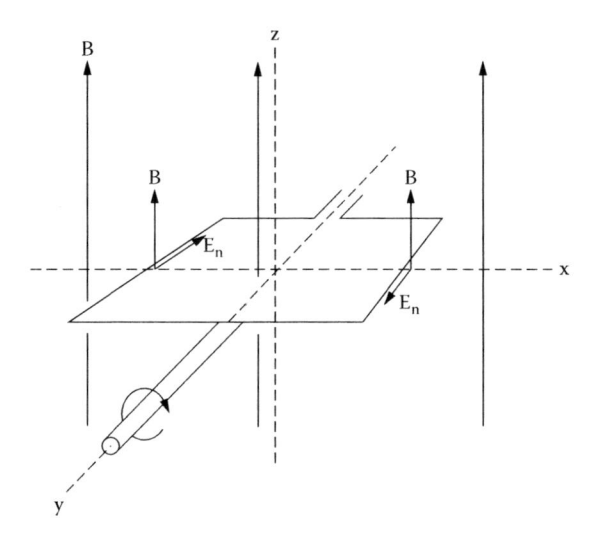

FIGURE 7.3 Rectangular loop rotated by outside force with angular velocity ω in a uniform external magnetic field (generator).

Generator: A loop of wire is moved (rotated) by an external force (Figure 7.3). The shaft power, $P = T \times \omega$ which, in this case, comes from the wind turbine rotor, directly or through a gearbox. The charges (electrons in the wire) are moving in an external magnetic field, and there is force on the charges. A coil may contain many loops of wire. A single coil is known as a single-phase generator; three coils of wire constitute a three-phase generator.

An external magnetic field can be produced by permanent magnets or electromagnets. For an electromagnet, a current in a coil produces a magnetic field, and an iron core in the coil will strengthen the magnetic field. The number of coils is referred to as poles. The current from a utility grid or a generator is used to produce the magnetic fields. Check the links at the end of the chapter to learn how an electric generator works.

7.1.1 Faraday's Law of Electromagnetic Induction

Another way of looking at electromotive forces is by Faraday's law of electromagnetic induction. The amount of magnetic flux, Φ_M, is equal to the strength of the magnetic field times the area:

$$\Phi_M = B \bullet A\cos(\theta) \tag{7.10}$$

where θ is the angle between **B** and **A**. The electromotive force is then equal to the negative change in magnetic flux with time:

$$\varepsilon = -\frac{\Delta \Phi}{\Delta t} \tag{7.11}$$

In generators and motors, the magnetic field and area can be kept constant and the angle between them may be changed by rotating a loop of wire. This yields an alternating voltage and current that vary like a sine wave.

Induction requires two coils. The changing magnetic flux in one coil causes a changing current in the next coil. A transformer works by induction. If the load is pure resistance, the voltage is in phase (0 phase angle) with the current. With a capacitor, the voltage lags the current by 90 degrees. With an inductor, the voltage leads the current by 90 degrees (Figure 7.4). All voltages in the figure are set at an angle of zero, and the current is shown in relation to the voltage (starting at a different angle for the sine curve). Check the links at the end of the chapter for information about the relation of voltage and current.

7.1.2 Phase Angle and Power Factor

The instantaneous voltage and current are given by

$$v = V_p \sin(\omega t), \; i = I_p \sin(\omega t + \varphi)$$

where V_p and I_p are the peak values; ω is the angular velocity in radians per second (2π times frequency); and the angle φ is the difference in degrees between the instantaneous voltage and current (sine wave for voltage and sine wave for current). For a resistor, the voltage and current are in phase and the average power over one cycle is

$$P = V \times I = V_p \sin(\omega t) \times I_p \sin(\omega t) = 0.5 \, V_p I_p \tag{7.12}$$

For capacitors and inductors, the voltage and current are 90 degrees out of phase and the average power is zero:

$$P = V_p \sin(\omega t) \times I_p \sin(\omega t + 90) = V_p \sin(\omega t) \times I_p \cos(\omega t) = 0$$

Electrical Issues

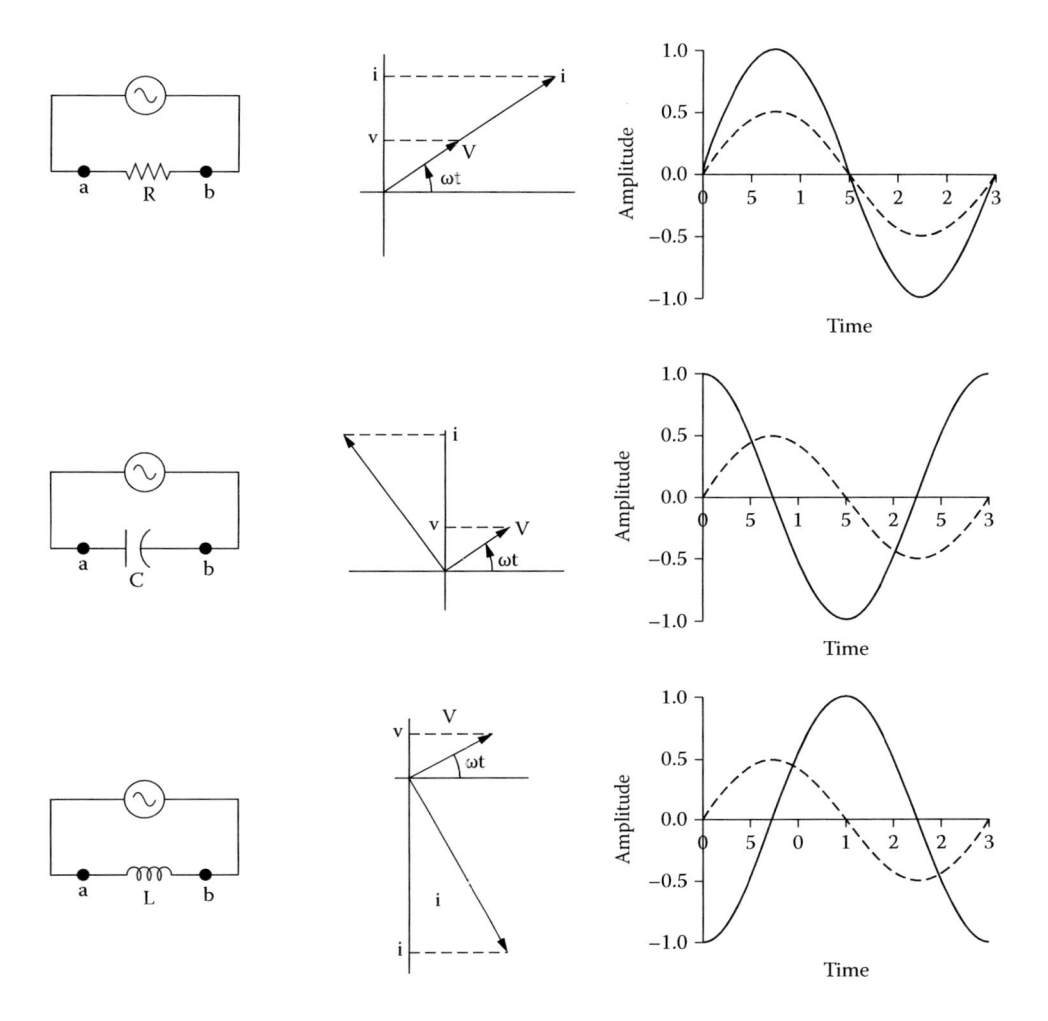

FIGURE 7.4 Current and voltage across a resistor, capacitor, and inductor, showing the phase relationship between voltage (dashed line) and current (solid line).

All real circuits involve inductance, capacitance, and resistance, so the current and voltage will not be completely in phase (Figure 7.5). The instantaneous power to an arbitrary AC circuit oscillates because both the voltage and current oscillate:

$$p = vi = V_p \sin(\omega t) * I_p \sin(\omega t + \phi) \tag{7.13}$$

The average power is found by integrating over one cycle:

$$P_{avg} = \frac{V_P I_P \cos\phi}{2}$$

However, parameters measured for AC circuits are average values of current and voltage that are given by the root mean square values:

$$V = (v \times v)^{0.5} = \{V_p \sin(\omega t) \times V_p \sin(\omega t)\}^{0.5} = V_p/2^{0.5} = 0.707\, V_p$$

The equation applies to a single-phase system. For three phases, the measured current is reduced by $3^{0.5}$. For a three-phase transfer of power, each leg transfers a current equal to

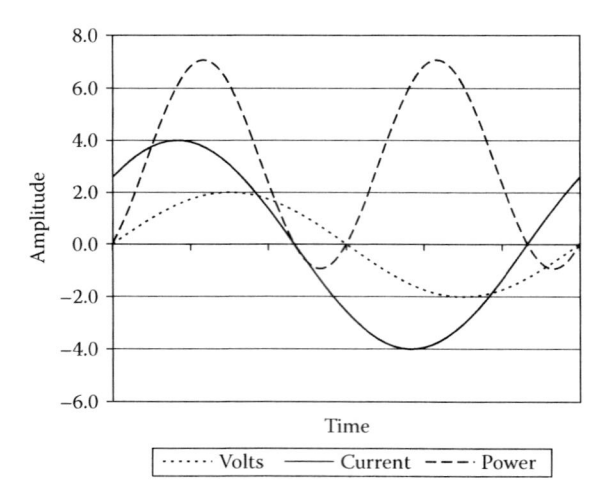

FIGURE 7.5 Instantaneous power in arbitrary AC circuit.

the coil current/1.73, and therefore the wire size needed is smaller. The real power generated or consumed is given by

$$P_{avg} = VI \cos \phi \quad (7.14)$$

where $\cos \phi$ is the power factor. Adding a number of induction generators to the utility line can change the power factor and reduce the actual power delivered—a concern of utility companies since a utility grid supplies the reactive power for the induction generators. Therefore, some wind turbines and most wind plants have capacitors added to the turbines or to the electric substations. There are a number of electrical conversion systems for wind turbines [1–8].

7.2 GENERATORS

The main classifications of generators are direct current, synchronous, and asynchronous (subdivided into induction generators and permanent magnetic alternators). The operation is constant or variable rpm, and as noted, constant rpm operation only reaches maximum power coefficient at one wind speed (Figure 6.14).

The variable rpm operation up to the rated wind speed is along the line of maximum power coefficient (Figure 5.13); however, above that wind speed not all the available power is captured.

Note: For this chapter, wind rotor will refer to the hub and blades, and rotor will refer to the rotating part of a generator.

The electrical conversion is achieved with constant wind rotor rpm with squirrel cage induction generators or synchronous generators, or with variable wind rotor rpm with doubly fed (wound rotor) induction generators, permanent magnet alternators, or direct-drive generators [9]. The variable frequency output is then converted to constant frequency. There is a trade-off between wind rotor efficiency and the cost and efficiency of conversion for variable frequency to constant frequency. An AC synchronous generator must be regulated to the correct rpm and synchronized with the grid. Induction generators essentially operate at constant rpm with a small variation known as slip. Induction generators are tied to the frequency of the grid since the grid supplies the reactive power for the field coils of the generator.

Rural electric grids may only have one phase, so a wind turbine connected directly to these utility lines would need a single-phase generator. If a wind turbine is connected through an inverter, the inverter can handle the phase.

Electrical Issues

A generator is composed of an armature (coil of wire around metal core) and a field. Power is taken from the armature. The field controls the power and consists of permanent magnets or an electromagnet (energized coil of wire). In the latter case, the generator contains two coils: (1) a stationary coil or stator and (2) a rotating coil or rotor. In a DC generator, the armature rotates and power is taken from a commutator by brushes. Brushes need maintenance so alternators are used. In an alternator, the field rotates and the variable AC output is converted to DC by a rectifier circuit. The DC is then converted into constant voltage and constant frequency by an inverter.

The advantage of a DC generator, permanent magnetic alternator, and doubly fed induction generator is the variable rpm (constant C_P) operation that is aerodynamically more efficient. For small wind turbines (to 10 kW; 50 kW under development), the elimination of a speed increaser is another advantage. Jacobs used a direct-drive, self-excited generator whose residual magnetization provides the initial voltage output. Feedback is used to increase the field and enhance power output. The generator output can be single- or three-phase. The Danish Wind Industry Association has a good explanation of types and operation of generators (see Links at the end of this chapter).

For HAWTs, the power is transferred to ground level through slip rings, or the power cord has enough slack to twist during yaw revolution. The second method has the desirable feature of eliminating the slip rings that are always potential problems for control signals and even for power transfer. However, strict observation schedules on length of the power drop cord or trip relay for yaw must be followed.

A number of wind turbines also use direct drive with permanent magnet alternators. Output is rectified to DC and then converted to AC by an inverter. Output is 120 or 240 V AC, single- or three-phase, for small wind turbines.

Synchronous generators and self-commutated inverters require a means of disconnect for safety during faults on a utility line because they are power sources. Induction generators at constant rpm operation drop offline during utility grid faults because the power for the field coils comes from the grid. For small wind turbines, synchronous generators will probably not be acceptable for interface with a grid, primarily because of the complications of controlling wind rotor rpm.

7.2.1 Induction Generator, Constant RPM Operation

Induction generators (Figure 7.6) are used for wind turbines because induction motors are mass produced, inexpensive, involve fewer operation and maintenance costs, and have simple controls. The induction motor or generator is brought up to synchronous speed and then connected to the utility line. All the features of synchronous generators for control of speed, excitation, and synchronizing are eliminated as the utility line provides these controls.

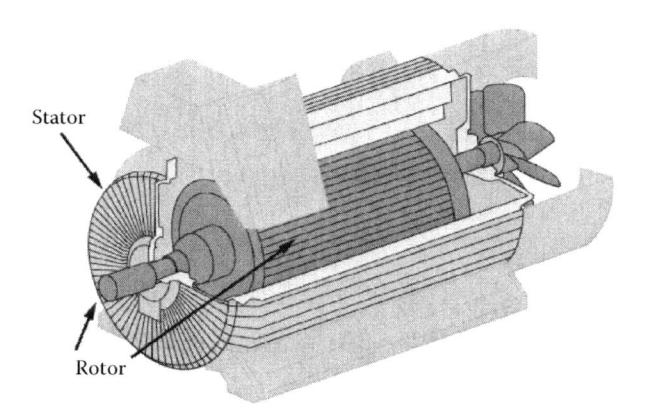

FIGURE 7.6 Cut-away drawing of induction generator.

The rotor is in the center of a four-pole stator whose magnetic fields are supplied by a three-phase utility grid. The rotor cage consists of a number of copper or aluminum bars connected by aluminum end rings. The rotor has an iron core consisting of thin insulated steel laminations with holes for the conducting bars. AC voltages across each pair of terminals create a rotating magnetic field and produce rotation in the center rotor; phase separation is 120 degrees (Figure 7.7).

The rotating magnetic field induces currents in a set of copper loops in the rotor, and magnetic forces on these current loops exert a torque on the rotor and cause it to rotate (as a motor). When it is forced to rotate past the synchronous speed (900, 1,200, or 1,800 rpm), it becomes a generator. The relationships of power, torque, efficiency, and rpm are shown in Figure 7.8.

Some large wind turbines had two generators, one for low wind speeds and the other for high speeds. A common design for newer machines is pole changing that allows them to run as a small or large generators, for example, 400 or 2,000 kW at two different rotational speeds. The use of one, two, or pole-switching generators depends on the energy produced and the extra cost of each option.

The switching mechanism must not allow a generator to operate below synchronous speed or it would turn into a gigantic fan. The control mechanism must measure rpm, with some leeway for wind speeds at the cut-in value, to turn the generator on and off, and at high wind speed, to cut out and restart the generator after the winds decline. Some wind turbines use the motor/generator for start-up because their blades do not have enough starting torque. When the winds become high enough (cut-in wind speed), the blades are turned by the motor/generator. As rpm increase due to wind power and when the motor/generator surpasses the synchronous speed, it now operates as a generator.

The time delay reduces on–off cycling when the winds are near the cut-in and cut-out speeds. In one case, a small (5-kW) downwind wind turbine would start in the upwind position because the winds shifted 180 degrees from the position when the turbine shut down. The upwind position is an unstable condition for a downwind rotor with coning. The control system indicated start-up but the blades were inefficient in that position and the wind turbine used 2 kW of power—it really was a big fan.

Induction generators (Figure 7.9) are the most common types for wind turbines from 25 kW to megawatts because the controls for synchronization to the line are simple, rugged, and mass produced. When a failure occurs on a utility grid, they automatically disconnect and present no safety problems. Induction generators decrease the power factor, and correcting capacitors are installed on individual wind turbines or at wind farms.

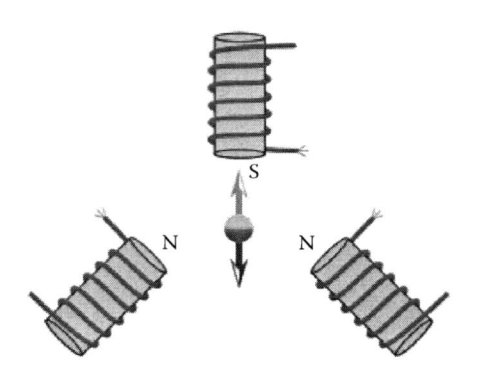

FIGURE 7.7 Three-phase AC generator. Rotating magnetic field produces AC voltages across each pair of terminals. Phase separation = 120 degrees.

Electrical Issues

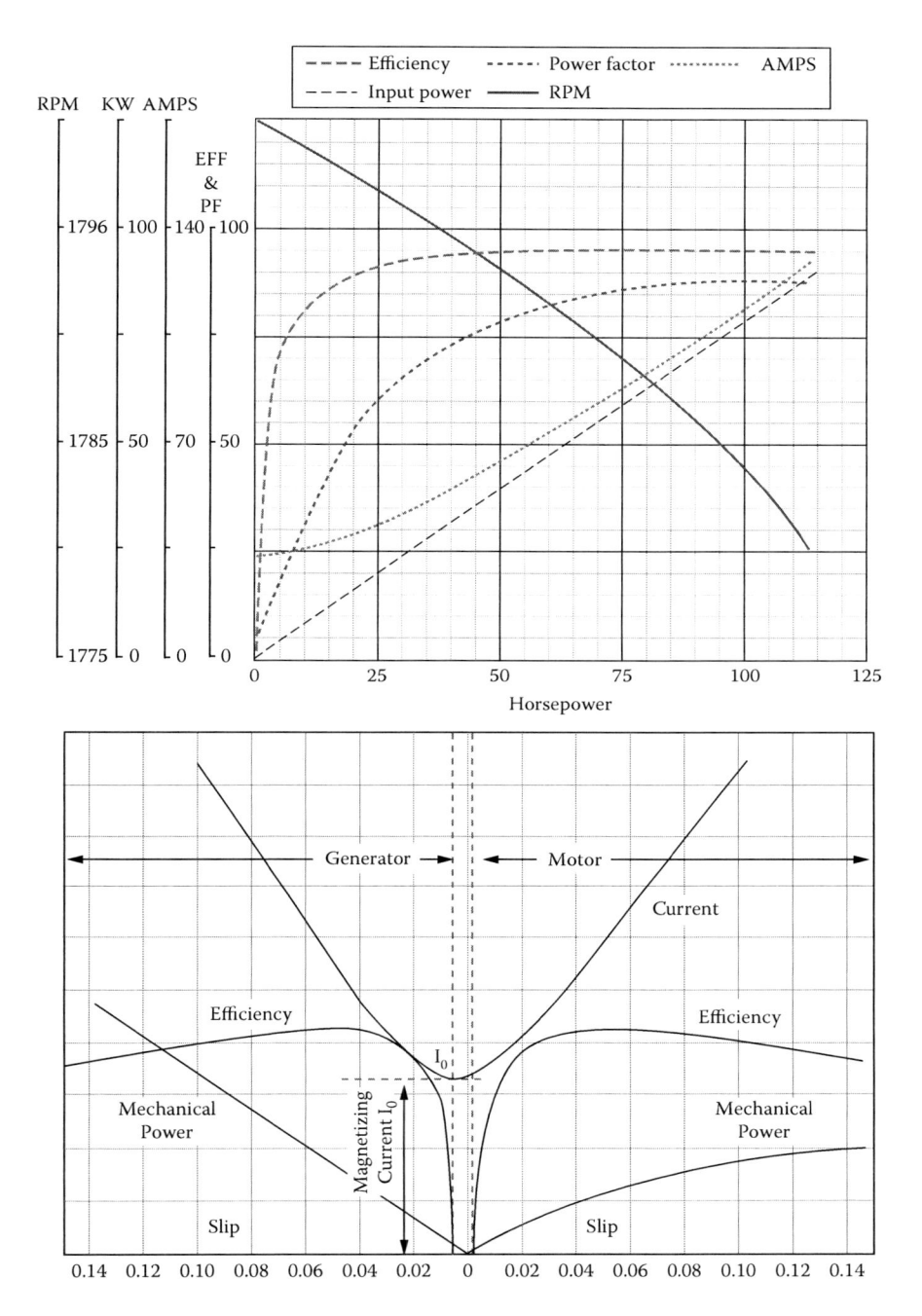

FIGURE 7.8 Operating characteristics of induction motor (420 V, 75 kW). The curves for induction generator are essentially mirror image as shown by bottom graph.

It is possible to have a resonance condition with inductance and capacitance; however, the variability of the wind ensures that induction generator output decreases rapidly when a fault occurs on a utility line. Remember, the induction generator is essentially a constant rpm operation for the rotor, fixed by the frequency of the utility grid. The wind rotor/generator combination reaches peak efficiency at only one wind speed.

FIGURE 7.9 Induction generator, 750 kW; stator; and slip rings for transferring power. (Photos courtesy of Wade Wiechmann.)

7.2.2 DOUBLY FED INDUCTION GENERATOR, VARIABLE RPM OPERATION

A standard (usually 1,500 rpm) doubly fed induction generator is connected to a gearbox. The stator is directly connected to the utility grid and the rotor of the generator is connected to a converter. A range of 60 to 110% of the rated rpm is sufficient for good energy production. At wind speeds above the rated speed, the blades are pitched to reduce aerodynamic efficiency. Variable blade pitch is also used for start-up, shut down, and over-speed.

7.2.3 DIRECT-DRIVE GENERATOR, VARIABLE RPM OPERATION

This type of generator has no gearbox and operates at the same rpm as the wind rotor, 10 to 25 rpm for megawatt wind turbines. These generators are very large (Figure 7.10), and the output is converted to constant frequency and voltage by power electronics. Again, control is by pitch of the blades. The trade-off is between no gearbox and large size with power electronics. More large direct-drive wind turbines are on the market.

7.2.4 PERMANENT MAGNET ALTERNATOR, VARIABLE RPM OPERATION

This is also a direct-drive system with no gearbox and is common on small wind turbines. The usual control for high winds is by furling using a tail. However, Southwest Wind Power has a downwind unit that uses electrodynamic braking for high winds and shutdowns. Some larger permanent magnet alternators are available. For example the Clipper 2.5-MW unit has four 660-kW generators, Goldwind produces 1.5- and 2.5-MW permanent magnet direct-drive units. General Electric manufactures a 2.5-MW unit. The advantages of permanent magnet excitation are lower losses, lower weight, and lower cost. A disadvantage is that the excitation cannot be controlled.

7.2.5 GENERATOR COMPARISONS

All the above generators have been used in wind turbines. The trade-off factors are (1) cost, size, and weight, (2) suitability for grid frequency, (3) blade noise, (4) energy production, (5) reliability and maintenance, (6) power quality, and (7) grid faults [9]. Many manufacturers have changed from constant to variable rpm operation because of energy production and smoother power due to inertia of the wind rotor. In the final analysis, the choice for the electric conversion depends on energy produced (annual) and the cost per kilowatt hour from the wind turbine.

7.2.6 GENERATOR EXAMPLES

At rated power, generators are very efficient; however, at low power levels the efficiency decreases. Therefore, some wind turbines utilized two generators, one of which was for lower wind speeds. The Vestas V47 had 200-kW and 660-kW generators. Another method is to change numbers of

Electrical Issues

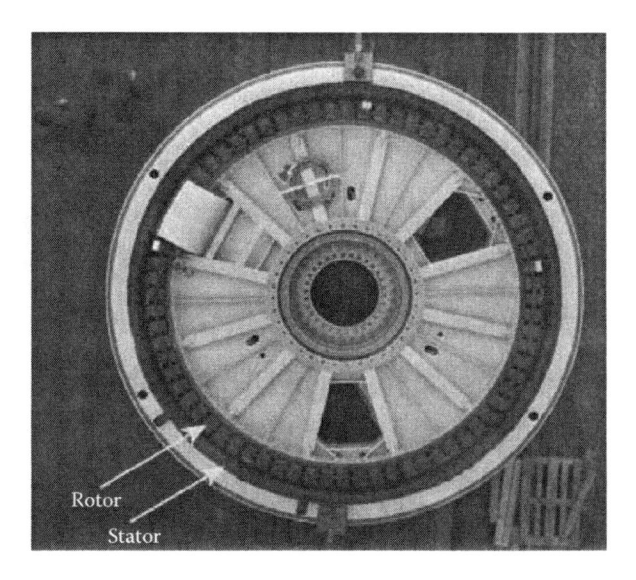

FIGURE 7.10 Ring generator for gearless Enercon, E66. Size can be estimated by comparison to two men in upper left corner. (Photo courtesy of Thomas Schips.)

poles, for example, six poles for low wind speed and four poles for higher speeds. The Bonus generator was rated for 260 and 1300 kW.

The generator for the MOD-5B was rated at 3.2 MW and was a variable speed (1,330 to 1,780) wound-rotor induction type. A cycloconverter system maintained a constant frequency output. The Westinghouse 600-kW wind turbine had a synchronous generator, and frequency was controlled by the variable pitch of the blades. A power control algorithm limited high instantaneous power output (spikes caused by wind gusts) by derating the maximum power by 10% when a power spike exceeded 800 kW.

Large wind turbines can be operated at variable rpm and maximum C_P. This means low rpm generators with large numbers of poles. Project Eole located at Cap Chat, Canada, was a large VAWT rated at 4 MW. Because it was a direct-drive system, the generator was quite large, 12 m in diameter with 162 poles. The output was rectified to DC and then inverted back to 60 Hz AC. The unit only operated for around 10,000 hours, and power output was limited to 2.5 MW.

Enercon, a German manufacturer, developed large-ring generators to eliminate the gearboxes on large wind turbines. The output is rectified and then converted to constant frequency. Over 1 million units from 300 kW to megawatt size have been installed. In 2007, Enercon built a 6-MW unit (now available as 7.58 MW), 126 m in diameter (E126).

The Sandia VAWT test bed (34 m diameter, rated at 500 kW) located at the Agricultural Research Service facility in Bushland, Texas, was designed as a variable speed, constant frequency system. It included a load-commutated inverter, AC-adjustable speed drive, and a synchronous motor/generator rated at 625 kW. Such systems are currently operated in industrial applications. Power electronics and inverters allow wind turbines to operate at either constant or variable rpm.

Jay Carter Sr. developed a wind turbine with the same rotor, hub, and drive train. It has two induction generator options: six poles, 30 kW (wind rotor 60 rpm) rating for medium wind speed regimes, and four poles, 50 kW (wind rotor 90 rpm) rating for good wind speed regimes (Figure 7.11).

Higher voltage generators are used in some wind turbines. A Spanish manufacturer developed a geared wind turbine with a brushless synchronous generator and a full converter.

FIGURE 7.11 Left: generator, gearbox, and Jay Carter Sr. Right: stator and rotor of generator, 50 kW.

TABLE 7.1
Wire size, Copper, 480 V Three-Phase, 2% Voltage Drop

Load, amps	Type Insulated	Overhead Bare, Covered	30	46	60	76	91	107	122	137	152	168	183
5	12	10	12	12	12	12	12	12	12	12	12	12	12
7	12	10	12	12	12	12	12	12	12	12	10	10	10
10	12	10	12	12	12	12	12	10	10	10	10	8	8
15	12	10	12	12	12	10	10	10	8	8	8	8	6
20	12	10	12	12	10	10	8	8	8	6	6	6	6
25	10	10	12	10	10	8	8	6	6	6	6	4	4
30	10	10	12	10	8	8	6	6	6	6	4	4	4
35	8	10	10	10	8	6	6	6	4	4	4	4	4
40	8	10	10	8	8	6	6	4	4	4	4	3	3
45	6	10	10	8	6	6	6	4	4	4	3	3	2
50	6	10	10	8	6	6	4	4	4	3	3	2	2
60	4	8	8	6	6	4	4	4	3	2	2	2	1
70	4	8	8	6	4	4	4	3	2	2	1	2	1
80	4	6	8	6	4	4	3	2	2	1	1	1	0
90	3	6	6	6	4	3	2	2	1	1	0	0	0
100	3	6	6	4	4	3	2	1	1	0	0	0	00
115	2	4	6	4	3	2	1	1	0	0	00	00	000
130	1	4	6	4	3	2	1	0	0	00	00	000	000
150	0	2	4	3	2	1	0	0	00	00	000	000	4/0
175	00	0	4	3	1	0	0	00	000	000	4/0	4/0	4/0
200	000	00	4	2	1	0	00	000	000	4/0	4/0	250	250
250	250	00	3	1	0	00	000	4/0	4/0	250	250	300	300

The sizes of the wires connecting a generator to a grid depend on the current and distance to the connection. For small wind turbines, manufacturers recommend wire sizes for different wire runs; however, see Table 7.1.

7.3 POWER QUALITY

Wind turbines and especially wind farms, which in reality are wind power plants, must provide the power quality [10,11] to ensure the stability and reliability of the system and meet the quality needs of customers on the grid. The four types (Figure 7.12) of connections depend on the electrical

Electrical Issues

conversion, generator, and connection (direct or partial and full converter). Induction generators require reactive power from a grid, and capacitor compensation is often used at the wind turbines or at a substation.

The power output of variable rpm wind turbines is smoother (less flicker) than the output of constant rpm wind turbines because rapid changes in the power are smoothed out by rotor inertia. If a converter is large enough, variable rpm wind turbines can also be used for voltage and frequency control. Power electronic converters produce harmonics that may need to be filtered.

The voltage at each wind turbine on a wind farm varies independently, and turbines may be shut down to correct faults (see Section 5.6.2) or perform maintenance. Capacitor compensation may lead to harmonics and self-excitation, with constant rpm, induction generators [10]; however, the wind speeds are so variable it is improbable that self-excitation will last very long. Fluctuations in voltage and frequency must be kept within ranges acceptable to the utility at the point of connection to the grid (Figure 7.13).

Faults on a utility grid will also cause a reaction from the wind turbines. A wind farm was monitored for one year [11], and 215 faults were noted. At the monitoring node, the voltage drop and spike in current described the fault (Figure 7.14). Most fault events occurred far from the wind farm, and most were cleared within ten cycles. Therefore, voltage ride-through capability of the wind turbines is important. For a doubly fed induction generator, the rotor currents increase very rapidly and should be disconnected from the grid within milliseconds to protect the converter. When constant rpm wind turbines come back online they need a lot of reactive power, which impedes voltage restoration.

The loss of generation from a wind farm during a fault varies from 0 to 100% of its capacity. In terms of loss of generation, the benefit of wind power generation is the amount of power disconnected from the wind farm, as the loss of a single generator in a wind power plant may be less than 1% of total generation. During the year of monitoring, only 1% of all the faults caused high generation losses ($P_{gen} > 0.8 \, P_{rated}$). Note that many engineers use the *wind plant* term because a wind farm really functions as a wind-powered electric generation plant.

7.4 ELECTRONICS

Electronics are used extensively in the control of the wind systems. Controllers generally consist of one or more computer processing units (CPUs) and programmable logic controllers (PLCs) that may be hardwired for wind turbine control and operation. Control systems run the gamut from simple controls for battery storage to supervisory control and data acquisition (SCADA) units for individual wind turbines, village projects, wind–diesel systems, and entire wind farms.

Electronics for power conversion and control are major parts of any wind turbine system, and solid-state inverters allow variable frequency output to be connected to a utility grid. Induction generators on constant rpm units may require a soft start to reduce mechanical stress and reduce the interactions between the utility grid and the wind turbine during connection.

7.4.1 CONTROLLERS

A controller monitors the condition of a wind turbine, collects statistics on its operation, and controls switches for different operations and functions. A controller contains one or more CPUs.

The simplest controller senses the voltage levels of batteries during full charge and discharge and may display the information on light indicators. As the battery bank voltage approaches the regulation level, a wind turbine is furled manually or the controller may switch power to a regular load or a dump load. If no load is available, the wind turbine may be brought to a slow rpm. The controller may include an electrical braking mode used for parking the turbine before climbing or lowering the tower to work on the turbine. Equalization of the batteries (restoring them to a high

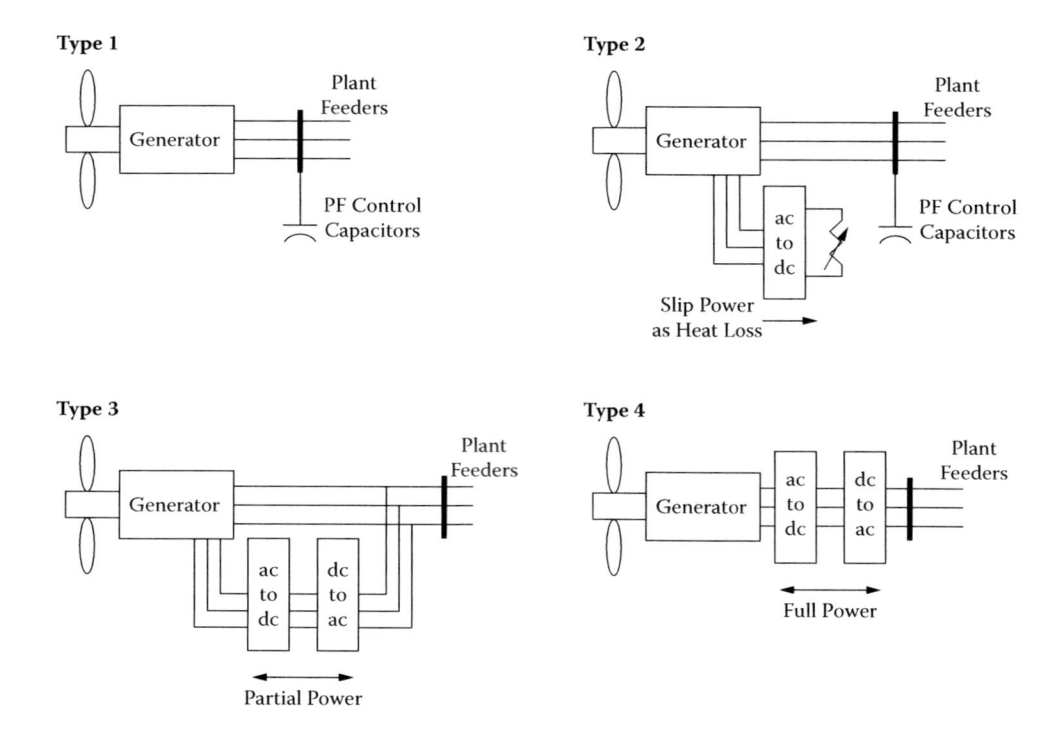

FIGURE 7.12 Four types of dynamic models, wind turbine connection to grid.

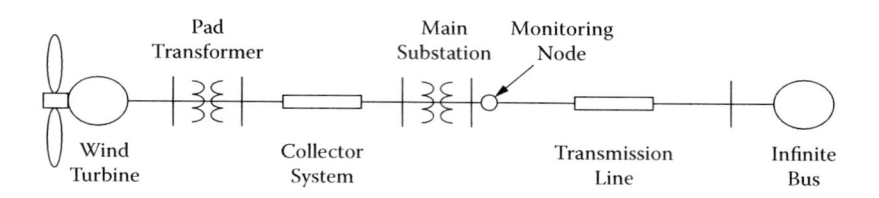

FIGURE 7.13 Typical network topology of large wind farm.

rate of charge) must be performed monthly. Water levels should be checked and distilled water added as needed.

Control of turbines for furling is accomplished mechanically. If a unit is connected to a grid through an inverter, the power output of the wind turbine is converted to DC and a disconnect switch is mandatory. Southwest Wind Power has a unique wind turbine. The DC rectifier, controller, and inverter are all inside the nacelle (see Figure 6.17). The controller regulates an electromagnetic brake for shutdown and to limit rotor rpm in high winds. The connections are the disconnect switch to the grid and a wireless two-way remote to turn the unit on and off.

One option is a wireless remote display to show performance in real time and collect kilowatt hour data for days, months, and years. The remote may be connected to a personal computer (PC) for monitoring turbine performance, and software can generate a power curve for the wind turbines.

For wind turbines with induction generators connected to a grid, the controller (Figure 7.15 and Figure 7.16) has more sensors and functions, for example, measurement of wind speed and rpm to determine switches for start-up if needed, connection of the generator, and control for shutdown and over-speed. The controller will also have sensors for faults and any fault will shut down the wind

Electrical Issues

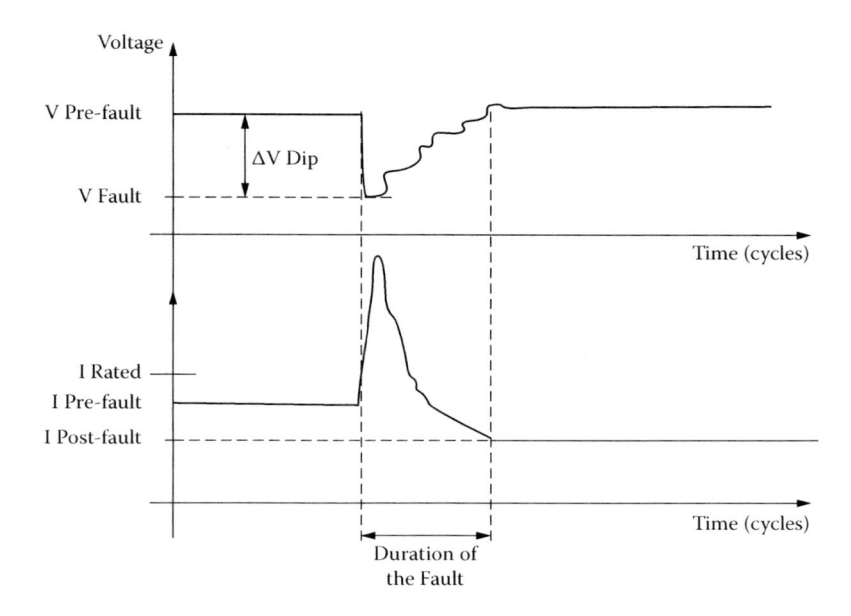

FIGURE 7.14 Voltage and current at connection to wind farm after fault on utility line.

FIGURE 7.15 Disconnects and controller for 50-kW wind turbine.

turbine. The controller may provide communication to an external personal computer on site or far away. Additionally, the computer may be able to change the parameters of the controller.

Large wind turbines require monitoring of 100 to 500 parameters and may require two controllers—one in the nacelle and one at the bottom of the tower, which is connected with fiberoptics on new turbines. Some models include a third controller in the hub for pitch control. CPUs and sensors for safety or operation-sensitive areas are duplicated for redundancy. The controller communicates status and operating conditions of the turbine and provides fault alarms and service requests to the operator, owner, or service contractor.

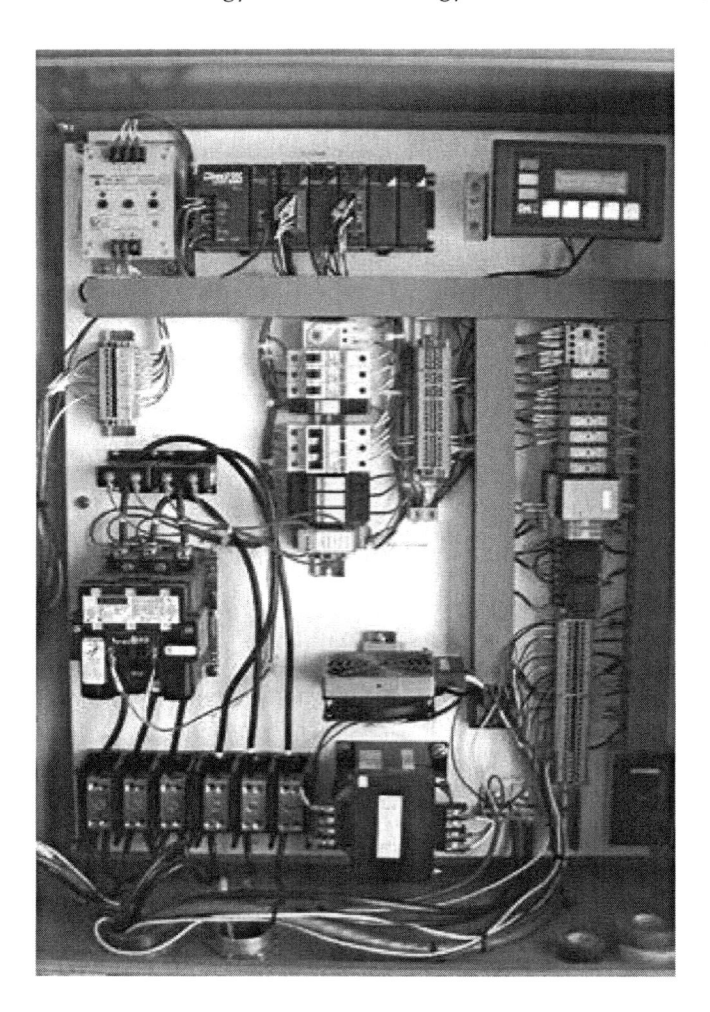

FIGURE 7.16 Controller for 50-kW wind turbine with induction generator (constant rpm operation).

Statistics are collected at the computer to provide a baseline for each wind turbine. Finally, for wind farms, supervisory control and data acquisition (SCADA) are components of control systems [12, Sec. 1.4]. Several companies produce SCADA systems for wind farms. Because wind farms can utilize different wind turbines, operational information on each wind turbine is compared to a baseline database to alert operators of potential problems. At an on-site or remote control room, operators can monitor each wind turbine on a farm and turn them on and off. In cases of high winds when transmission lines were full, wind farm output had to be curtailed. Control systems shut down the turbines within a short period.

A management system can integrate wind farm SCADA, on-site meteorological wind forecasting, and market price data to enable operators to maximize energy production and the resulting income. Control strategies are proprietary because they are intended to maximize energy production within wind regimes and minimize turbine system wear and tear during very high winds.

7.4.2 Power Electronics

Power electronics convert the variable-frequency and voltage power from the generator to the utility grid, constant frequency and voltage within the ranges set by the utility to ensure power quality. Converters are classified by AC to AC without a DC link (output voltages are chopped from input

Electrical Issues 163

voltages) and by AC to AC with a DC link (input voltages are converted into intermediate DC voltages that are stored and then converted to the output voltages).

An overview presents the types of three-phase AC–AC converters [13]. The power electronics of large wind turbines allow them to operate more efficiently. A common system for doubly fed wound induction generators involves a converter connected to the rotor of the generator that directly controls currents in the rotor windings so the mechanical and electrical rotor frequencies are decoupled. Only a fraction (20 to 40%) of the rated generator power passes through the converter. The operational speed range of the generator depends only on the converter rating.

7.4.3 INVERTERS

A number of inverters are on the market, but only a few manufacturers produce inverters for wind turbines. Inverter designs need to accommodate the very different inputs of wind turbines. Inverters for photovoltaic devices have less stringent operating requirements. There are inverters for hybrid systems where power is taken from the battery storage. For wind turbines with permanent magnet alternators, the output is rectified to DC, and the inverter converts DC to the constant voltage and frequency of the grid.

Since a wind turbine is controlled mechanically, the inverter controls the electrical aspects of synchronization of phase and power transfer. Some wind systems use battery storage before the inverter and require a different inverter design. The early inverters used short-length, square wave pulses with proper timing on the cycle to input power to a grid. The square wave pulses added harmonics to the output. Later inverters improved and had efficiencies over 90% under 75% load with 2% harmonic distortion. At low winds and loads, inverter efficiencies will decrease.

The field test of a wind turbine (permanent magnet alternator, three-phase, 10 kW) connected to a grid through an inverter (single-phase, 10 kW) indicated a problem with the inverter in wind speeds of 13 m/sec and greater [14]. Less power was delivered because the inverter entered a pause mode. If the pause happens too many times within a certain period, the inverter quits functioning and must be reset manually.

The main safety function of the inverter is to disconnect the wind turbine from the utility line when a fault occurs on the utility line. This prevents hot wires because the generate will continue to operate. A disconnect (may be fused) is needed between the wind turbine and inverter along with a fused disconnect between the inverter and utility grid.

7.5 LIGHTNING

Lightning is always a problem for electronics, especially for wind turbines connected grids. Lightning strikes on a grid will send spikes over long distances. A wind turbine is generally the tallest lightning rod around, so lightning protection via a path to ground is imperative. Manufacturers' instructions on grounding and number and connection of copper rods (size and length) must be implemented along with other measures for protecting controllers and inverters, from varistors to blow-out cans. Even then lightning can still cause problems by damaging controllers, electrical systems, blades, and generators. Furthermore, the induced electromagnetic fields may damage the pitch control systems inside a hub. Damage due to lightning is a very costly repair because blade and generator replacement may require a crane.

A 1995 German study estimated that 80% of wind turbine insurance claims paid covered damages caused by lightning [15]. Mean annual thunderstorm days and lightning flash density data show regions of the U.S. where wind turbines are subject to the greatest risk from lightning. Lightning was monitored at wind farms in the Turbine Verification Project [16] by collecting data on direct strikes on wind turbines and utility line surges.

The estimated average number of strikes per turbine per year ranged from 0.04 for California to 0.43 in Nebraska. The study information also includes the repair costs [17]. Lightning protection for wind turbines has improved but lightning is capricious and sometimes even the best protection is not sufficient. Blades should have internal lightning conductors running all the way to their tips. One example of lightning protection added after installation was implemented after surges on the utility line damaged the controller. The solution was an underground copper grid connecting all the guy wires plus the turbine tower.

7.6 RESISTANCE DUMP LOAD

If a wind turbine uses resistive loads for over-speed control, the resistors must be outside. During loss of load and high winds at the AEI Wind Test Center, the resistors inside the control shed, along with the controller and inverter, became so hot the control shed caught fire and burned to the ground. Luckily, the fire burned the insulation off the power wires from the wind turbine. They shorted together and shut the wind turbine down before it was destroyed.

LINKS

ABB. www.abb.com. wind power generators and www.abb.com/product/us/9AAC100348.aspx?country=US
Danish Wind Industry Association. www.windpower.org/en/tour.htm
How Stuff Works. Electric motor. www.howstuffworks.com/motor1.htm
How Stuff Works. Generator. http://science.howstuffworks.com/electricity2.htm
Lessons on electricity. www.sciencejoywagon.com/physicszone/07electricity/
Walter Fendt. Direct current electrical motor and generator. ww.walter-fendt.de/ph14e/www.walter-fendt.de/ph14e
Wikipedia. Grid tie inverter. http://en.wikipedia.org/wiki/Grid_tie_inverter

REFERENCES

1. T.S. Jayadev. 1976. *Induction generators for wind energy conversion systems*. AER-75-00653. Available from NTIS.
2. G.L. Johnson and H.S. Walker. 1977. Three-phase induction motor loads on a variable frequency wind electric generator. *Wind Engineering. 1*, 268.
3. D. Curtice et al. 1980. *Study of dispersed small wind systems interconnected with a utility distribution system*. RFP-3093/94445/3533/80/7. Available from NTIS.
4. M. Hackleman. 1975. *The Home-Built Wind-Generated Electricity Handbook*. Saugus, CA: Earthmind.
5. J. Barble and R. Ferguson. 1954. Induction generator theory and application. *AIEE*, February, p. 12.
6. L.L. Freris. 1990. *Wind Energy Conversion Systems*. Englewood Cliffs, NJ: Prentice Hall, chap. 9.
7. D. Eggleston and F. Stoddard. 1987. *Wind Turbine Engineering Design*. New York: Van Nostrand Reinhold, chap. 14.
8. G. Johnson. 1985. *Wind Energy Systems*. Englewood Cliffs, NJ: Prentice Hall. Available as PDF files from Gary Johnson, www.eece.ksu.edu/~gjohnson
9. H. Polinder et al. 2004. *Basic operation principles and electrical conversion systems of wind turbines*. www.elkraft.ntnu.no/norpie/10956873/Final%20Papers/069%20–%20electrical%20conversion%20system
10. E. Juljadi et al. 2006. *Power quality aspects in a wind power plant*. IEEE and Power Engineering Society. www.nrel.gov/wind/pdfs/39183.pdf
11. E. Juljadi et al. 2008. *Fault analysis at a wind power plant for one year of observation*. IEEE and Power Engineering Society.
12. European Wind Energy Association. 2009. *Wind Energy: The Facts*. London: Earth Scan.
13. R. Erickson, S. Angkititrakul, and K. Almazeedi. 2006. *A new family of multilevel matrix converters for wind power applications*. Final report, July 2002–March 2006. NREL/SR 500-40051. www.nrel.gov/wind/pdfs/40051.pdf
14. NREL. 2003. *Wind turbine generator system safety and function test report for Bergey Excel-S with Gridtek-10 inverter*. NREL EL-500-33963. www.nrel.gov/wind/pdfs/33963.pdf
15. R. Kithil. 2008. Lightning hazard reduction at wind farms. www.lightningsafety.com/nlsi_lhm/wind1.html
16. T. McCoy, H. Rhoads, and T. Lisman. 2000. Lightning activities in the DOE-EPRI turbine verification program. Paper presented at *Proceedings of Windpower Conference*. www.nrel.gov/docs/fy00osti/28604.pdf

17. B. McNiff. 2002. *Wind Turbine Lightning Protection Project, 1999–2001*. NREL/SR-500-31115. www.nrel.gov/docs/fy02osti/31115.pdf

PROBLEMS

1. What is the voltage drop across a 100-ohm resistance if the current is 2 amps?
2. How much power is lost as heat through that resistance?
3. The maximum power rating of the Carter 25 is 30 kW, and it has a single-phase 240-V generator. What is the maximum current produced? Remember the difference between root mean square values and peak values.
4. What are the peak voltages for 110-, 240-, and 480-V AC?
5. If the phase angle of a 240-V AC, 20-amp circuit is 20 degrees, how much is the power reduced from maximum?
6. What is a three-phase generator?
7. What is the angular velocity for 60-Hz frequency?
8. The synchronous point on an induction generator is 1,200 rpm. If the generator is rated at 500 kW, what is the shaft torque into the generator?
9. Look at Figure 7.8. At what slip is the efficiency maximized for generator operation?
10. If a 25-kW (rated) wind turbine has a three-phase, 480-V generator, what minimum size wire will be needed for each phase to connect the wind turbine to a load that is 50 m away? Remember, you need to count the length of wire down the 25-m tower. Peak power can be 30 kW. Calculate maximum current and reduce it by a factor of 1.7 since the system is three-phase. Each leg (wire) of the three-phase system carries 1/3 of the current. Use Table 7.1.
11. A 100-kW, three-phase, 480-V generator is connected to a transformer within 10 m of the base of a wind turbine. Peak power can be 120 kW. Remember to count the wire down the 30-m tall tower. What minimum size wire is needed for each phase? Calculate current and reduce it by a factor of 1.7, since the system is three-phase. Each leg of the three-phase system carries part of the current. Use Table 7.1.
12. What is power factor? What affects the value of the power factor for a wind farm?
13. List two advantages and two problems of induction generators, constant rpm, and stall control.
14. List two advantages and two disadvantages of doubly fed induction generators, variable rpm, and pitch control.
15. What happens at a wind farm when faults occur on the utility line?
16. Why are SCADAs used at wind farms?
17. What type of wind turbines use power electronics? Why?
18. List three functions of a wind turbine controller.
19. What is the function of an inverter? What types and sizes of wind turbines use inverters?
20. Lightning strikes a blade and essentially destroys it. If you have Internet access, use Reference 17 and estimate the cost to replace the blade.

8 Performance

It is important to remember that the load is part of a wind energy conversion system (Figure 8.1). The most common application is the generation of electricity—a good match of the characteristics of the rotor and the load. The other major application for wind power is pumping water, which is a poor load match when the rotor is connected to a positive displacement pump (constant torque device). However, a farm windmill is well designed to pump low volumes of water with a positive displacement pump, even though it is inefficient.

Overall, performance of a system is measured by annual energy production and annual average power for its wind regime. Compromises on efficiencies for system components should be combined to maximize annual energy production within the initial and life cycle costs. The last two factors may oppose each other, for example, reducing the initial costs may increase life cycle costs. The comparison will cover wind turbines that generate electricity.

Power curves and power coefficients have been measured experimentally, and peak efficiencies are around 0.40 for vertical axis wind turbines (VAWTs) to 0.50 for horizontal axis wind turbines (HAWTS; Figure 6.14). For constant rpm operation, for example, by an induction generator, the rotor will operate at peak efficiency at only one wind speed (Figure 6.13). Also, for a variable speed rotor, the efficiency will decrease above rated wind speed as power output is limited to the rated value.

To increase generator efficiency, some units have two generators. One operates at low wind speeds and the other at high wind speeds. The Vestas V27 utilized a 50/225 kW asynchronous generator with synchronous speeds of 750/1,000 rpm. Another possibility for increasing generator efficiency is to change the number of poles of the generator between low and high wind speeds. The Mitsubishi rated for 1 MW has an induction generator rated at 250/1,000 kW, with wind rotor speeds of 21/14 rpm.

8.1 MEASURES OF PERFORMANCE

Capacity factor: Capacity factor is the average power and is equivalent to an average efficiency factor.

$$CF = \text{average power/rated power} \tag{8.1}$$

In general, capacity factors are calculated from kilowatt hours produced during a certain period: power = energy/time. The time periods vary; however, the most representative period would be one year, although capacity factors for a month and a quarter have been reported. Capacity factors of 0.3 are good, an 0.4 is excellent, and an 0.10 factor is too low.

For wind sites and wind farms with class 4 and above winds, annual capacity factors should be 0.35 or greater. During windy months, the capacity factors can exceed 0.50. Capacity factors are somewhat arbitrary because of the different sized generators that have the same rotor diameters. For February 2002, Lake Benton I, Minnesota, reported a capacity factor of 0.49, and Lake Benton II reported 0.60. The difference was caused by the larger diameters of the wind turbines at Lake Benton II. Both systems had the same size generators but Lake Benton II's turbines swept a larger area.

Availability: The availability is the percentage of time a unit is available to operate and it serves as a measure of reliability. For prototypes and early production models, the availabilities

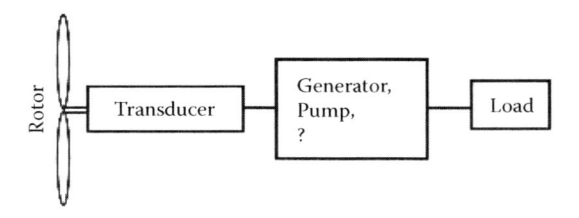

FIGURE 8.1 Wind energy conversion system.

were 0.50 or even lower. Third-generation models have availabilities of 0.95 to 0.98. Manufacturers may define availability differently, so be careful when comparing availabilities of different wind turbines. Reliability and operation and maintenance affect system performance.

Connect time: The connect time (energized hours) is the amount of time or percent of time a unit actually generates power. In the Texas Panhandle, a typical unit should generate power around 60% of the time. This is a large number and can be put into perspective by comparing wind turbines to automobiles. Suppose your car went 160,000 km (100,000 miles) without maintenance. At an average speed of 50 km/h, that is only 3,200 h of operation, which is equivalent to just over a half a year of operation for a wind turbine.

Lifetime: Wind turbines are designed for twenty- to thirty-year lifetimes. This can be achieved by considering the lifetimes of the components [1] and by providing preventive maintenance. Some components such as bearings in gearboxes may have to be replaced within a turbine lifetime. Twenty-five years of operation for a wind turbine would be equivalent to 8 million km for a car, so major repairs will be required.

Jamie Chapman, formerly of U.S. Windpower, noted that, "Estimated minimum standards for non-routine maintenance are one down tower per five years and one up tower per year." "Down tower" means that the nacelle or rotor had to be removed—a major problem. Some first-generation wind turbines had many problems and were replaced within five years or dismantled. Others required major retrofits. Some of the early wind farms in California began replacing the 50- to 100-kW wind turbines with megawatt-size turbines (repowering), starting in 1998. The smaller wind turbines were then refurbished for the distributed market.

Design of generators and gear trains is well known. Loads produced by a rotor are the major unknown factors, especially stochastic loads caused by the turbulent character of winds. As the industry matured, engineers designed blades, gearboxes, and generators specifically for wind turbines. Airfoils for horizontal-axis wind turbines have been designed to provide improved performance and increased energy production.

Reliability: Most of the first-generation wind turbines [2] lacked reliability and quality control. Prototypes generally failed within months. Lack of reliability leads to larger maintenance and operation costs after installation. Manufacturers and dealers were caught in a bind because retrofit programs in the field were very expensive. The most successful wind farms utilize reliable wind turbines and follow good operation and maintenance programs.

If a dealer has to service a small wind system more than once during the first year of warranty, he probably lost money. Typical service charges are $60/hr or more. A large service area means a dealer spends most of his time on the road. Increases in gasoline prices will increase service charges.

Specific output: The most important factors for determining the annual energy production are the wind regime and the rotor swept area. One way to compare wind turbines is by annual specific output (kilowatt hours per square meter). Stoddard [3] tabulated data for wind turbines in California, where the best value was 1,000 kWh/m² without considering the wind regime. Comparing the averages of a large number of units in similar locations will give a good estimate of performance. Calculating the ratio of annual kilowatt hours/weight of rotor or weight on top of tower gives an idea of a goal for cost comparisons. As for any mature industry, costs are based primarily on the ratio of cost/weight of material. Another specific output is kilowatt hours/kilowatts, but this ratio is not as useful.

The wind turbines manufactured in Denmark were more massive and captured over 50% of the California wind farm market from 1982 to 1985 because they were rugged and reliable. However, after five years, a major problem developed with deterioration of fiberglass-reinforced plastic (FRP) blades at the roots due to fatigue. The repair and replacement market for blades was estimated at $80 million.

8.2 WIND STATISTICS

WindStats Newsletter [4] contains reports and wind energy production tables covering thousands of wind turbines. It reports location, manufacturer, kilowatt rating, swept area, tower height, estimated annual energy production, monthly and quarterly energy production, quarterly capacity factor, specific output (kWh/m^2 and kWh/kW), annual production for the previous one or two years, and installation data. The newsletter also published information about reliability. The information for wind turbines in Denmark is available online. The quarterly reports of the California Energy Commission [5] and the monthly values in *WindStats* provide better information on load matching to utilities.

The Sindal Report [6, free download] is a quarterly publication about wind power market trends along with individual production data for more than 6,000 wind turbines in Denmark and Sweden. Data are similar to *WindStats* and cover land and offshore locations. Graphs of monthly averages by quarter are presented for energy (kWh), capacity factors, specific output (kWh/m^2 and kWh/kW) for a range of turbine sizes. For some megawatt wind turbines for 2012, capacity factors ranged from 33 to 47% with the offshore wind farms having the largest values. Specific outputs range from 1,100 to nearly 1,400 kWh/m^2 (Table 8.1). For these megawatt wind turbines, energy production for 2012 ranged from 6 to 13 GWh. Of course energy production depends primarily on wind regime (offshore production will differ somewhat) and rotor area, and secondarily on rated power.

8.3 WIND FARM PERFORMANCE

Capacity factors are somewhat arbitrary and depend on rated power of a wind turbine, which can be the same for different size rotors or different based on rated wind speed. Of course, the most important issue is annual energy production that depends primarily on rotor area and wind regime. Some energy production statistics published by the Federal Energy Regulatory Commission (FERC) are

TABLE 8.1
2012 Performance Data for Megawatt Wind Turbines Cited in Sindal Reports

Manufacturer	Number of Units	Rating (MW)	Land (L) or Offshore (O)	Energy (GWh)	CF (%)	Specific Output (kWh/ m^2)
Enercon	34	2.3	L	6.1	33	1,386
Siemens	74	2.3	L	7.1	36	1,146
Siemens	229	2.3	O	9.4	47	1,398
Siemens	15	3.6	L	13.0	42	1,296
Vestas	73	3.0	L	8.5	32	1,311
Vestas	8	3.0	O	8.9	34	1,324
Vestas	15	3.1	L	10.7	40	1,084
Vestas V80	80	2.0	O	160		
Bonus	72	2.3	O	165.6		

actual reported values for all power plants in the U.S. In many countries, electricity production figures are calculated from average capacity factors and installed capacity.

Since non-hydro renewables, especially wind power, have increased dramatically in installed capacity and energy production, national energy agencies such as the Energy Information Administration (EIA) in the U.S., now report those values. The EIA also reports international data. Of course, other sources of world production data are international energy, national energy, and national wind energy associations.

Capacity factors have improved with the newer and larger wind turbines, so it is expected that wind farms installed in good to excellent wind regimes since 2000 will demonstrate better capacity factors (35 to 50%) than the older installations. The early wind farms in California had average capacity factors below 20%. Availability and capacity factor are related. Both measurements will be low if wind turbines have operational problems. For example, during the first year of operation at Horns Rev, an offshore wind farm in Denmark, the capacity factor was only 26%; however, the next year it reached the expected value.

At the Scroby Sands offshore wind farm (30 wind turbines, 60 MW) in the U.K., energy production was limited in the first year of operation. Numerous mechanical problems led to the replacement of twenty-seven intermediate speed and twelve high-speed gearbox bearings along with four generators. The capacity factor for the first year was 29%, rather than the predicted 40%. Another example was a thirty-eight-turbine, 80-MW wind farm that encountered software and blade malfunctions. One year after installation, thirteen turbines were still not operational, but all thirty-eight turbines were operational in the second year.

8.3.1 CALIFORNIA WIND FARMS

The California Energy Commission (CEC) instituted a program in 1984 for regulating reporting performance of wind systems [5]. All California wind projects producing more than 100 kW that sell electricity to power purchasers must report quarterly performance. The quarterly reports must indicate turbine manufacturers, model numbers, rotor diameters, kilowatt ratings, number of cumulative and new turbines installed, projected output per turbine, output for each turbine model, and output of entire project.

Annual reports are compilations of data from the four quarters and contain summary tables reflecting resource areas. The reports do not provide information on every wind energy project in California. Non-operating wind projects and turbines that do not produce electricity for sale, such as those installed by utilities, government organizations, and research facilities, are not required to file reports. Wind performance report summaries are available from 1985 through 2005 [5].

Only small wind turbines (10 to 18 m diameters, 5 to 100 kW) were available in the early 1980s. At the end of 1985, the largest installed capacity was U.S. Windpower's 181 MW, followed by Fayette's 146 MW. The wind farms produced 0.65 TWh, which was 45% of that predicted by the plant operators. Average capacity factor was 13%—much lower than the 20 to 30% cited in technical reports. Foreign (and newer) wind turbines had a capacity factor of 17%. The ten largest manufacturers represented 80% of the installed capacity and four of those accounted for 53%. The average installed cost of the 10,900 wind turbines was $2,000/kW (range of $700 to $2,300).

By 1990, 1,500 MW had been installed in California and produced 2.68 TWh—enough to power the residential needs of San Francisco [5]. Kenetech, formerly U.S. Windpower, still produced the largest number of units and largest installed capacity. The sizes of wind turbines increased from 100 to 750 kW. By 2010, California wind farms had a capacity of 3 GW and generated over 6 TWh per year—around 3% of total electric production and new wind turbines were capable of generating megawatts.

The annual capacity factor is an average from operational wind turbines (Figure 8.2). In 1990 the better projects had capacity factors in the twenties. For the third quarter, Kenetech had a value of 40% and Bonus achieved 39%. Fayette had a capacity factor problem. At one time, it had the

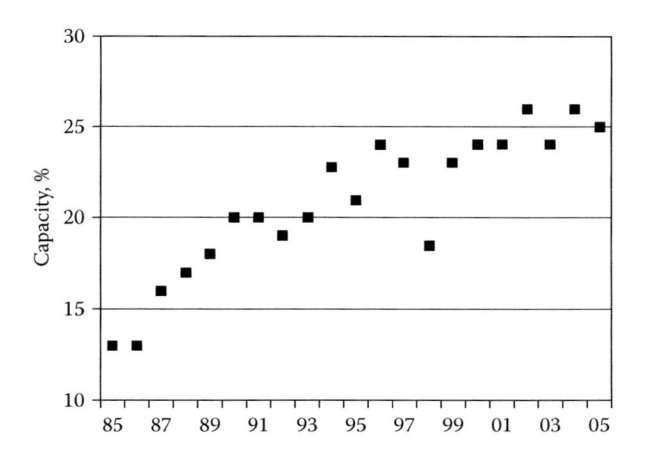

FIGURE 8.2 Average capacity factor for wind turbines in California. (*Source:* California Energy Commission.)

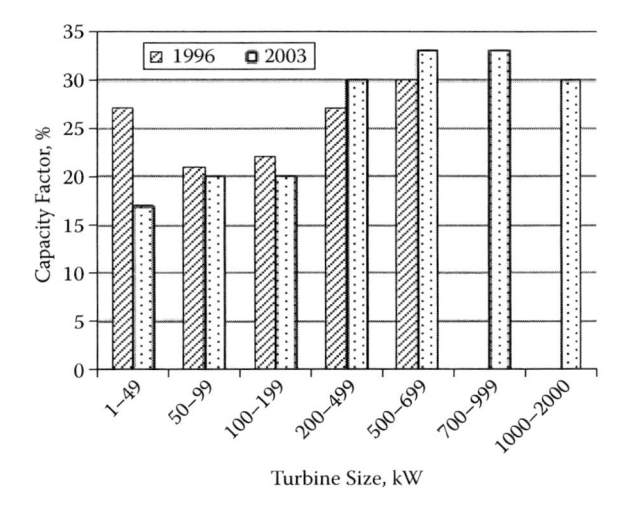

FIGURE 8.3 Capacity factor by range of wind turbine sizes in California.

second largest installed capacity, but its capacity factor was a very low 5% because its 90-kW, 10-m turbines were overrated. The vertical axis wind turbines of Flowind also had low capacity factors (10%). Its annual capacity factor increased to 30% with the new, larger wind turbines (Figure 8.3). The specific kWh/m^2 output varied from low values to over 1,000 kWh/m^2 (Table 8.2).

In the 1990s, the older wind turbines, primarily in the range of 50 to 100 kW (55% of MW capacity installed), were cannibalized for parts and uneconomic wind turbines were dismantled. A number of trends are noted. Wind turbines became larger (megawatt ratings), capacity factors improved, and reliability increased. Also, the drop in production in 1997 was caused when older, smaller units were taken out of production and replaced with bigger turbines in 1998.

Specific output (Figure 8.4) increased when poorly performing units were taken out of service and newer wind turbines installed. The larger specific output shows the type of performance that can be expected with good wind turbines in an excellent wind regime. Turbines will show annual variations of both capacity factor and specific output because of differences in yearly wind regimes and differences between locations. Wind is site-specific and variations must be expected.

TABLE 8.2
Specific Output, kWh/m², for Wind Turbines (Most, but not all Manufacturers) in California, 1989

Turbine	Dia m	Rated kW	# Units	Capacity MW	Per Turbine kWh	Per Turbine kWh/m²
Fayette	10	90	1,363	123	41,000	522
Bonus 65	15	65	644	42	113,000	640
Vestas 15	15	65	1,330	86	53,000	300
Micon 60	16	60	531	32	95,000	473
Nordtank 60	16	60	152	9	170,000	846
Micon 65	16	65	126	8	184,000	916
Nordtank 150	16	65	375	24	100,000	498
Vestas 17	17	100	1,071	107	145,000	639
US Windpower	18	100	3,419	342	220,000	865
Micon 108	20	108	967	104	230,000	732
Bonus 120	20	120	316	38	276,000	879
Carter 250	21	250	24	6	250,000	722
Nordtank 150	21	150	164	25	240,000	693
Flowind 19	21	250	200	50	142,000	410
Danwin 23	23	160	151	24	390,000	939
Vestas 23	25	200	20	4	434,000	885
WEG MS2	25	250	20	5	560,000	1,141
Mitsubishi	25	250	360	90	486,000	991
DWT 400*	35	400	35	14	1,000,000	1,040
* Estimated kWh					average	756

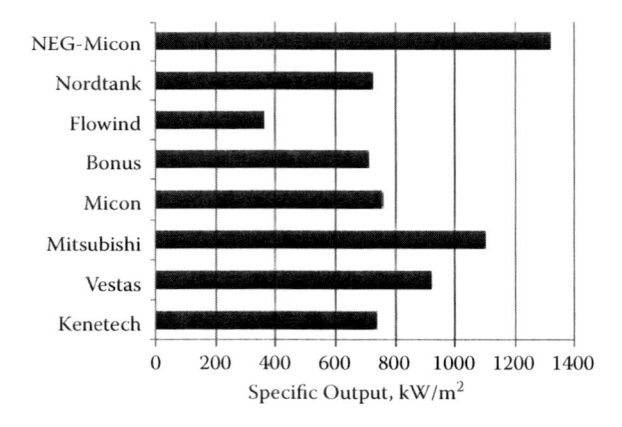

FIGURE 8.4 Specific outputs of manufacturers with largest installed capacities in California. NEG-Micon systems do not include older Micon units ranging around 100-kW capacity.

8.3.2 WIND FARMS IN OTHER STATES

Wind farms in the U.S. generated an estimated 140 TWh in 2012, and some capacity factors were over 40% [7]. The capacity factors (Figure 8.5) and specific outputs (Figure 8.6) were analyzed for five wind farms (Table 8.3) in the Southern High Plains. The same types of wind turbines were compared, but all turbines at White Deer and some at Fluvana had smaller rotor diameters and hub

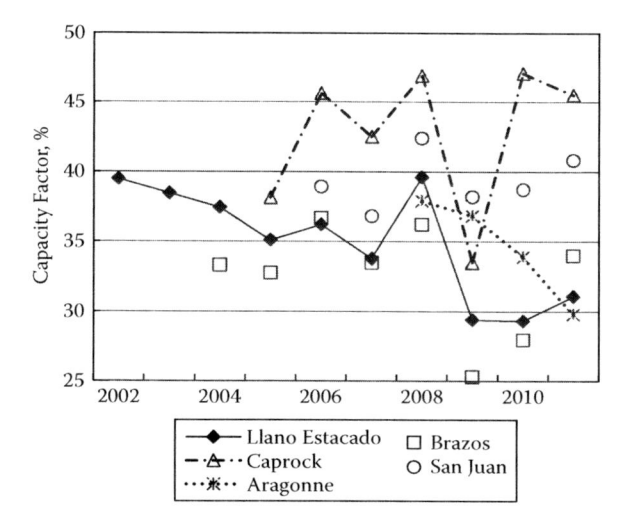

FIGURE 8.5 Annual capacity factors for wind farms in Texas and New Mexico.

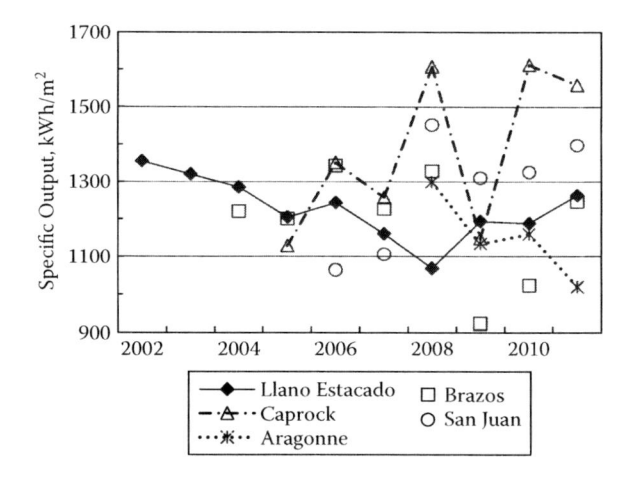

FIGURE 8.6 Annual specific outputs for wind farms in Texas and New Mexico.

TABLE 8.3
Installed Capacities of Mitsubishi 1-MW Systems at Various Wind Farms

Location	Wind Farm	Capacity (MW)	Rotor Diameter (m)	Hub Height (m)
White Deer, TX	Llano Estacado	80	56	60
Fluvana, TX	Green Mountain, Brazos	60	56	60
		100	61.4	69
San Jon, NM	Caprock	80	61.4	69
Elida, NM	San Juan Mesa	120	61.4	69
Pastura, NM	Aragonne Mesa	90	61.4	69

heights. Capacity factors ranged from less than 30% to over 45%. The largest annual specific output was 1,606 kWh/m². The yearly variation was consistent across the region (2009 was a low wind year). However, the downtrend in capacity factor at Llano Estacado, Brazos, and Aragonne may have been caused by a decline in reliability.

Manufacturers now offer wind turbines with various sized rotors for different wind regimes. The estimated energy output of Texas would improve with an 8% increase of wind turbine rotor diameter rather than increasing hub height from 75 m to 100 m. For 2009 through 2011, the Wildorado Wind Ranch in Texas had an average capacity factor of 46%. Data for Figures 8.5 and 8.6 were obtained from the FERC (see Links at the end of this chapter; you must know the name of the reporting entity). The Electric Reliability Council of Texas (ERCOT) reported that wind power set a new record of 26% of system demand on the morning of November 10, 2012.

Wind farms in the Turbine Verification Program had to provide public data on performance through the Electric Power Research Institute (EPRI). The twelve Zond turbines near Fort Davis, Texas, had a capacity factor of 0.16 and a specific output of 568 kWh/m² over three years. These turbines were rated at 500 kW and had aileron control. Eleven Zond turbines with full-span pitch control near Searsburg, Vermont, had a capacity factor of 0.25 and specific output of 884 kWh/m² over two years. Part of the difference was because of the control method and some due to differences in wind regimes.

The Sandia National Laboratory maintains the Continuous Reliability Enhancement for Wind (CREW) database for the analysis of wind plant operations. The wind turbine reliability benchmark reports [8] include operating performance at a system-to-component level and identify opportunities for technology improvement. At present, the database represents a small portion of U.S. wind farms (10 plants, 800 to 900 turbines, and 180,000 turbine days of operation). However, the database is very useful as it provides first benchmarks on performance. Key metrics improved from 2011 to 2012 (Table 8.4) and components affected are shown in Table 8.5. The availability time accounting (Figure 8.7) showed that the units were generating 59% of the time and unscheduled maintenance took 1% of the time.

8.3.3 OTHER COUNTRIES

By the end of 2012, the installed wind capacity in Europe reached 106 GW and produced 230 TWh—about 6% of the gross final consumption. A major difference from the 2008 data was that in 2011, most countries in Europe had some wind capacity. Thirteen countries had capacities over 1 GW, and in twenty-one countries, wind produced at least 1% of total electric consumption.

The five largest wind power producers are Denmark (25.9%), Spain (15.9%), Portugal (15.6%), Ireland (12%), and Germany (10.6%). At the end of 2011, as in previous years, Germany had the

TABLE 8.4
Key Metrics (%) for 2011 and 2012

Metric	2012	2011
Operational availability	97.0	94.8
Utilization	82.7	78.5
Capacity factor	36.0	33.4
Mean time between events (hr)	36	28
Mean downtime (hr)	1.6	2.5

Source: Continuous Reliability Enhancement for Wind (CREW) Database, Sandia National Laboratory, Albuquerque, NM.
Operational availability = generating + reserve shutdown, wind + reserve shutdown, other.

TABLE 8.5
Average Events per Turbine per Year by Component[a]

Component	Number per Turbine per Year
Wind turbine (other)[b]	90
Electric generator	40
Controls	12
Power distribution	10
Gearbox	8
Braking system	5
Structure and enclosures	4
Yaw	3
Hydraulic control	2
Drivetrain	1

[a] Some events automatically reset; others need intervention.
[b] Wind turbine (other) condition occurs mainly when technician has turbine in maintenance/repair mode.

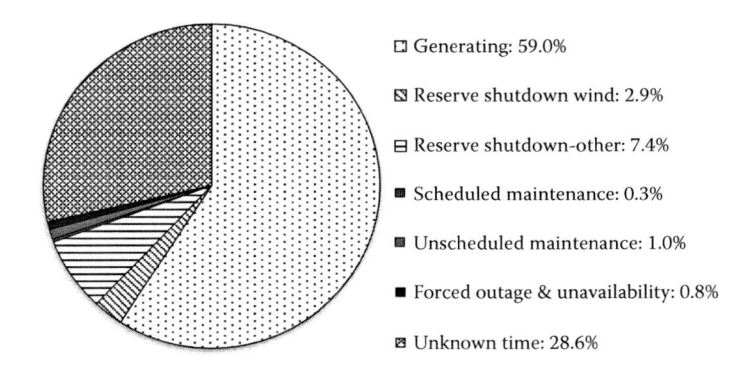

FIGURE 8.7 Availability time accounting. Substantial portion of unknown time attributable to pilot program and beta testing. (*Source:* Continuous Reliability Enhancement for Wind (CREW) Database, Sandia National Laboratory, Albuquerque, NM.)

largest installed capacity (29.1 GW), followed by Spain with 21.7 GW, average capacity factor of 26%, and generation of 41.6 Ton Wh. On a night in September 2012, wind power produced 64.2% of Spain's electric demand. Offshore wind farms in Europe produced around 14 TWh of electricity from an installed capacity of over 5 GW in 2012.

The average capacity factors for the U.K. were 26% for land and 30% for offshore (2006 through 2010 data), and estimated annual energy production (2012 data) was 18.7 TWh. The U.K. now leads in offshore wind farms with over half the Europe capacity and maintains the largest offshore wind farm, London Array at 630 MW.

Many countries supply national energy data and statistics, for example, wind power produced 6.1 TWh (2011) from installed capacity of 2.8 GW in Sweden, and a 25% capacity factor in Spain. The average capacity factor for New Zealand wind farms (623 MW) in 2011 was 40% and they provided around 4% of the electricity consumption.

China is now the leading country in installed wind capacity (76.4 GW in 2012), and wind produced around 150 TWh (1.2%) of the country's electricity. However, transmission problems resulted in an average rate of curtailment of 16%, and the largest curtailment was 23% in East Inner

Mongolia. Of course, curtailment reduced capacity factors for those wind farms. India wind capacity is over 18 GW (2012), ranks fifth in world, and accounts for around 1.6% of electric generation.

Denmark was an early proponent of wind power and has reached the stage at which its non-offshore cumulative capacity has essentially leveled off. Its total installed capacity and number of turbines increased through 2002. From 2001 through 2003, 1,300 small wind turbines and those with poor siting were replaced with larger turbines, so the installed capacity still increased until 2003, then leveled off at 3,130 MW although the number of turbines decreased from 6,400 to 5,267 (January 2007).

This meant the average power per wind turbine increased from 1 MW to around 2 MW, which included the large wind turbines installed offshore. The second stage, mostly offshore installations, increased the capacity to 3,871 MW by end of 2011, and wind energy supplied 26% of the country's electric consumption. The average capacity factor was 25% for the smaller wind turbines; the offshore wind farms have capacity factors of 45 to 50% [6].

The first-year performance of the Nysted offshore wind farm was a wind turbine availability of 97% and a wind farm availability of 96% [9]. The energy production of around 50 GWh per month was within the predictions for the wind regime for the first half of 2004. The monitoring system revealed increased vibration levels in the gearboxes. The gearboxes were designed for easy gear changes and two gearbox bearings were replaced in every wind turbine. A nacelle crane was used and average downtime was 48 hr per turbine. The problem was solved and the 2011 capacity factor was 50%.

8.4 WAKE EFFECTS

Vortices are generated by the tips of blades, the trailing edges of blades, and the tower, and they increase the turbulence of a wake. The tips of airfoils and trailing edges are designed to reduce the vortices and also reduce the noise accompanying some vortices. The three primary methods of wake and array loss research have been numerical modeling, wind tunnel simulations, and field measurement. A database of literature on wind turbine wakes and wake effects through 1990 is available [10].

The wake is expanding and the wind speed is reduced downwind, so if there are multiple wind turbines, how far apart should they be placed? The wakes from wind turbines create turbulence and along with wind speed deficits result in array losses reflected by reduced annual energy production. Therefore, the placement of wind turbines in a wind farm is a trade-off between energy production and cost of installation.

Downwind units will produce reduced energy, so the question is how much reduction of spacing within a row and between rows will increase production. In fairly flat areas, the rows will be placed perpendicular to the predominant wind direction, within-row spacing would be two to four rotor diameters, and between-row spacing would be five to ten or more rotor diameters.

Offshore wind farms generally utilize larger spacing; for example, Horns Rev in the North Sea off the coast of Denmark has a seven-rotor diameter spacing (within and between rows). The physical factors controlling wake interference are downwind spacing, power extracted by the wind turbines, turbulence intensity, and atmospheric stability. Wind turbine wakes develop in fairly well defined regions at different downwind distances, and wake geometry models show this information [11]. Field tests on single and multiple wind turbines measured the velocities and power deficits downwind. The wake effects were still noticeable at ten rotor diameters downwind from a rotor.

Wind turbines have close spacing between rows at wind farms in San Gorgonio Pass, California, due to the high cost of land. Energy production was reduced for the second row and even more for the third row that experienced wake effects from both the first and second rows. Field measurements of wake effects inside wind farms have generally been limited to two to four rows of wind turbines. Energy deficits of 10 to 15% in row 2 and 30 to 40% in row 3 have been reported for densely packed wind farms. Measurements of wake deficits downwind of large arrays indicate that the losses may be larger and extend farther downwind than expected. Energy deficits of 15% were estimated 5 km downwind from a 50-MW array [12]. In California, early wind turbines were small (25 to 100 kW). Larger wind turbines on taller towers were interspaced within rows later.

It is more difficult to predict output and array losses without an extensive wind measurement program for a wind farm. The exceptions are offshore wind farm where ocean waves provide data on wind speeds at 10 m height determined from satellite data (see Section 4.4).

High-resolution data are used to estimate the wind resources of the Danish Seas. Comparisons have been made of those data with met tower data taken offshore. Ocean wind maps covering the Horns Rev wind farm (400-m grid cells) in the North Sea and the Nysted wind farm (1.6-km grid cells) in the Baltic Sea were used to quantify the wake effect [13].

The Nysted wind farm has eighty-two turbines (82.4 m diameter, 2.3 MW) with a 9 × 8 array. The distance between turbines within a row is 480 m and between rows is 850 m (5.8 D × 10.3 D). The velocity deficit is around 10% at 0 to 3 km downwind, and the wind recovers to 2% of the upstream values at around 5 to 20 km downstream, which depends on the ambient wind speed, atmospheric stability, and number of operating turbines [14]. The recovery is faster for unstable than for near-neutral conditions. In calm winds, the turbines are clearly visible in ocean wind speed maps.

The influence of wake effects on energy production [15] was estimated using data from met towers northwest and east of the Horns Rev wind farm and from the SCADA database that contains all observed data for each wind turbine. The Horns Rev wind farm has eighty turbines (80 m diameter, 2 MW) with an 8 by 10 array and a distance of 560 m between the turbines (7D spacing). For most selected cases, the wind turbines were operating at high wind speeds. An analytical model links the small- and large-scale features of the flow in the wind farm with equidistant space between units within a row and between rows.

For wind perpendicular to the row, a large power drop occurs between rows 1 and 2 (around 30%), and a smaller, almost linear power drop occurs between subsequent rows. From row 2 to row 9, the power drop is around 10 to 15%. For winds along the diagonal, the spacing is 9.3 D; however, this covers only three lines with eight turbines. At wind speeds of 9 to 10 m/sec, a large power drop (25 to 35%) is noted from line 1 to line 2, a slight drop from line 2 to line 5, and then essentially a constant drop from line 5 to line 8.

The atmospheric conditions were just right to show the wakes of wind turbines at Horns Rev (Figure 8.8). Notice that the downwind wind turbines are in the wakes of the wind turbines from each of the previous rows. Compare this photo with the photo of the Nysted wind farm (Figure 9.15), also offshore of Denmark.

In the final analysis of performance, the main issues are energy production, return on investment, and value of the energy, which should include externalities. Capacity factors give indications about the wind regime and the relation between rotor area and generator size. However, the main

FIGURE 8.8 Wind turbine wakes at Horns Rev, Denmark. (Horns Rev 1 owned by Vattenfall. Photographer Christian Steiness.)

measures of performance should be annual energy production and average specific output (kilowatt hours per square meter) per turbine type and model. Wind class should also be included as a check on comparisons of performances of wind turbines.

8.5 ENERTECH 44

A long-term performance test of an Enertech 44 [16] provided monthly values of energy production, connect time, availability, and wind speed. The variations of power by month and year are shown in Figure 8.9. Connect time (time during which a unit is connected to a grid) is about 60% (Table 8.6) or over 5,000 hr per year. From 1989 to 1996, when the unit was rated at 40 kW, it averaged 78,000 kWh per year.

The prototype wind turbine (induction generator, constant rpm, stall control) was installed at the Agricultural Research Service site in Bushland, Texas, in May 1982. All three models had 13.4-m diameter rotors. The original turbine was a 240-V single-phase induction generator with a rated capacity of 25 kW. The gearbox and generator were changed to a three-phase, 480-V, 40-kW induction generator (Table 8.7) in 1984. Later that year, a gearbox and three-phase, 480-V, 60-kW induction generator were installed. In July 1988, the gearbox was replaced with the previous 40-kW gearbox, making the rated power closer to 50 kW.

The availability was good, even though the unit was a prototype, and several component failures occurred. The downtime was estimated at 1% for routine maintenance and service, 1% for repair of component failures, and 1% for weather-related events, mainly icing. Additional downtime was for replacing gears in the gearbox and installing different generators. The year 1992 was a low year for wind power. The unit was down over 2 months while a yaw bearing was replaced in 1995, down for 1.5 months for a major oil leak in 1996, and down for 2.5 months as a soft start was installed to reduce the loads on the motor/generator.

After that, the unit was connected part of the time to a wind–diesel test bed (village grid), so it would not have the same connect time and energy production. The unit was down 0.5 month in 1999 due to control failure caused by lightning. A report on the reliability is available from the Agricultural Research Service. It shows causes for all downtimes for twenty years of operation. The unit was operational until 2012.

Capacity factors are higher for small generators, but annual energy production is better with larger generators. However, the energy differences between the 40- and 60-kW generators were not significant. The power curves (Figure 8.10) include all efficiencies, from wind to electric output.

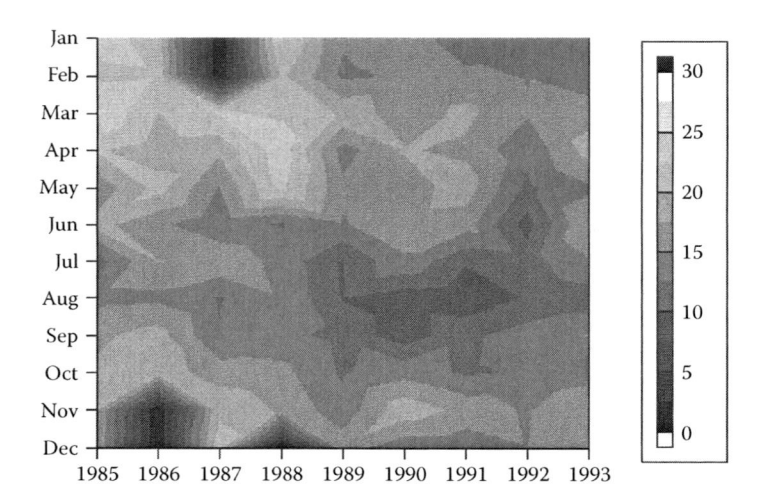

FIGURE 8.9 Average power in kilowatts (legend on right) by month for Enertech 44.

Performance

TABLE 8.6
Enertech 44 Wind Turbine, Fixed Pitch, Induction Generator

Year	Operating Time (hr)	Connect Time (%)	Energy (kWh)	Capacity Factor (%)	Availability (%)	Wind Speed (m/s)	Rated Power (kW)
82	3218	63	48092	40	99.9	5.7	25
83	5567	63.6	63710	29	92.6	6.0	
84	4611	52.6	72295		86.3	5.9	40
85	4662	55.5	91732	17	94.9	5.6	60
86	4121	47.1	77522	15	82.1	5.7	
87	3850	44	65638	12	81	5.6	
88	3971	45.3	71643		77	5.6	50[a]
89	5893	67.3	83452	19	99.4	5.3	
90	5831	66.6	86592	20	97.5	5.6	
91	5705	65.1	82390	19	96.6	5.9	
92	5641	64.6	73510	17	98	5.4	
93	5754	65.9	88363	17	96.4	5.7	
94	5769	66.4	79392	18	95.7	5.6	
95	4099	46.8	51931	12	72.8	5.7	
96	4991	56.8	76470	17	86.8	5.8	
97	4608	52.6	56958	13	75.4	5.5	Hybrid
98	4944	56.4	68885	16	93.2	5.5	
99	4487	51.2	65147	15	93.3	5.7	
2000	4241	48.3	66589	15	85.3	5.7	
Average						5.7	

[a] Generator, 60 kW; gear box, 40 kW.

The same information is presented by power coefficient curves (Figure 8.11). In other words, winds above 12 m/sec are insufficient for the larger generator to offset the differences in generator efficiency at lower wind speeds. Furthermore, a larger generator and gearbox would increase the cost.

8.6 BERGEY EXCEL

A Bergey Excel wind turbine installed at the AEI Wind Test Center in August 1991 operated until 2010, when the test center was moved from its location adjacent to the campus. The specifications were three phases, 240 V, permanent magnet alternator, rated at 10 kW. The variable voltage, variable frequency was converted to DC, then inverted to 60 Hz for connection to a utility line. Power and wind speed were sampled at 1 Hz and averaged over 15 min. The time sequence data were then averaged over 1 month for each 15-min period to calculate an average day for the month. As expected, the power (Figure 8.12) varied widely by season and time of day. Spring 1992 was a below-average wind period.

Power curves indicate performance, and when compared to the manufacturer's curve, the measured power curve (Figure 8.13) at the site was lower, even when corrected to standard density [17]. This means that the energy production would be lower than that predicted from the manufacturer's power curve. Part of the reduction is due to the efficiency of the inverter, especially at high wind speeds (see information on inverters in Chapter 7).

Power curves for shorter periods will indicate performance of a wind turbine when compared to baseline experimental power curves at a site. Of course, there is some data scatter, especially at high wind speeds with few data points. However, low power curves indicate a problem. Power curves for

TABLE 8.7
Performance, Enertech 44/40 kW, 44/60 kW, Bushland, Texas, Apr September 1986 (Anemometer at 10 am)

Dates	Number of Days	Operating Time (hr)	Connect Time (%)	Energy (kWh)	Availability (%)	Average Speed (m/sec)
44/40						
Mar 20–Apr 1					Shakedown	
Apr 2–Apr 30	29	571	82	11,148	100	7.4
May	31	568	76	9,078	99.7	6.4
Jun	30	511	71	8,281	100	6.3
Jul	31	430	58	5,017	100	5.0
Aug	31	302	41	2,443	99.7	4.1
Sep	30	461	64	7,240	100	5.8
Oct	31	412	55	6,260	100	5.3
Total	213	3254	64	49,467	100	5.8
44/60						
Nov 1–30, 1984	17				Shakedown	
Dec 1984	31	366	49	7,877	87.3	5.6
1985	365	4897	56	91,732	94.9	5.7
Jan–Sep 1986	273	3824	58	72,905	100	5.8
Total	686	9087	57	172,514	97	5.7
Nine-Month Breakdown – 1986						
Jan	31	450.8	61	9,790	100	5.7
Feb	28	342	51	7,578	100	5.5
Mar	31	441.7	59	8,803	99.9	5.9
Apr	30	466	65	8,635	99.8	6.5
May	31	430	58	9,103	100	6.2
Jun	30	360.9	50	5,638	100	5.1
Jul	31	518.1	70	9,839	100	6.4
Aug	31	369.5	50	5,293	100	5.3
Sep	30	445	62	8,226	100	6
Total	273	3824	58	72905	100.0	5.8

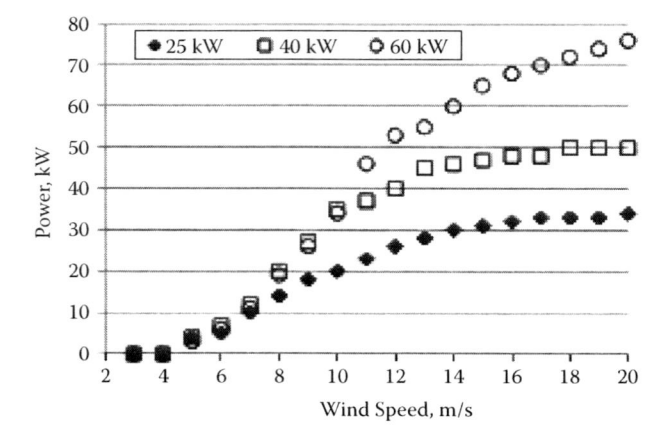

FIGURE 8.10 Power curves for Enertech 44 with different size generators.

Performance

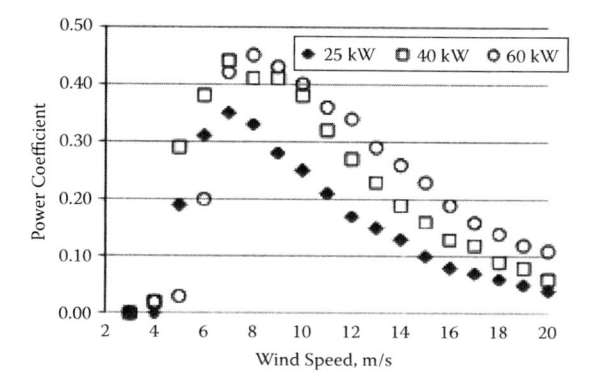

FIGURE 8.11 Power coefficient curves for Enertech 44 with different size generators.

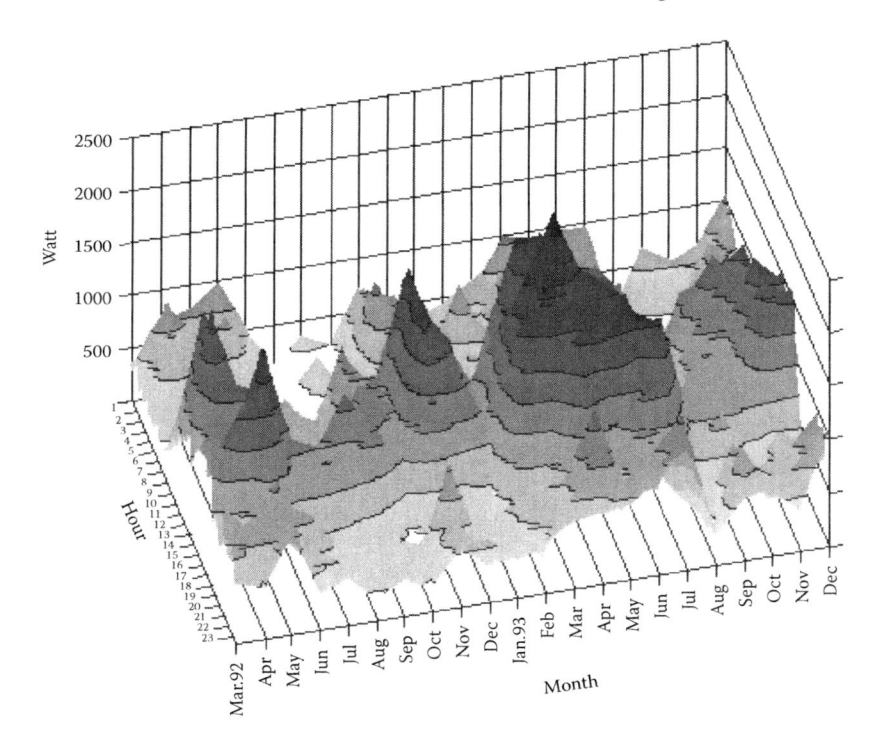

FIGURE 8.12 Power for Bergey 10-kW wind turbine installation for average day, by month from March 1992 through December 1993.

each month were plotted and then averaged to obtain a baseline curve (Figure 8.14). Notice that the system definitely had a problem in November.

8.7 WATER PUMPING

Water pumping by windmills is an old technology. Most changes of farm windmill technology have not been commercial. The electric-to-electric system for pumping larger volumes of water for villages and small irrigation plots [18] was designed and prototypes have been tested. Now such systems are available commercially. Windmill performance for water pumping can be estimated by a flow curve for water and a wind speed histogram to estimate the amount of water pumped by month or year.

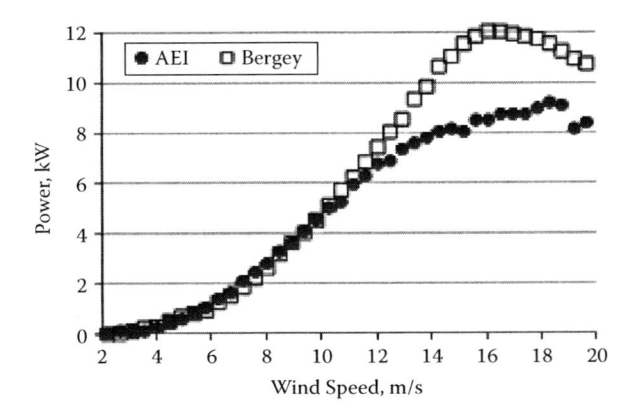

FIGURE 8.13 Power curve data for Bergey Excel power curve measured at AEI Wind Test Center, 1997 through 1999.

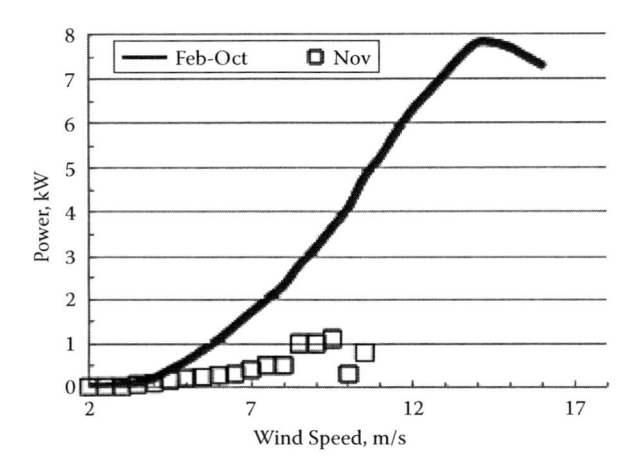

FIGURE 8.14 Power curves for Bergey, 10 kW, at Leroy, Texas.

8.7.1 FARM WINDMILLS

The farm windmill designs have not changed since the 1930s. These windmills are well designed to pump small volumes of water for livestock and residential use. They are comparable to drag devices because of the large solidity (close to 1). The wind rotor has a peak efficiency of 15 to 18% at a tip speed ratio around 1. The wind rotor efficiency is higher than the overall efficiency because pump efficiency limits system performance.

Since a windmill is connected to a positive displacement pump, the rotor needs a lot of blades to achieve high starting torque. For a mechanical farm windmill with a positive displacement pump, the water flow rate is directly related to the number of strokes per minute. Overall average annual efficiency (wind to water pumped) is around 5 to 6% [19]. The curve for water flow is similar to the efficiency curve. Tables are available to estimate performance of farm windmills for different wind regimes, for example, Table 8.8. The same information is shown in Figure 8.15; however, the strong wind data from Table 8.8 were not plotted since they were close to fair wind data.

Performance tests of eight farm windmills [20] showed little difference between the four units that had reciprocating pumps, two of which had conventional reduction gears and two did not.

Performance

TABLE 8.8
Estimated Water Pumped by Farm windmill

Depth (m)	Pump Diameter (cm)	Light Wind (3 to 4.5 m/sec) Cubic m/hr	Fair Wind (5 to 7.5 m/sec) Cubic m/hr	Strong Wind (>8 m/sec) Cubic m/hr
9	3.6	8.1	12.5	13.7
17	2.7	4.6	7.1	7.8
24	2.2	3.2	4.9	5.4
38	1.8	2.0	3.1	3.4
49	1.6	1.6	2.4	2.6
67	1.3	1.1	1.8	2.0
79	1.2	0.9	1.5	1.6
91	1.1	0.8	1.2	1.4
110	1.0	0.6	1.0	1.1
140	0.89	0.5	0.7	0.8
171	0.84	0.5	0.7	0.8
183	0.78	0.4	0.6	0.6
Depth (ft)	Pump Diameter (in.)	Gal/hr	Gal/hr	Gal/hr
30	8.0	2145	3300	3630
55	6.0	1220	1875	2060
80	5.0	845	1300	1430
125	4.0	540	830	910
160	3.5	420	640	700
220	3.0	300	470	520
260	2.75	250	385	425
300	2.5	210	325	360
360	2.25	170	260	285
460	2.0	125	190	210
560	1.875	120	180	200
600	1.75	100	150	165

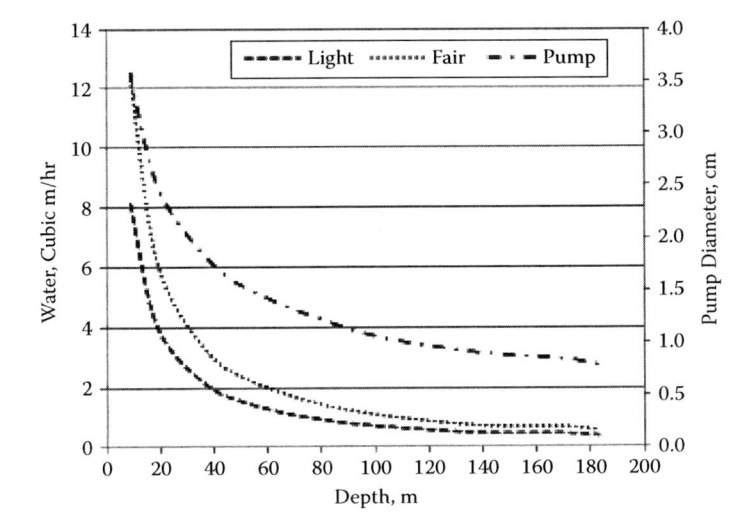

FIGURE 8.15 Pump diameter to use for depth to water and amount of water that a farm windmill would pump in light and fair winds.

The windmill equipped with a Moyno pump performed well, but the three airlift units performed poorly. The advantage of no moving parts in the well was offset by the lower efficiency of the pump and air compressor.

8.7.2 Electric-to-Electric Systems

A very promising development is a combination wind and electric-to-electric water pumping system [21]. The wind turbine alternator is connected directly to a motor that is connected to a centrifugal or turbine water pump. This system is a better match of the characteristics of the wind turbine rotor and the load. The overall annual efficiency is 12 to 15%—double the performance of a farm windmill. The water flow is higher at the higher wind speeds for the wind–electric system shown in Figure 8.16) because of more wind power in this region. This is also a region where a farm windmill is furled, limiting the power output.

A farm windmill and a 1.5-kW wind–electric system [22] using a submersible pump are essentially the same size and the costs are almost the same. The wind–electric system pumped twice the amount of water from the same depth (Figure 8.17). However, during the low wind month of August,

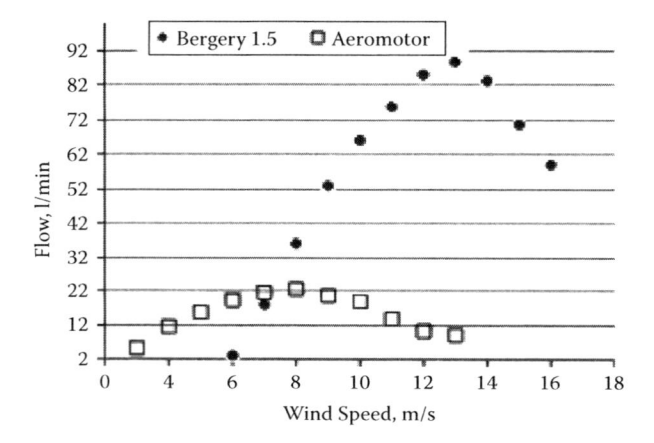

FIGURE 8.16 Water flow rates for mechanical, multiblade windmill (Aeromotor) and Bergey 1.5-kW electric-to-electric water pumping system.

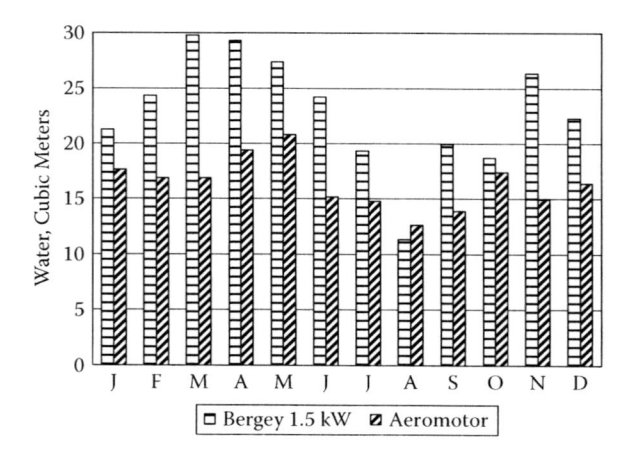

FIGURE 8.17 Predicted annual water pumped by 1.5-kW wind–electric water pumping system and farm windmill.

the farm windmill pumped more water. Another issue is that larger wind–electric systems can pump enough water for villages [23] or low-volume irrigation [24].

8.8 WIND–DIESEL AND HYBRID SYSTEMS

Around 1.4 billion people (about 70% of the populations of developing countries) do not have access to electric power because they are too distant from transmission lines of conventional electric power plants. Extension of the grid is too expensive for most rural areas, and if extended, must be subsidized heavily. These people depend on wood, charcoal, biomass, or dung for cooking and heating, mainly collected by women and girls.

For remote villages and rural industries, diesel generators represent the power standard. Remote electric power was estimated at 10.6 GW in 1990. Much of that was produced by 133,816 diesel generators ranging in size from 5 to 1,000 kW, with power rating estimated at 9.1 GW. These estimates are low. In 2012, Brazil produced 3 GW and India produced 22 GW of village power from diesel systems. In the U.S., remote power and back-up generators produced 5.6 GW.

Many islands, especially those with low populations, rely on diesel generation sets to produce electricity. For example, the Maldives population of 316,000 relies totally on diesel sets (62-MW capacity). Canada utilizes more than 800 diesel generating sets with a combined installed rating of over 500 MW. Diesel generators are inexpensive to install but they are expensive to operate and maintain. Major maintenance is needed every 2,000 to 20,000 hours, depending on the size of the system. Most small village systems generate power only in the evenings. Wind–diesel [25] is an alternative because of the high costs for generating pure diesel power in isolated locations. By 1986, more than a megawatt of wind turbine capacity was combined with existing diesel systems.

The grid of the Kotzebue Electric Association (KEA) in Alaska uses five diesel generators with a combined capacity of 11.04 MW. The annual average load is about 2.5 MW, with a peak around 3.9 MW, and the minimum load is around 1.8 MW. Loads are greatest during the winter months for heating and lighting. KEA maintains a high reserve capability to prevent loss of power during winters. Critical electrical and heating loads include the regional hospital, airport, and water system.

Typically, KEA runs one generator continuously during the winter and maintains the others as back-ups. In 2012, the association used around 5.3 million L (1.4 million gal) of diesel fuel, with an average efficiency of 3.8 kWh/L (14.6 kWh/gal). The fuel cost for the diesel generators was estimated at $0.23/kWh since the delivered diesel cost was $0.80/L ($3.39/gal). Fuel costs represent about 60% of the operational cost. KEA receives its annual fuel supply during the short summer season when its river is navigable by barge. The fuel is stored in two, 3.78 million liter (1 million gallon) steel tanks.

The KEA wind farm project added wind turbines to an existing diesel plant [26]. As of 2012, the farm consisted of nineteen turbines after two new EWT (900 kW) wind turbines increased the capacity to 2.94 MW. The farm is a high penetration system and storage and/or dump loads are needed to allow excess power to be absorbed and then released during peak loads or used for thermal applications. KEA plans to install a 500- kW/3.7 MWh flow battery and electrical dump loads.

The first three turbines were installed in July 1997, and seven more were added in May 1999. The ten wind turbines (Atlantic Orient, 66 kW, 15 m diameter) are located on a relatively flat plain 7 km south of Kotzebue and 0.8 km from the coast (Figure 8.18). The site is well exposed to the easterly winter winds and the westerly summer winds, with an annual average wind speed of 6.1 m/s. The cost of energy for the wind turbines was estimated at $0.13/kWh for the first two years of operation.

The ten wind turbines should reduce the annual fuel consumption by about 340,000 L, which is about 6% of normal requirements. At the 1998 fuel cost of $0.25/L, the fuel reduction saves KEA and its member-owners around $84,600 each year. In addition to direct fuel cost savings, KEA's costs for storage and meeting pollution control requirements associated with diesel fuel will decrease.

In 2000, the ten wind turbines produced 1.1 MWh of electricity, saving 265,000 L of diesel fuel. The wind turbines were shut down during part of the summer due to construction on the distribution system, so availability was only 85% during that period. KEA added two more AOC turbines

FIGURE 8.18 Wind turbines at Kotzebue Wind Farm. Starting at left foreground; Northern Power, 100 kW; Vestas, 65 kW; some Atlantic Orient wind turbines (downwind), 50 kW; and EWT, 900 kW. (Photos courtesy of Rich Stromberg, Alaska Energy Authority.)

in spring 2002. Because of the cold weather and high-density air, KEA had to change the control system to reduce peak power output.

A Northern Power wind turbine (100 kW), three more 50-kW units and one remanufactured 65-kW Vestas V15 were installed, and the association had seventeen wind turbines at the site. In 2007, they generated 667,580 kWh of energy, resulting in a savings of 172,240 L of diesel fuel. Installing foundations in permafrost and operating in cold climates presented problems not found at lower latitudes. After the price for diesel fuel escalated to $1.25/L in 2008, wind–diesel generation became more economical.

With the addition of the two EWT wind turbines (900 kW) in 2012, the wind farm will generate around 4 million kWh per year so the annual displacement of diesel is around 1.05 million liters. With diesel cost at $0.80/L, the savings to KEA will be $800,000 per year and will increase when diesel cost increases.

A number of prototype and demonstration hybrid systems combining wind and photovoltaic generation have been installed, but performance for most projects has been poor. In the past, hybrid systems [27] experienced high failure rates from faulty components, poor maintenance, and inadequate support by system suppliers after installation. One problem with Atlantic Orient (Entegrity) wind turbines in Nome, Alaska, is that ice chunks sliding off the blades bent the tip brakes and caused loss of over-speed control and damage to the turbines. Hybrid systems will be covered in more detail in Chapter 10.

8.9 BLADE PERFORMANCE

A smart rotor blade [28] would include active aerodynamics control with spanwise distributed devices: trailing edge devices and camber control, micro tabs, boundary layer control (suction, blowing, synthetic jets, vortex generators), and structural integration. Blade performance has been

Performance

evaluated through research and field experiments analyzing lift and drag data for airfoils and changing attack angles. The experiments studied the effects of surface roughness, boundary layer control, flow visualization, pressure taps, and vortex generators.

Data from pressure taps on a blade were used to obtain lift, drag, and pitching momentum coefficients during normal operation and dynamic stall [29]. The blade was a new NREL S809 thin airfoil, constant chord, no twist, on a three-blade downwind rotor (10 m diameter) with constant rpm and variable pitch. Dynamic stall occurred at a 30 degree yaw angle and during high angles of attack.

8.9.1 Surface Roughness

Performance will be reduced by the airfoil sensitivity to blade roughness. Just as for wings on airplanes, ice reduces performance drastically (Figure 8.19), to the point where a rotor will not turn. Also, falling chunks of ice from large blades present safety hazards. If icing is a major problem, heated blades may be economic solutions. Black blades have been used on some wind turbines to assist thawing when the sun comes out.

Accumulation of surface debris on the leading edges of blades from insects, grease, dust, and air pollution causes energy losses, as noted by wind farm operators. A 60-kW wind turbine at an Agricultural Research Service installation showed a monthly energy loss of over 20%. Energy losses of 40% were observed when the wind speeds were above 13 m/sec [30]. Insects on the blades (Figure 8.20), like on the windshield of a car, can reduce performance by 30% or more.

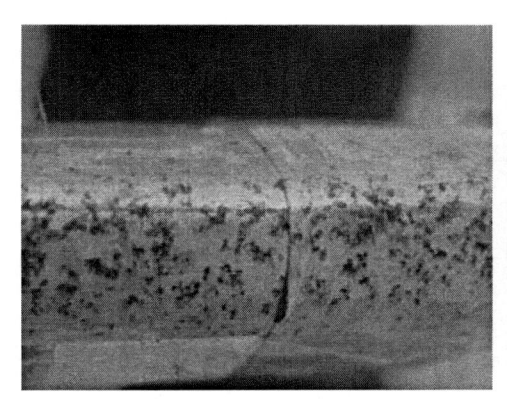

FIGURE 8.19 Left: bugs on leading edge of PM blade. Right: graph paper on ground shows shadow of leading edge bugs that protrude 1 to 3 mm.

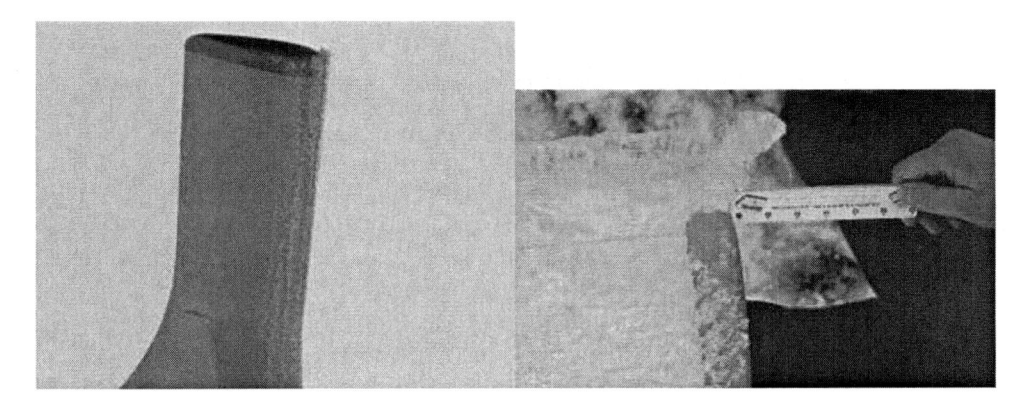

FIGURE 8.20 Ice on Carter 25 blade. Blades with this much ice did not rotate.

The impacts of insects on leading edges can be severe; but data showing amounts and heights of contamination are difficult to obtain [31]. Adhesive tape was wrapped around the leading edges of blades at equally spaced radial locations. Strips were collected and scanned by laser profilometry. The results showed that grit can adequately model surface roughness for wind tunnel and field testing. An artificial scale for roughness (light, medium, and heavy) was developed by NREL based on testing of wind turbines in California. This corresponded to using #80 rock tumbler grit at approximately 100 to 150, 250 to 300, and 500 to 600 particles per 5 cm^2.

Power was measured for two 24-hr periods, with the data averaged over 5 min for the Enertech 44. Measurements were made on dirty blades and made again after a rain cleaned the blades (Figure 8.21). Insects on the blades reduced the peak power about 20 kW (Figure 8.22) or 40%. The power curves for the data show the same information (Figure 8.23). This is another reason for wind farms to have baseline power curves for all turbines. They can then compare weekly power curves to baselines for performance checks.

California wind farms wash the blades after an insect hatch (Figure 8.24) to improve energy production. Active stall wind turbines attempt to compensate automatically for reductions in power output at wind speeds above the rated speed. Since insects on blades reduce aerodynamic efficiency, the active stall will compensate for this by changing the pitch angle toward zero degrees.

The compensation by active stall was evaluated on a NEG-Micon 72C wind turbine (72 m diameter, 1,500 kW) [32]. No reduction in the power curve was noted because the active stall provided complete compensation for moderately contaminated blades. Slight reductions occurred around the knees of the power curves for severely contaminated blades. There was a slight reduction in power curves in the lower wind region for extremely contaminated blades. However, the power still reached nominal rates at a wind speed above the rated one and a significant reduction in high winds occurred beyond that point. The compensation for blade roughness is another reason for using active stall over passive stall. The effect of blade roughness on energy production is the reason airfoils with less sensitivity to blade roughness have been designed specifically for wind turbines.

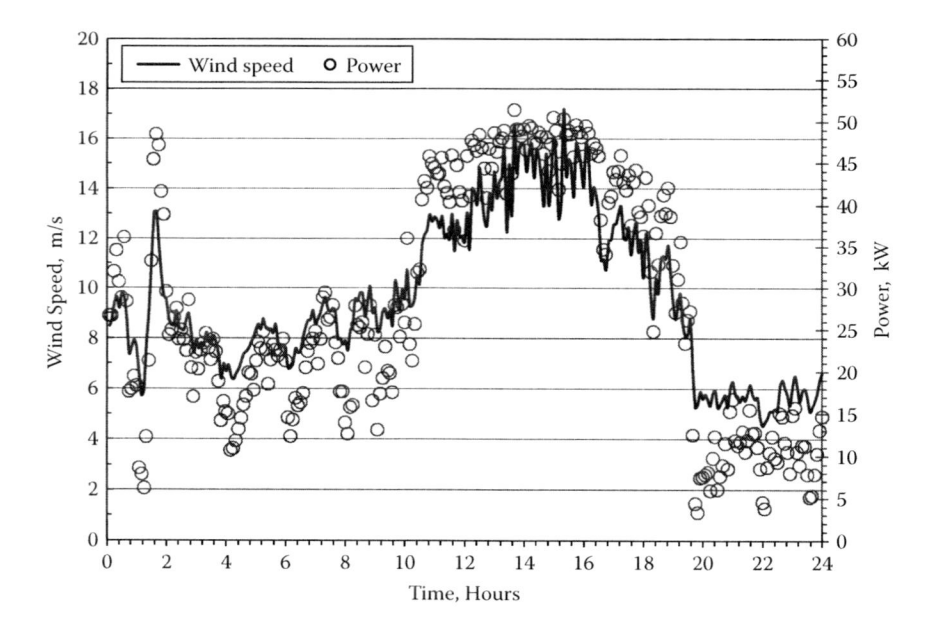

FIGURE 8.21 Enertech 44 performance with clean blades after rain on April 17, 1986. Maximum power is close to the 50-kW rated power.

Performance

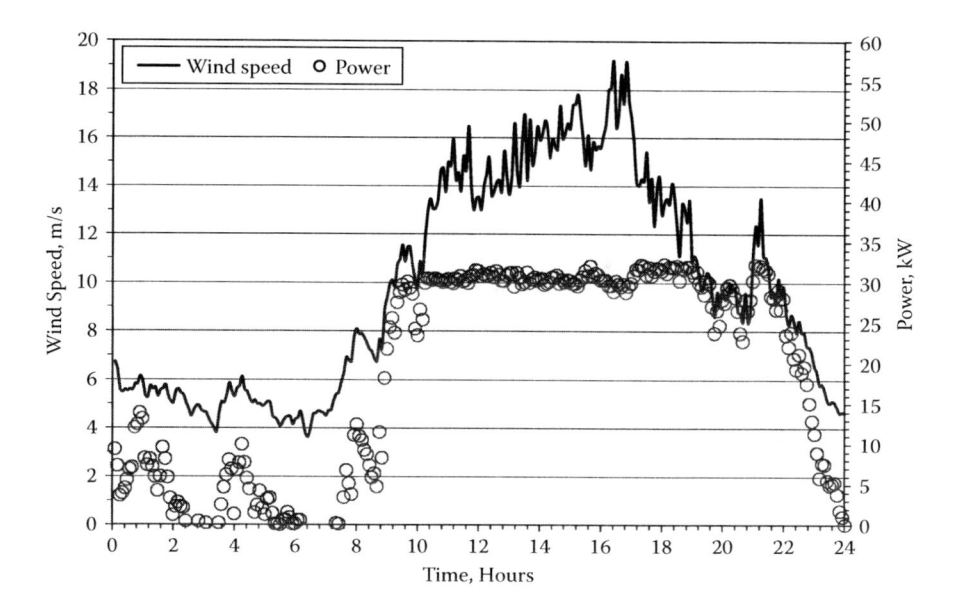

FIGURE 8.22 Enertech 44 performance with dirty blades on April 13, 1986. Maximum power leveled off around 32 kW.

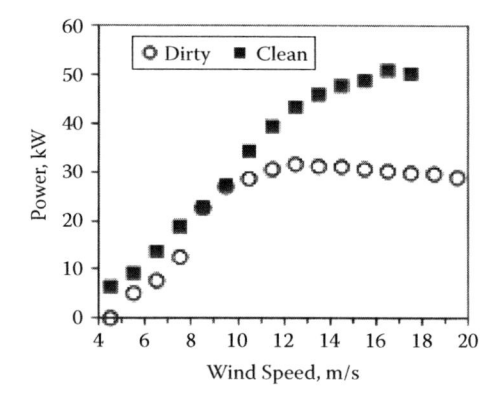

FIGURE 8.23 Enertech 44 power curves from data in Figures 8.19 and 8.20.

8.9.2 BOUNDARY LAYER CONTROL

Boundary layer control describes all the methods that can be used to reduce skin friction drag by controlling the transition to turbulent flow, reducing the development of turbulent flow, and preventing separation of laminar and turbulent flows. Boundary layer control is intended to keep the flow attached further along the chord, thereby increasing lift and reducing drag and preventing dynamic stall—a hysteresis loop of lift caused by changing high angles of attack on blades that creates high loads.

One method of boundary layer control is using suction or blowing air through holes in a blade. Suction can prevent laminar and turbulent separation by removing flows of low momentum. The pressure difference needed for suction or blowing can be obtained from the centrifugal force acting on the air inside a blade or a pump can supply the difference.

FIGURE 8.24 Spraying from tower to clean blades at San Gorgonio Pass in California. Spray is powered by the truck on the ground.

8.9.3 VORTEX GENERATORS

A vortex generator mixes faster moving laminar flow with the boundary layer and delays flow separation from the blade and stall. Typically, counter-rotating pairs of vortex generators with ±20 degree angles of incidence at 10% chord are installed on the inner portions of a blade (Figure 8.25) that are thicker and more prone to dynamic stall. Vortex generators were installed on the MOD-2 and the MOD-5 wind turbines, and the performance improved [33].

A Carter 25 wind turbine has an optimal blade with a large amount of twist and taper at the root. When vortex generators were tested on the unit, the maximum power increased, but power below the rated wind speed was reduced because of the added drag of the vortex generators. In other words, the inner portion of the blade did not enter stall and did not need the vortex generators.

In general, vortex generators improve blade performance by 4 to 6%. A unique concept is the air-jet vortex generator that showed improved performance over vane vortex generators in wind tunnel tests. The air-jet device was installed on a 150-kW wind turbine and increased the maximum power. The potential benefits were not conclusive, probably due to the placement of the jets on the outer part of the blade rather than on the inboard section. Production blades now have vortex generators (Figure 8.26).

8.9.4 FLOW VISUALIZATION

The performances of blades, rotors, and towers can be checked by flow visualization: smoke, tufts, stall flags, pressure-sensitive liquid crystals, and oil streaks. Tufts (small pieces of string, frayed on the end) are driven by frictional drag and stall flags are pressure driven. A stall flag responds to separated flow with an optical signal that exceeds a tuft signal by a factor of 1,000 [34]. Smoke shows the stream flow for airfoils in wind tunnels, the generation of tip vortices from the ends of blades, and their propagation downstream [35]. Smoke released from tethered smoke generators was used to observe the evolution of tip vortices from the MOD-2 [36]. The vortex became unstable when it passed through the wake of the turbine tower.

Blades on downwind turbines pass through the wake of a tower and the change in attack angle and flow across the blade also generate noise. Flow visualization was used to study the flows [37]

Performance

over the blades of an Enertech 21 (6.4 m diameter, 5 kW), with and without tip brakes; a Carter 25 (10 m diameter, 25 kW); and an Enertech 44 (13.4 m diameter, 50 kW). All three were downwind, constant rpm wind turbines. A video camera and a 35-mm camera were mounted on a boom attached to the root of a blade. Tufts and oil flow revealed the nature and many of the details of the flows, such as laminar separation bubbles, turbulent reattachment, and complete separation over part or most of the blade.

Full or partial reattachment due to tower shadow was observed on each unit (Figure 8.27). Spanwise, flow was observed near the leading edge of the Enertech 21, and most of the blade was in stall at high wind speeds. The tip brakes on the Enertech units are important in retaining attached flow near the tip. The oil streak pattern after 4 min in winds from 7 to 15 m/sec on the Enertech

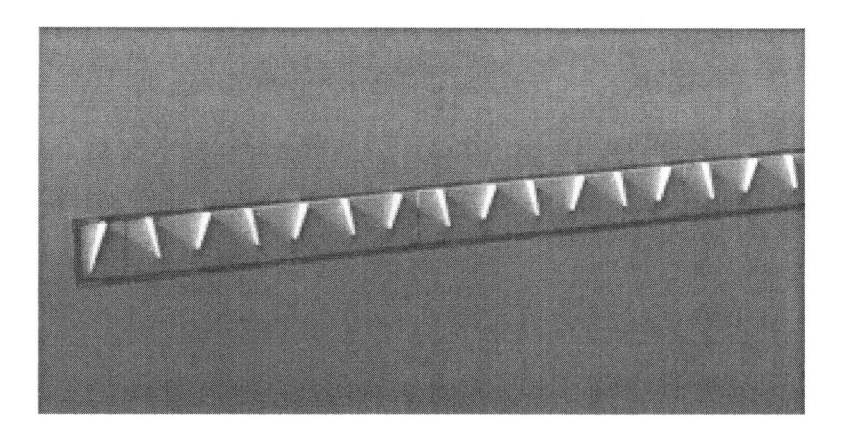

FIGURE 8.25 Shape and orientation of vortex generators on blade of GE wind turbine (77 m diameter, 1.5 MW).

FIGURE 8.26 Vortex generators on inner portion of blade of GE wind turbine (77 m diameter, 1.5 MW).

FIGURE 8.27 One blade of Enertech 21 over one revolution. Shaded areas show representative pattern of attached flow. Note strong reattachment due to tower shadow.

44 blade showed that flow was separated completely below 0.5 blade length. However, the flows on the highly twisted and tapered Carter 25 blade were attached in medium winds. The flow data showed a turbulent-type edge separation beginning at about half the radius and progressing forward. Pressure-sensitive liquid crystals were tried, but field results were not good because precise lighting is required to observe color changes.

It should be noted that vortices alternating on each side will be shed by cylinders in wind flow that can induce vibration in the cylinder. On the VAWT 34-m test bed, a spiral staircase on the torque tube eliminated these vortices.

8.10 COMMENTS

Wind turbine and wind farm performance data (annual, quarterly, monthly, or by period of peak demand) determine economic viability and aid comparisons of turbines. The main performance factors are the amount of energy produced and the cost of the energy compared to other energy sources. Of course, electricity is the major application; water pumping is secondary. Capacity factors in good to excellent wind regimes should range from 30 to 50%, and annual specific outputs should be over 1,000 kWh/m^2. For wind farms, availabilities of 98% and turbine lifetimes of twenty-five or more years should be the norm if good preventative maintenance programs are followed.

LINKS

Federal Energy Regulatory Commission Electricity quarterly reports. http://eqrdds.ferc.gov/eqr2/frame-summary-report.asp
Global Wind Energy Council. www.gwec.net
International Energy Agency. www.iea.org
New Zealand Wind Energy Association. www.windenergy.org.nz
Performance for Vestas 47 and V80. www.hullwind.org
RenewableUK. Wind and marine energy. www.renewableuk.com

Swedish Energy Agency. www.energimyndigheten.se/en/
Windicator. www.windpower-monthly.com/wpm:WINDICATOR
Wind in power. European Wind Energy Association. www.ewea.org/fileadmin/files/library/publications/statistics/Wind_in_power_2011_European_statistics.pdf

REFERENCES

1. R.N. Clark. 1983. *Reliability of wind electric generation.* ASAE Paper 83-3505.
2. W. Pinkerton. 1983. Long term test: Carter 25. In *Proceedings of Wind Energy Expo and National Conference,* p. 307.
3. F. S. Stoddard. 1990. *Wind turbine blade technology: a decade of lessons learned, 1980–1990, California Wind Farms Report.* Alternative Energy Institute, West Texas A&M University.
4. *WindStats Newsletter.* www.windstats.com
5. California Energy Commission. Wind Performance Reporting System. PDF reports: www.energy.ca.gov/wind; Electronic Wind Reporting System. http://wprs.ucdavis.edu
6. Sindal. http://www.sindal-lundsberg.com/cms/
7. B.D. Vick, R.N. Clark, and D. Carr. 2007. Analysis of wind farm energy produced in the United States. In *Proceedings* of *Windpower Conference.* CD.
8. V. Peters, A. Ogilvie, and C. Bond. 2012. Wind plant reliability benchmarks. Sandia National Laboratory. Report 2012-7329P. http://energy.sandia.gov/wp/wp-content/gallery/uploads/Sandia-CREW-2012-Wind-Plant-Reliability-Benchmark-Presentation.pdf
9. P. Volund, P.H. Pedersen, and P.E. Ter-Borch. 2004. The 165-MW Nysted offshore wind farm first year of operation: performance as planned. www.2004ewec.info/files/24_0900_pervolund_01.pdf
10. W. Cleijne. 1990. *Literature database on wind turbine wakes and wake effects.* Report 90-130. Apeldoorn, The Netherlands: TNO.
11. P.B.S. Lissaman. 1994. Wind turbine airfoils and rotor wakes. In D.A. Spera, Ed., *Wind Turbine Technology.* New York: ASME Press, p. 283.
12. D.L. Elliott. 1991. Status of wake and array loss research. In *Proceedings of Windpower Conference,* p. 224.
13. C.B. Hasager et al. 2007. Wind resources and wind farm wake effects offshore observed from satellite. Risoe National Laboratory, Wind Energy Department. www.risoe.dk/rispubl/art/2007_79_paper.pdf
14. M.B. Christiansen and C.B. Hasager. 2005. Wake effects of large offshore wind farms identified from satellite SAR. *Remote Sensing of Environment,* 98, 251.
15. M. Méchali et al. Wake effects at Horns Rev and their influence on energy production. http://johnstonanalytics.com/yahoo_site_admin/assets/docs/WakeEffectsWindTurbines.21172331.pdf
16. R.N. Clark and R.G. Davis. 1993. Performance of an Enertech 44 during 11 years of operation. In *Proceedings of Windpower Conference,* p. 204.
17. K. Pokhrel. 2001. Performance of renewable energy systems at a demonstration project. Master's thesis, West Texas A&M University.
18. V. Nelson, N. Clark, and R. Foster. 2004. *Wind Water Pumping.* CD. Alternative Energy Institute, West Texas A&M University (also available in Spanish).
19. R.N. Clark. 1992. Performance comparisons of two multibladed windmills. In *Proceedings of 11th ASME Wind Energy Symposium,* Vol. 12, p. 147.
20. J.A.C. Kentfield. 1996. The measured field performances of eight different mechanical and air-lift water-pumping wind-turbines. In *Proceedings of Windpower Conference,* p. 467.
21. J.W. McCarty and R.N. Clark. 1990. Utility-independent wind electric water pumping systems. In *Proceedings of Solar Conference,* p. 573.
22. B.D. Vick, R.N. Clark, and S. Ling. 1999. One and a half years of field testing a wind–electric system for watering cattle in the Texas Panhandle. Paper presented at Windpower Conference. CD.
23. M.L.S. Bergey. 1990. Sustainable community water supply: a case study from Morocco. In *Proceedings of Windpower Conference,* p. 194.
24. B.D. Vick and R.N. Clark. 1998. *Ten years of testing a 10 kilowatt wind-electric system for small scale irrigation.* ASAE Paper 98-4083.
25. R. Hunter and G. Elliot. 1994. *Wind Diesel Systems: A Guide to the Technology and Its Implementation.* Cambridge, UK: Cambridge University Press.
26. Kotzebue Electric Association. http://kea.coop/news/renewable-energy.php

27. V.C. Nelson et al. 2001. *Wind hybrid systems technology characterization: report for NREL.* West Texas A&M University and New Mexico State University. http://solar.nmsu.edu/publications/wind_hybrid_nrel.pdf
28. H.E.N. Bersee. 2007. *Smart rotor blades and rotor control.* The Netherlands: Delft University of Technology. www.upwind.eu/pdf/070510%20WP1B3%20Harald%20Bersee.pdf
29. C.P. Butterfield et al. 1991. Dynamic stall on wind turbine blades. In *Proceedings of Windpower Conference*, p. 132.
30. R.N. Clark and R.G. Davis. 1991. Performance changes caused by rotor blade surface debris. In *Proceedings of, Windpower Conference,* p. 470.
31. E.M. Moroz and D.M. Eggleston. 1992. A comparison between actual insect contamination and its simulation. In *Proceedings of Windpower Conference*, p. 418.
32. C.J. Spruce. 2006. Power performance of active stall wind turbines with blade contamination. Paper presented at European Wind Energy Conference. http://proceedings.ewea.org/ewec2006/allfiles2/0672_Ewec2006fullpaper.pdf
33. R.E. Wilson. 1994. Aerodynamic behavior of wind turbines. In D.A. Spera, Ed., *Wind Turbine Technology.* New York: ASME Press, p. 215.
34. G.P. Corten. 2001. Flow separation on wind turbine blades. http://igitur-archive.library.uu.nl/dissertations/1950226/inhoud.htm
35. L.J. Vermeer. 2001. *A review of wind turbine wake research at TUDELFT.* AIAA-2001-0030. ww.windenergy.citg.tudelft.nl/content/research/pdfs/as01njv.pdf
36. H.T. Liu. 1983. Flow visualization in the wake of a full-scale 2.5 MW Boeing MOD-2. wind turbine. www.stereovisionengineering.net/mod-2.htm
37. D.M. Eggleston and K. Starcher. 1989. A comparative study of the aerodynamics of several wind turbines using flow visualization. In *Proceedings of Eighth ASME Wind Energy Symposium*, Vol. 7, p. 233.

PROBLEMS

1. From Table 8.2 calculate annual specific output (kWh/kW) for two different wind turbines.
2. From Table 8.2, calculate capacity factor for Fayette, Vestas 23, and Bonus 120.
3. From Table 8.4, what is the average capacity factor for 1989 through 1996?
4. Calculate the specific output (kWh/m^2) in (a) 1985 for Enertech 44/60 and (b) 1990 for Enertech 44/40.
5. From Table 8.5 calculate: (a) specific output (kWh/m^2) and (b) capacity factor for 7 months for Enertech 44/25.
6. From Table 8.5 calculate (a) specific output (kWh/m^2) and (b) capacity factor for 1985.
7. From Table 8.5 calculate specific output (kWh/m^2) for May and August 1984. Does specific output depend on the wind?

Information for problems 8–10. Today, Carter is not manufacturing wind turbines, and Vestas V27 is not in production.

8. For the Carter 300, calculate (a) specific output (kWh/m^2), (b) $IC/kW, (c) specific output (kWh/kg), and (d) kWh/$IC.
9. For the Vestas V27, calculate (a) specific output (kWh/m^2), (b) $IC/kW, (c) specific output (kWh/kg), and (d) kWh/$IC.
10. Estimate the annual capacity factor for the Carter 300 and Vestas V27.
11. Go to the Vestas web page (www.vestas.com). (a) For the Vestas V52 (1.65 MW), estimate the annual specific output (kWh/m^2) for a good wind regime. (b) For the Vestas V90 (3 MW), estimate the annual specific output (kWh/m^2) for a good wind regime.
12. For a farm windmill, what is the approximate pump diameter if the water depth is 40 m? Approximately how much water could be pumped in a light wind?
13. For a farm windmill, what is the approximate pump diameter if the water depth is 20 m? Approximately how much water could be pumped in a light wind?

Performance

14. For a farm windmill, what is the approximate pump diameter if the water depth is 100 m? Approximately how much water could be pumped in a fair wind?
15. For a farm windmill (use Figure 8.14 for flow data), estimate water pumped for 1 month that has an average wind speed of 5 m/sec. Use Rayleigh distribution (1 m/sec bin width).
16. Check the Internet to see which companies sell wind–electric water pumping systems.
17. For a wind–electric water pumping system (use Figure 8.14 for flow data), estimate water pumped for one month that has an average wind speed of 5 m/sec. Use Rayleigh distribution (1 m/sec bin width).
18. For an annual average wind speed of 6 m/sec, compare the predicted annual energy production of the Enertech 44 for 25- and 60-kW wind generators. Use Figure 8.8 for power curves and use Rayleigh distribution (1 m/sec bin width).
19. The Electronic Wind Performance Reporting System is available online [5]. For the latest year available, what is the statewide energy production for California? Which manufacturer had the largest installed capacity? Which manufacturer had the largest number of turbines installed?
20. By approximately what percent will bugs on blades reduce the power?
21. Select a wind farm near your home town or city. What is the installed capacity? How much electricity did it produce last year? If values are not available, estimate from installed capacity and capacity factor.
22. Are there any village power systems in your country? If the answer is yes, determine whether performance data are available for one system. What is the size of the system and how much energy is produced annually?
23. Are there any wind–diesel systems in your country? If the answer is yes, determine whether performance data are available for one system. What is the size of the system and how much energy is produced annually?
24. List two types of boundary layer controls for wind turbines. Briefly explain each type.
25. Which type of wind turbines would perform best with heavy insect contamination on the blades? Why?
27. Go to the Sindal report. For the latest quarter available, for wind turbines 3 MW and above, which units have largest capacity factor on kWh/kW land? Offshore?
28. From the latest benchmark report of the Sandia Laboratory CREW database, what is the generating time? What is the capacity factor? Note year for data.

9 Siting

The crucial factor in siting a wind turbine or wind farm (also called wind park or wind plant) is the annual energy production and how the value of the energy produced compares to other sources of energy. Many data from meteorological stations worldwide are of little use in predicting wind power potential and expected energy production from wind turbines.

9.1　SMALL WIND TURBINES

For small wind turbines, a measuring program may cost more than the turbine; therefore, other types of information are needed. Many countries are developing wind maps to aid development of wind farms. These maps can be used as guides to determine regions with enough wind for small wind turbines. Also, wind maps for countries and large regions obtained from numerical models have sufficient resolution for determining general areas for siting of small wind turbines. In the U.S., Wind Powering America [1] provides residential scale wind speed maps at 30 m for every state.

An annual average wind speed of around 4 m/sec and greater is considered suitable for small wind projects. Tower heights for small wind turbines range from 10 to 35 m. Since small wind turbines are located close to loads, local topography will influence the estimations of wind speeds and siting decisions. If a location is on exposed terrain, hills, or ridges, wind speeds will be higher than speeds in a valley. In complex terrain, some sites will be adequate for small wind turbines and some will be too sheltered.

One of the factors in the settlement of the Great Plains of the U.S. was the farm windmill that provided water for people and livestock. Therefore, if farm windmills are used or were used in the past in a region, the wind is sufficient for the use of small wind turbines in the region. Another possibility is to install met towers to compile reference data for a region. Generally, this is done by regional or state organizations or governments, not by individuals interested in siting small wind turbines.

Small wind turbines can be cost effective as stand-alone systems using the general rule that the average wind speed for the lowest wind month should be 3 to 4 m/sec. General maps of wind power or wind energy potential for small wind turbines have been developed for large regions (Figure 9.1) [2]. These gross wind maps will be replaced by national wind maps developed for determining wind energy potential for wind farms. Finally, if wind farms already exist in an area, the wind is sufficient for small wind turbines.

It is obvious that a small wind turbine should be located above (10 m if possible) obstructions and away from buildings and trees [3]. Towers for small wind turbines should be a minimum of 10 m and preferably 20 m high; higher towers generally capture more energy (Figure 9.2). Again, the trade-off is the extra energy versus the cost of a taller tower. Towers of 35 m height are sometimes used.

As a general rule for avoiding most of the adverse effects of building wakes, a turbine should be located (1) upwind of the prominent wind direction or maybe the prominent wind direction of low wind months at a distance more than two times the height of the building, (2) downwind a minimum distance of ten times the building height, or (3) at least twice the building height above ground if the turbine is immediately downwind of the building. The above rule is not foolproof because the size of the wake also depends on the building's shape and orientation to the wind (Figure 9.3).

Downwind from a building, power losses become small at a distance equal to fifteen times the building height. However, a small wind turbine cannot be located too far away from the load

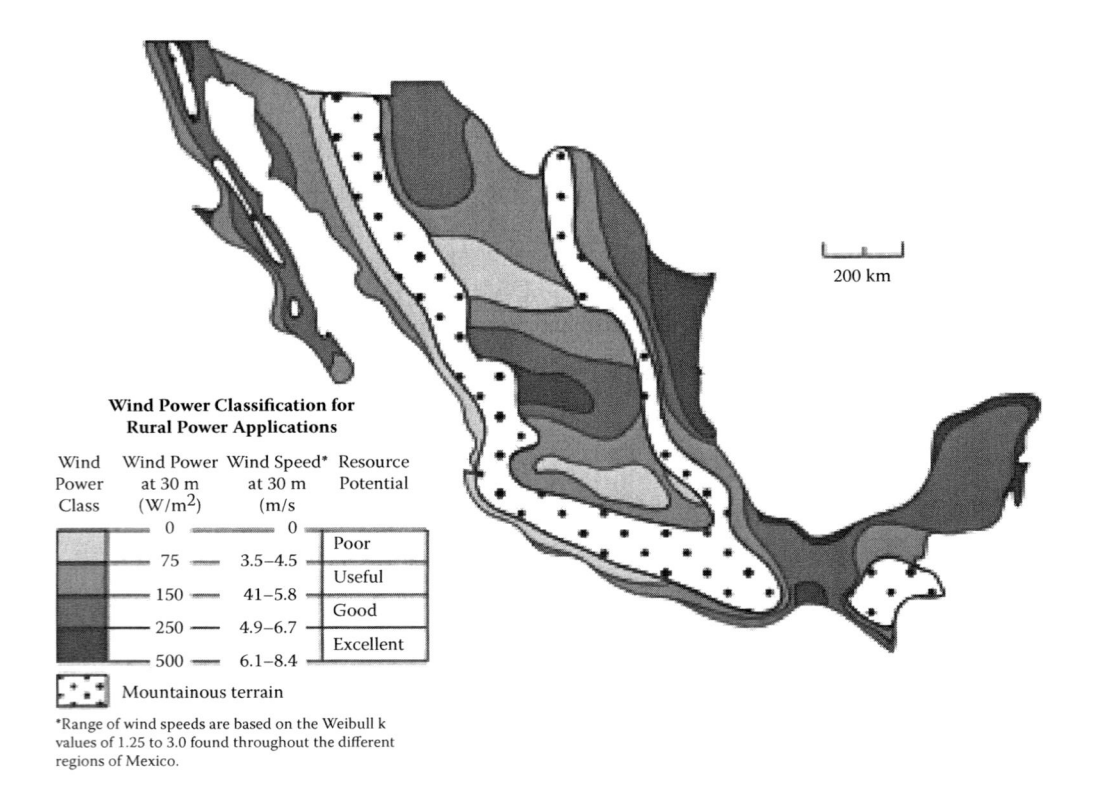

FIGURE 9.1 Wind power map for rural applications In Mexico. Notice difference in definition of wind power class and height at 30 m.

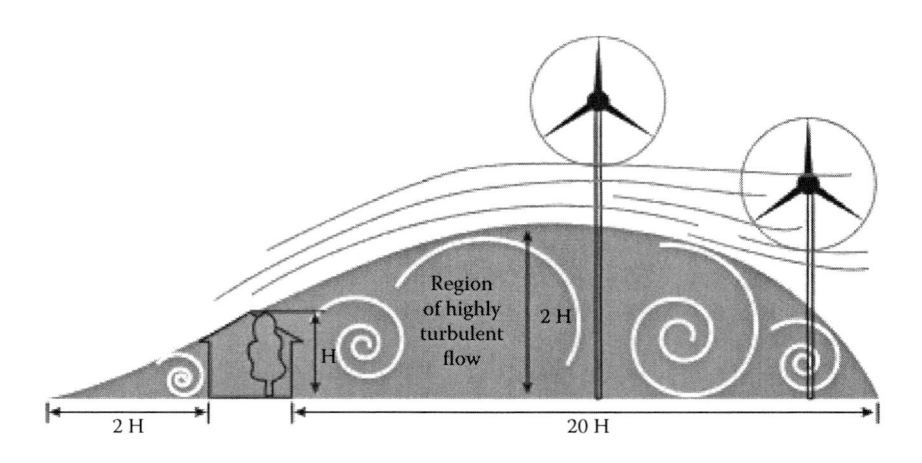

FIGURE 9.2 Height of small wind turbine near obstacles of height *H*.

because the cost of wiring over distance is prohibitive. Also, more losses in wires will occur if DC rather than AC transmits power from the wind turbine to the load. In general, small wind turbines should not be mounted on occupied buildings because of possible noise, vibration, and even turbulence. Tower heights for very small wind turbines vary from stub poles on sailboats to short (3 to 5 m) towers, and some are even mounted on buildings. Paul Gipe wrote numerous articles on all aspects of wind energy [4], and two of his books are about small wind systems [5,6].

Siting

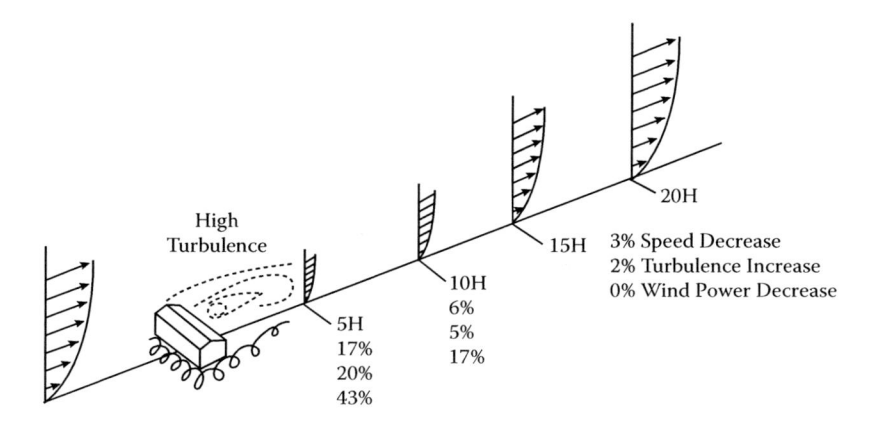

FIGURE 9.3 Estimate of speed, power decrease, and turbulence increase for flow over building. Estimates shown are for building height *H*. (Source: M.N. Schwartz and D.L. Elliott. 1995. In *Proceedings of Windpower Conference*. With permission.)

Is there such a concept as wind rights if a neighbor erects a tall structure that obstructs the flow of wind to your turbine? From a visual standpoint, a wind turbine in every backyard in a residential neighborhood is much different from a photovoltaic (PV) panel on the roof of every home.

The American Wind Energy Association [7] and the Canadian Wind Energy Association [8] have online information about small wind turbines including information on siting. A guide for small wind turbines available from the National Renewable Energy Laboratory (NREL) [9] also contains similar information about siting.

RenewableUK, formerly the British Wind Energy Association, maintains a section on small and medium wind turbines [10] that includes information on the national wind speed database, small wind turbine technologies, planning, and case studies. An interactive map for wind speeds at 10, 25, and 45 m is available online [11], and the RenSmart Site Planner estimates energy production, yearly value, and pay-back time for wind and solar systems. National wind energy associations in other countries probably have sections on small wind turbines.

A number of designs were developed by architects, inventors, and even people selling wind systems (most not built or tested) to integrate wind turbines into building structures in urban areas. The designs usually touted the increase of wind speed caused by the building. However, in the real world, incorporating wind turbines into buildings is a difficult choice because of noise, vibration, and safety concerns. In some concepts of installations on buildings, the wind turbines must be mounted perpendicular to the predominant wind direction because the wind turbines are fixed in yaw.

According to Dutton et al., the estimated energy production is in the range of 1.7 to 5.0 TWh in the built environment (turbines in urban areas, turbines mounted on buildings, and turbines integrated into buildings) in the U.K. [12]. The technical feasibility and various configurations are also discussed. There is an Internet site for urban wind turbines [13]. Available downloads include the *European Urban Wind Turbine Catalogue; Urban Wind Turbines: Technology Review* (companion text to the European Union's *UWT Catalogue); Urban Wind Turbine Guidelines for Small Wind Turbines in the Built Environment*; and *Windy Cities: Wind Energy for the Urban Environment*. The wind turbine guidelines include images of wind flow over buildings and example projects.

A newspaper in Clearwater, Florida, installed a stacked Darrieus unit next to its building. The unit consisted of three Darrieus turbines, 4.5 m in diameter, 6 m tall, 4 kW each (Figure 9.4). Fortis mounted three wind turbines (5 m diameter, 2 kW rather than the nominal 5 kW) on a factory and office building and experienced a small problem with vibration at high wind speeds due to the flexibility of the roof. The Aeroturbine has a helical rotor mounted in a 1.8 × 3 m frame rated at 1 kW [14].

FIGURE 9.4 Three stacked Darrieus 4-kW wind turbines next to building. Notice man on top. (Photo courtesy of Coy Harris, American Wind Power Center and Museum.)

A building in Chicago mounted eight units horizontally on top of a building (Figure 9.5) although other buildings mounted units vertically. Two 6-kW wind turbines mounted on the roof of a civic center in the U.K. were described in a case study [15]. A different concept is mounting a number of small wind turbines on the parapets of urban and suburban buildings [16]. The horizontal axis wind turbine had a rated power of 1 kW and was mounted in a modular housing measuring approximately 1.2 ×1.2 m). Fourteen wind turbines installed on a corner of the Energy Adventure Aquarium building (Figure 9.6) in California constitute a kinetic sculpture.

The most spectacular structure featuring integrated large wind turbines is the Bahrain World Trade Center. The two 240-m towers with sail silhouettes have three cross bridges that carry wind turbines [17]. The turbines are 29 m in diameter, rated at 225 kW, and are predicted to generate around 1,100 to 1,300 MWh per year—11 to 15% of the energy needed by the buildings. The aerodynamic design of the towers funnels the prevailing onshore Persian Gulf breezes into the paths of the wind turbines.

9.1.1 Noise

Although zoning is an institutional issue, the regulations will affect the potential for erecting small wind turbines and may specify turbine size, tower height, required space surrounding the tower, noise restrictions, and even visual concerns of neighbors. The noise from a small wind turbine is around the level of noise in an office or in a home. Noise from a small wind turbine is rarely a problem because the level drops by a factor of 4 at a distance of 15 m, and is generally masked by background noise.

Siting 201

FIGURE 9.5 Eight helical horizontal axis wind turbines, 1 kW each, on top of building. (Photos courtesy of Kurt Holtz, Lucid Dream Productions.)

A sound study with a 10-kW wind turbine (wind speeds at 9 to 11 m/sec) showed levels of 49 to 46 dBA for the running turbine and at a distance of 15 m from the turbine, respectively. Essentially no difference was found at distances of 30 m and more. However, if a wind turbine rotor is downwind, some sound is made every time the blade passes the tower. Even if the sound is at the same level as background noise, it can be annoying. In California, noise from a wind turbine must not exceed 60 dBA at the closest inhabited building.

9.1.2 Visual Impact

The State of Vermont has a scoring system for possible adverse visual impacts of small wind turbines [18] from the vantage points of private property (neighbors' views) and public views (roads, recreation facilities, and natural areas). The considerations for neighbors' views are:

1. What is the position of the turbine in the view?
2. How far away is the turbine seen?
3. How prominent is the turbine?
4. Can the turbine be screened from view?

FIGURE 9.6 Twelve 1-kW wind turbines mounted on parapet of building. (Photo courtesy of AeroVironment.)

TABLE 9.1
Vermont System for Scoring Visual Impacts of Small Wind Turbines

	Neighbor View				Public View	
	1	2	3	4	5	6
Points	View Angle (degrees)	Distance (m)	Prominent	Screened	Vista	Duration (sec)
0	> 90	> 900	Below tree tops	Complete	Degraded	0
1	0–45	450–900	At horizon line	Multiple trees	Common	<15
2	50–60	150–450	Above horizon line	Single tree, 1/2 to 2/3	Scenic	<30
3	60–90	<150	Above tallest mountain	No screening	Highly scenic	>60

For public views, two additional factors must be considered:

5. Is the turbine seen from an important scenic or natural area?
6. What is the duration of the view?

Each factor is rated by a point system (Table 9.1), with a total of 12 points for the residential view and 18 for the public view. If the score (Table 9.2) is below the significant range, the wind turbine is unlikely to have a visual impact unless it is near or within a scenic view. The score is only a general indicator for visual impacts of small wind turbines. Wind turbines will be visible, at least from some viewpoints because they will tower above surrounding trees.

In the midwestern plains of the U.S. that have few trees, small wind turbines are noticeable from 1 to 3 km—the same as trees around a farmhouse. Comparable structures such as cell phone towers, light towers at highway interchanges, radio towers, and towers for utility transmission lines have comparable heights. The difference is that those towers do not have moving rotors.

TABLE 9.2
Score Sheet for Determining Visual Impacts of Small Wind Turbines

Impact	Neighbor View	Public View
Negligible	0–3	0–3
Minimal	3–6	3–9
Moderate	6–9	9–14
Significant	9–12	14–18

9.2 WIND FARMS

Long-term wind data are critical for siting wind farms. Data should be collected at a potential site for two to three years, after which other questions arise. What is the long-term annual variability? How well can we predict the energy production for a wind farm? The siting of turbines over an area the size of a wind farm, about 5 to 20 km² is termed micrositing. The turbines should be located within a wind farm to maximize annual energy production and yield the largest financial return. Array losses have to be considered in the siting process.

In general, there will be a number of landowners and a developer who will lease an amount of land based 20 hectares (ha) or 50 acres per megawatt of planned production. Not all the land will be used for turbines, and in many cases, developers lease land for further expansion. Actual values after construction will be from 12 to 18 ha per megawatt. Negotiation with a large number of landowners can present some difficulties, for example, one lease of 1640 ha involved 120 landowners.

9.2.1 Long-Term Reference Stations

To determine whether historical data from a site are adequate to describe long-term wind resources at another site, a rigorous analysis should be done. Simon and Gates [19] recommend that the annual hourly linear correlation coefficient be at least 0.90 between the reference site and off-site data. Remember to consider wind shear if the heights are different at the two locations. If the two sites do not exhibit similar wind speed and direction trends and lack similar topographic exposures, they will probably not have sufficient correlation value.

Long-term reference stations should be considered at all locations with wind power potential everywhere in the world. These stations should continue to collect data even after a wind farm is installed. The data will improve siting of wind farms and also provide reference sites for delineating wind resources for single or distributed wind turbines in the region. As wind turbine sizes increased, hub heights became higher. Because wind speed increases with height in most locations, reference stations are needed to collect data at least at 50 m, and if possible to 100 m.

9.2.2 Siting for Wind Farms

The number of met stations and duration of data collection to predict the energy production for a wind farm vary depending on the terrain and the availability of long-term base data in the vicinity. In general, numerical models of wind flow will predict wind speeds to within 5% for relatively flat terrain and 10% for complex terrain, which means an error in energy of 15 to 30%. Therefore, a wind measurement program is imperative before a farm is installed. However, if a number of wind farms are already in the region, one year of data collection may suffice.

For complex terrain, one met station per three to five wind turbines may be needed. Since wind turbines for wind farms are now megawatt size, one met station per two wind turbines may be required in complex terrain. For somewhat homogeneous terrain as in the U.S. plains, a primary tall met station and one to four smaller met stations may suffice. The tallest met station should be installed at a representative location, not at the best point of a wind farm.

Contour maps are used for locations of wind turbine pads and roads. In general, the wind turbines will be located at higher elevations within the wind farm area. The U.S. Geological Survey has topographical maps that can be downloaded. Topozone (now a subscription service) has interactive U.S. topography maps (at different scales) available online [20]. These maps are very useful for selecting met tower locations, micrositing, roads, and other physical aspects of wind farms.

The key factors for array siting for the Zond wind farms [21] in Tehachapi Pass were an extensive anemometer data network, the addition of new stations during the planning period, a timeframe of one year to refine the array plans, a project team approach to evaluate the merits of siting strategies, and the use of initial operating results to refine the rest of the array. A large number of met stations were needed because the spatial variation of wind resources over short distances on a complex terrain was greater than expected. The energy output from 2 projects consisting of 98 wind turbines and 342 wind turbines was within 3% of the predicted value. This experience shows it is possible to estimate long-term production from a wind plant with acceptable accuracy for the financial community. One of the key factors was an extensive network of met towers.

The money spent on micrositing is a small fraction of project cost, but the value of the information gained is critical for estimating energy production accurately. Many problems with low energy production are the results of poor siting.

Wind turbines have become larger, with rotor diameters from 60 to 150 m and hub heights of 60 to 100 m. Few data show conditions above these heights, but NREL had a program for tall tower data [22]. The problem is that all tower data collected by wind farm developers are proprietary. Because of wind shear, wind turbines are located at higher elevations on rolling terrain and mesas and on ridges on complex terrain. In the past, turbulence was considered a big problem for siting at the edges of mesas and ridges. However, taller towers allow placement of wind turbines on the edges that are perpendicular to the predominant wind direction. Consider wind turbines on mesas in Texas. The north edge of the mesa would have increased winds from northern storms in the winter due to the rise in elevation. The southern winds in summers allow room for expansion of the wake. Turbulence data for these sites are proprietary, primarily because turbulence affects operation and maintenance.

9.3 DIGITAL MAPS

Digital maps are useful as they give a general overview of wind resource, provide confidence in the data, and information about land use, transmission lines, and other features can easily be displayed on the same maps. NREL created a higher resolution digital wind map for the U.S. and is in the process of updating the map by state using terrain enhancement and geographic information systems (GIS). NASA's World Wind is an open source virtual glove similar to Google Earth, but the maps do not cover wind resources.

The Wind Site Assessment Dashboard (formerly windNavigator), based on GoogleMaps®, is an interactive tool that includes wind resource maps and world data [23]. The map (2.5 km resolution) provides wind speeds at 60, 80, and 100 m and a pointer to locate minimum and maximum mean annual wind speeds. Selectable area maps at 200 m resolution (PDF or GIS data set) can be purchased. Satellite, hybrid, and terrain views are available for the entire world. The SmallwindExplorer interactive map is available to the public online [24]. Mean wind speed data for heights of 24.4 (80 ft), 30.5 (100 ft), and 36.7 (120 ft) m are available.

A similar interactive wind resource map (map, satellite, hybrid, and terrain views) and data for much of the world, are available [25]. FirstLook, has wind speed data for 20, 50, and 80 m and with Wind GIS Data Layers, resolution is at 90 m. In addition, a solar resource map and prospecting tools are available. Remember, wind speed maps are useful indicators of wind energy and wind power maps are the next step.

9.4 GEOGRAPHIC INFORMATION SYSTEMS

A geographic information system (GIS) is a computer system capable of holding and using spatially oriented data. A GIS typically links different data sets or it displays a base set over which overlays of other data sets are placed. Information is linked as it relates to the same geographical area. A GIS is an analysis tool, not simply a computer system for making maps.*

The two general bases of representing data are raster and vector. In raster-based data, every pixel has a value. Vector-based data are represented mathematically—endpoints for lines and lines for polygons. Each pixel can represent an attribute and the number of attributes depends on the number of bits: 16 to 256 colors or shades of gray. Therefore, pixels and vectors can have different attributes and are linked to a database that may be queried. A GIS allows a user to associate information with a feature on a map and create relationships that can determine the feasibilities of various locations, for example, a hierarchical system for locating anemometer stations for wind prospecting.

An overlay is a new map with specific features placed on top of a base map. An overlay is one form of a database query function. The overlay and base maps can be raster or vector images. The number of overlays is generally limited only by the amount of information that can be presented with clarity.

The main types of terrain data are the Digital Elevation Model (DEM) and the Digital Line Graph (DLG). They are available at different scales, for example, the DLG at 1:2,000,000, 1:100,000, and 1:24,000. Depending on the scale, the DLG data show highways, roads (even trails), lakes and streams, gas and utility transmission lines, and other features. The problem is that the data may have been taken from fairly old maps and may be incomplete. The DEM shows terrain height to 1 m on a latitude–longitude grid with a resolution of 3 arc seconds [pixels around 90 m × 90 × cos(latitude) m]. NREL coupled the DEM database with software to produce shaded relief maps of 1 degree × 1 degree.*

A technique of terrain enhancement [26] was used to identify windy areas in the Midwest. In the flat or rolling terrain found in most of the Midwest, the two most important factors influencing wind speed are terrain elevation and surface roughness. The wind map (normalized from PNL digital map) was adjusted to an average elevation and average surface roughness in a circle (12-km radius) around that point. The U.S. Geological Service Terrain Elevation Data was the base map consisting of average elevations in 1 km² grid cells rounded to the nearest 6 m. Terrain exposure was determined by subtracting actual elevation from the average elevation for each 1 × 1 km grid cell. Then a power correction factor was calculated by

$$\frac{P}{P_a} = \frac{\left(\ln\left[\frac{H_h + E}{z_o}\right]\right)^3}{\left(\ln\left[\frac{H_h}{z_o}\right]\right)^3} \tag{9.1}$$

where:
P_a = average power/area from normalized wind map.
H_h = hub height, 50 m.
E = exposure, m.
z_o = roughness length (crop land, 0.03 m; crop land and mixed woodland, 0.1 to 0.3 m; forest, 0.8 to 1.0 m.

Care must be taken when using P_a. Do you use the bottom or the middle of the wind class? Do you limit the number of wind class changes to one, especially for mountainous terrain?

---

* PC versions of GIS are available from IDRISI (www.clarklabs.org) and ArcGis (www.esri.com/products, U.S. phone: 909-792-2853). Mention of these products does not imply endorsement.

9.5 WIND RESOURCE SCREENING

As an example, wind resource screening for the Texas Panhandle is presented [27,28]. The DEM (3 arc seconds resolution) and DLG data were used. The original DEM data were in blocks of 1 degree × 1 degree. Data for utility transmission lines (69 kW and higher) were input by hand. Two GIS systems (IDRISI and PC ARC INFO) for personal computers were used. IDRISI has built-in functions that enhance its use for wind resource screening: slope, hill shading, aspect, and orthographic projection. A data sheet showing bin sizes, maximums, and minimums accompanies these functions.

The Panhandle of Texas is part of the Southern High Plains, with rolling hills in the East and flat plains above the caprock. The elevation rises from 450 m in the Southeast to 1,460 m in the Northwest. The Canadian River goes from west to east across the Panhandle. The other notable feature is Palo Duro Canyon. The graphs can be viewed in color or gray scale, with colors selectable up to 256.

At 256 colors, a DEM map for the entire Texas Panhandle would display contours 4 m apart. The base map (Figure 9.7) is the DEM data for the Panhandle. Most of the images were created using sixteen values. The elevation data of the base map can be analyzed by various commands in IDRISI. Instead of the whole area, subsets of the data can be analyzed in the same manner to view more detail. Resolution is limited by the cell size of the original data.

The Panhandle has a large wind energy potential since it has class 3 and 4 winds over the whole area. On the flat open plains covering much of the Panhandle, almost 100% will fall into the same wind power class. In this region, wind speed increases with height; therefore, modest elevation may increase wind power dramatically. Terrain exposure affects areas above and below the average elevation. A 150 km radius was used to determine an average elevation and the maximum change from this average was 190 m (Figure 9.8). An orthographic projection with an overlay of terrain elevation shows more clearly the areas of higher elevation. On the basis of terrain exposure, a revised wind map was calculated. Some of the regions with positive exposure were put into a higher wind class by this process and low areas were assigned a lower wind class.

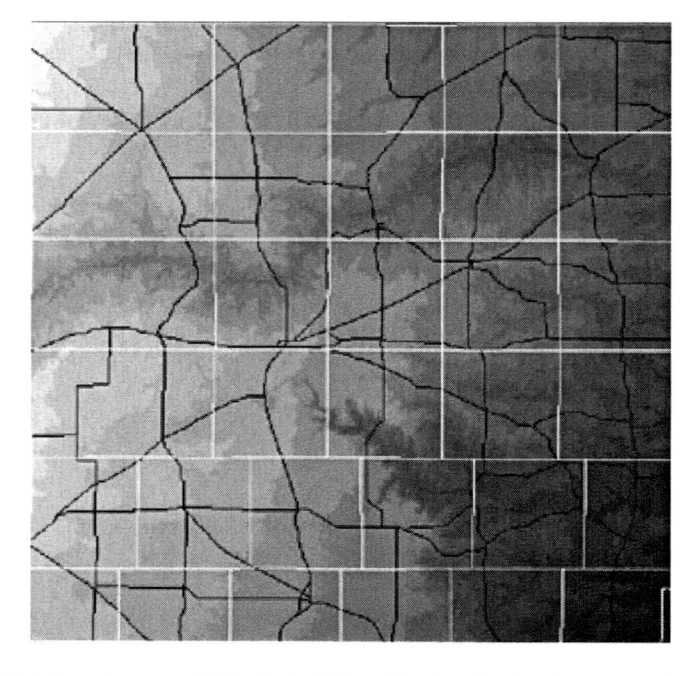

FIGURE 9.7 Digital elevation map (16 shades) of Texas Panhandle showing county boundaries and major highways. Contour lines are 62 m apart.

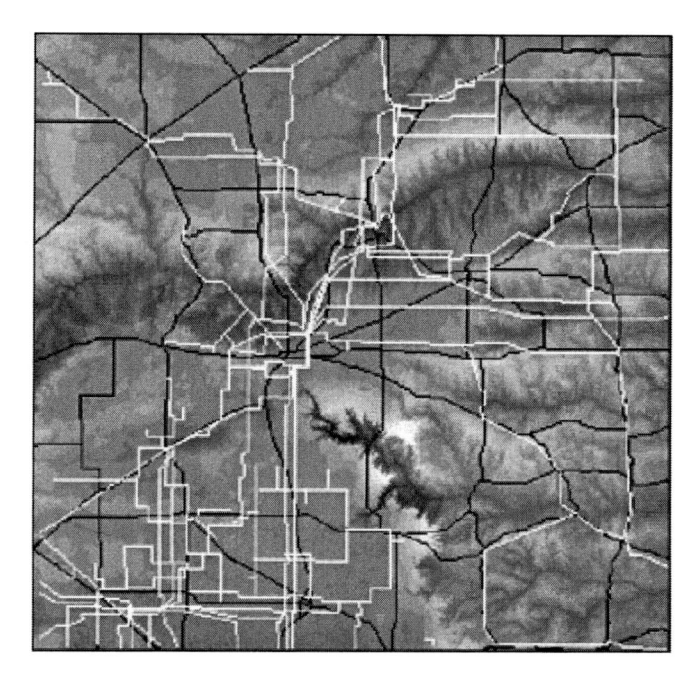

FIGURE 9.8 Terrain exposure for Texas Panhandle showing major highways and transmission lines. Light areas have better exposure (range of 16 levels from –195 m to +168 m height).

GIS was used to screen wind resources based on the criteria of wind power class, terrain type, proximity to transmission line, slope, and aspect. Within these criteria, classes or levels can be selected to exclude or limit an area's suitability for wind plants. A map was generated for the following screening parameters:

Wind class 3 and above
Slope of 0 to 3 degrees
Aspect from 155 to 245 degrees for area where slope exceeds 1 degree
Multiples of 8 km from transmission line (69 kV and above)
Excluded lands: parks, roads, urban, lakes, wildlife refuges

The maps were combined to generate a map of the possible areas for wind farms by wind class. Within 8 km of transmission lines, the total area was 28,600 km²—around 37% of the land in the Panhandle.

9.5.1 Estimated Texas Wind Power (Pacific Northwest Laboratory)

The Pacific Northwest Laboratory (PNL) estimated the capturable wind power for Texas at 50 m height as 134,000 MW from class 3 and above winds and 28,000 MW for the class 4 winds that blow primarily in the Panhandle. The PNL estimate was based on treating total power intercepted over a given land area as a function of the number of wind turbines, rotor swept area, and available power in the wind. Environmentally sensitive lands, urban areas, and terrains in valleys and canyons were excluded. The following formula was used to calculate the power intercepted by the rotor areas of wind turbines:

$$P_i = P_a A_t N \qquad (9.2)$$

where P_a = average wind power potential (W/m²); A_t = rotor area ($\pi D^2/4$); D = rotor diameter (m); and N = number of wind turbines. The calculation for the number of turbines that can be placed on a land area is

$$N = \frac{A_i}{S_r S_c} \quad (9.3)$$

where A_i = land area; S_r = spacing between turbine rows (D); and S_c = spacing within turbine row (D m^2). Note that $S_r S_c$ is the land area devoted to one turbine. In general, wind plants only remove 3 to 10% of the land from other productive uses and most of the removed land is used for roads. Some wind farm roads are only 5 m wide. The roads at another wind farm with 3-MW wind turbines are over 10 m wide.

If the cost of land is high, the land area for a single wind turbine will be smaller; but the wind plant output will be lower due to array effects. In California, some wind plants have turbine spacing of 2D within the rows and 5D to 7D to the next row. As a general rule, in the Plains area, 5 to 12 MW can be installed per square kilometer (4D × 8D spacing). For the edges of bluffs and on ridges, 6 to 15 MW can be installed per linear kilometer (2D to 3D spacing, one row only). With closer array spacing the megawatts per square kilometer would be larger and so would the array losses.

The average intercepted power can be calculated from Equation (9.2) or the intercepted power per unit land area can be calculated from

$$\frac{P_i}{A_t} = \frac{nP_a}{4 S_r S_c} \quad (9.4)$$

Remember, the calculation is for intercepted power, and capacity factors of 0.30 to 0.35 are used to estimate the capturable wind power.

9.5.2 Estimated Texas Wind Power (Alternative Energy Institute)

The same procedures of terrain enhancement and GIS were used to estimate the capturable wind power, also known as wind power potential, for Texas [29]. The selection criteria were class 3 or higher winds from a revised wind map showing terrain exposure, slope of 0 to 3 degrees, and exclusion of urban areas, highways, federal and state parks, lakes, wildlife refuges, and federal wetlands and land within 15 km of transmission lines carrying 115 kV or more.

The capturable annual power was calculated for the wind turbines with 50-m hub height, 10D × 10D spacing, 30% capacity factor, and no array losses (reasonable for large spacing). With these assumptions, the estimated annual capturable wind power was 157,000 MW (525,000 MW of wind turbines at 30% efficiency) with an annual energy production of 1,300 TWh. These results are somewhat larger than the estimates determined by PNL.

The estimates were further revised with data (at 40 and 50 m) from Alternative Energy Institute (AEI) and private meteorological sites [30]. The estimates were then used to update the wind map (1 km pixel size) for Texas (Figure 9.9). Class 3 and 5 lands were reduced from the previous estimate and class 4 lands increased. The selection parameters were the same, except for slopes (0 to 10 degrees) and areas within 16 km of electrical transmission line (≥69kV) for usable land for wind power (Figure 9.10).

The estimate for capturable wind power (Table 9.3) is larger also because a spacing of 7D × 9D was used and the capacity factor was 30% for class 3 lands and 35% for class 4 and above lands. The estimates show the large wind potential, 172,000 MW (500,000 MW of installed capacity). However, only a fraction will be installed because the total electrical generating capacity of Texas was 120,000 MW (11,500 MW wind) in 2012. Maps and estimates are available from AEI [31].

A number of wind farms have been built on mesas and terrain involving edges and bluffs. In one area of West Texas (Pecos, Upton, and Crockett Counties), 759 MW of wind farms were installed on mesas. Over 3,000 MW (installed from 2005 to 2009) in wind farms are sited along Interstate High 10 from Abilene to Roscoe and then northwest to Snyder along Highway 84. Some of these are on mesas with exposures from cliffs and bluffs on one side.

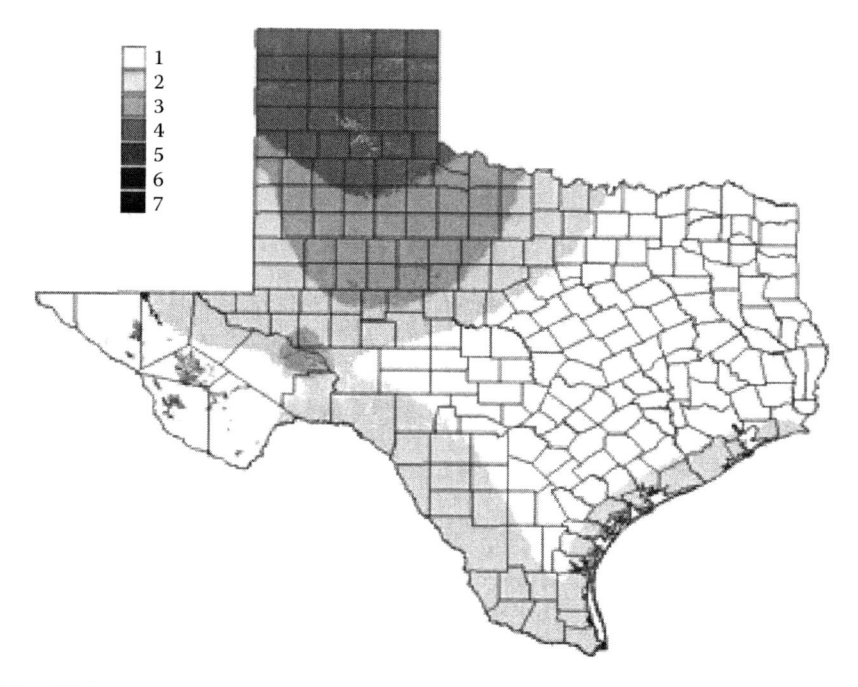

FIGURE 9.9 1995 Texas wind power map.

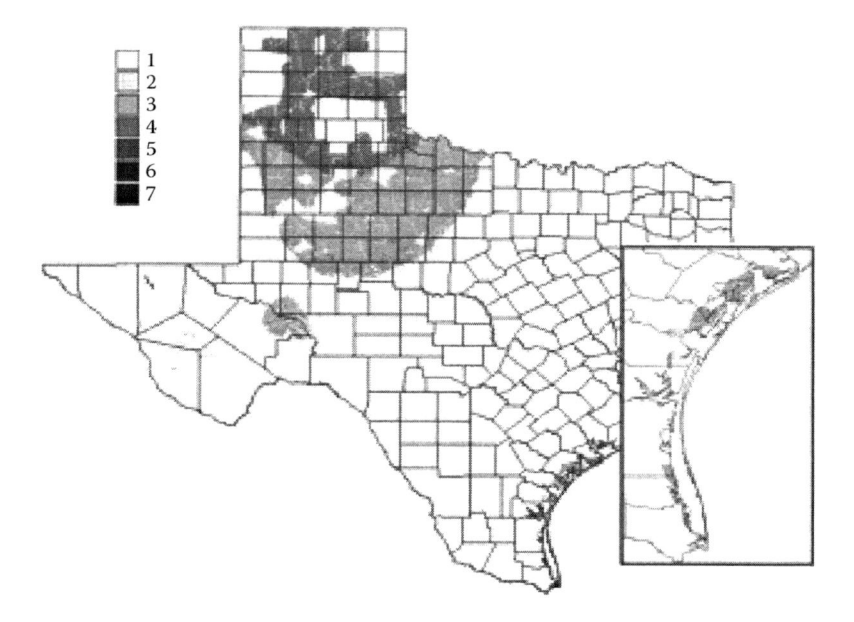

FIGURE 9.10 Texas land suitable for wind farms, 1995.

The limit of proximity to transmission lines has now changed and wind farms have been built within 40 km of major transmission lines. Also, the Texas Public Utility Commission promoted new transmission lines from West Texas and the Panhandle to connect with major load centers of the rest of the state. This will provide The Energy Reliability Council of Texas (ERCOT) a total of 18,000 MW of wind power—about 10,000 additional MW of wind capacity. Without the constraint of proximity to transmission lines, the estimate for the amount of intercepted wind power is

TABLE 9.3
Texas, Intercepted and Capturable Wind Power and Annual Energy Potential from Land that Satisfies the Screening Parameters

Wind Class	Area km²	Intercepted MW	Capturable Power MW	Energy TWh/year
3	69,299	302,365	90,170	795
4	41,391	232,196	81,269	712
5	42	288	101	1
6	54	471	165	1
7	2	22	8	
Total	110,788	535,342	172,252	1,509

850,000 MW with capturable wind power around 270,000 MW. If offshore winds are included, the estimate will be even larger.

9.5.3 Wind Power for United States

Similar estimates have been made for all the U.S. regions and states. Winds of class 4 and above [32] and access to transmission lines are the most common criteria. The *State Wind Working Group Handbook* contains articles and PowerPoint presentations by several authors [33].

9.6 NUMERICAL MODELS

Numerical models for predicting winds are becoming more accurate and useful, especially for areas of the world where surface wind data are scarce or unreliable. Models were derived from numerical models for weather prediction [34]. Remember that a small difference in wind speed can make a large difference in energy production. In the final analysis, surface wind data are still needed for wind farms.

MesoMap: This system was developed specifically for near-surface wind forecasting. It is a modified version of the Mesoscale Atmospheric Simulation System (MASS) weather model. MesoMap uses historical atmospheric data spanning twenty years and a fine grid (typically 1 to 5 km). It simulates sea breezes, mountain winds, low-level jets, changing wind shear due to solar heating of the earth's surface, effects of temperature inversions, and other meteorological phenomena. MesoMap does not depend on surface wind measurements although surface measurements are desirable for calibration.

The model provides descriptive statistics utilizing wind speed histograms, Weibull frequency parameters, turbulence and maximum gusts, maps of wind energy potential within specific geographical regions, and even the annual energy production data for wind turbines of any height for selected sites in a region.

WAsP: The Wind Atlas Analysis and Application Program software was developed by Denmark's Risoe National Laboratory to predict wind climate and power production from wind turbines. The predictions are based on wind data measured at stations in the region. The program includes a complex terrain flow model. WAsP was used to develop the European wind map (Figure 4.3) and is used by other governments and organizations across the world. Other models are available from links listed at the end of this chapter and elsewhere on the Internet.

9.7 MICROSITING

Wind maps, data compiled by meteorological towers, models, and other criteria are used to select wind farm locations. Other considerations for a wind farm developer are the type of terrain (complex to flat plain); wind shear; wind direction; and spacing of turbines based on predominant wind

Siting

direction and availability, land cost, and requirements such as roads, turbine foundations, and substations. Terrain may be classified as complex, mesa, rolling, or plain. Passes may be classified as one type or a combination. Spacing is generally stated as diameter D of a wind turbine, so larger turbines will be farther apart.

As turbines have become larger, are wind shear data from 25 to 50 m sufficient to predict wind speeds at 70 to 100 m heights? The first answer is yes, for those parameters, but it is not definitive if the inquiry concerns another location in the same region.

In complex terrain, such as mountains and ridges, micrositing is very important. On flat plains, the primary consideration is spacing between turbines in a row and spacing between rows. On mesas, the highest wind speed is on the edge of the mesa facing the predominant wind direction so turbines may be set in a single row In rolling terrains such as hills, wind turbines should be placed at higher elevations.

In California, the high wind classes arise from the rise of hot desert air and cooler air from the sea traveling through the passes. California has complex terrain at Tehachapi Pass, rolling terrain at Altamont Pass (east of San Francisco), and both ridges and flat terrain at San Gorgonio Pass near Palm Springs. The winds in the passes are predominantly from the west, so the turbine rows are primarily sited north–south. At San Gorgonio Pass, some wind turbines in rows were only 2D apart and rows were spaced 4 to 5D apart because of the high cost of leasing land. Where space is tight, turbines can be placed at different heights. As expected, the array losses are fairly large. Starting in 1998, smaller turbines were replaced with larger ones.

The wind farm near White Deer, Texas, has 80 1-MW wind turbines of 56-m diameter. The wind turbines have 4D spacing within rows and 15D between rows (Figure 9.11). North is at the top of the figure and the lines indicate roads at 1 mile (1.6 km). The buffer zone on the west is because the adjacent land was not under lease to the wind farm. Predominant winds are south–southwest during the spring and summer and from the north in winter. As lower winds occur in July and August, the rows are situated perpendicular to those predominant winds. The low spots are playa lakes that contain water only after rain so no turbines were installed in those locations. Only the west side of the wind farm is visible in the photo; there are more turbines to the east. Examples of wind farms in other terrain are shown in Figure 9.12 through Figure 9.14. Figure 9.15 shows an offshore wind farm for comparison.

The amount of land taken out of production depends primarily on the lengths and widths of wind farm roads. Values vary from 0.5 to 2 hectares (ha) per turbine. If county roads exist, the developer will use less land; however, the developer may have to improve the county roads to handle heavier traffic. Roads may be very expensive for a wind farm on a mountain ridge. The access road from the bottom to the top of the Texas Wind Project in the Delaware Mountains cost $1 million in 1993.

FIGURE 9.11 West side of wind farm in plains near White Deer, Texas. White lines are roads, to show one square mile, which is equal to 260 ha. (Photo courtesy of Cielo Wind Power.)

FIGURE 9.12 Wind farm in rolling terrain, Lake Benton, Minnesota. (Photo, courtesy of Wade Wiechmann.)

FIGURE 9.13 Wind farm on Southwest Mesa, near McCamey, Texas. Example of mesa with one row. (Photo courtesy of Cielo Wind Power.)

Siting 213

FIGURE 9.14 Wind farm in complex terrain, Northwest Spain.

FIGURE 9.15 Nysted wind farm in Baltic Sea, Denmark. (Photo courtesy of Siemens.)

Civil engineering aspects of a wind farm site include location of assembly area and construction of electrical substation and roads (length, width, and grade over complex terrain). Roads must allow wide turns by trucks hauling the long blades. Many sites erect batch cement plants on site, especially for construction on complex terrains of ridges and mesas.

A general rule of thumb is that around 5 to 10 MW/km^2 can be installed on land suitable for wind farms. However, on ridge lines at 2D to 3D spacing, the value would be around 8 to 12 MW km.

The kilometer measure is linear and the ridge is assumed to be more or less perpendicular to the predominant wind flow. As wind turbines become larger, the megawatts per square or linear kilometer will increase due to the energy output increase as the square of the radius. Most landowners lease blocks or areas of land, not just the places where turbines will be located. It was interesting in the Texas Wind Power Project that land leased for the wind farm included all land at the 1,453 m contour and above (elevation of ridges is 1,830 m). The landowner is now trying to determine whether land below the contour has any wind potential.

Satellite and aerial images are used in micrositing and are available from various sources; some are free. Flash Earth (www.flashearth.com) has the option of switching among sources, such as Google Maps, Microsoft VE, and others. The wind farms in the images are fairly distinct, primarily because of the roads at the sites and the areas around the wind turbines. Oil fields show the same pattern, but the roads are not as wide.

In some farming areas, round circles for irrigation sprinklers are very prominent; large circles represent section sprinklers (1 square mile, 260 ha), and small circles represent quarter-section sprinklers. The shadows of the wind turbines are more obvious than the wind turbines, and the angle of a shadow may be different from one part of the wind farm to an adjacent part because the images were taken at different dates and times. Images from different sources will also be taken at different dates and times. New wind farms will not appear in satellite images until the images are updated—more than a year may elapse between updates.

Micrositing techniques of wind farm developers are proprietary. However, satellite images show the layout of wind farms, and good information about siting may be obtained from the images and topographic maps. If the type and model of a wind turbine are known, the spacing can be estimated from an image. The image of Trent Mesa, Texas (Figure 9.16) shows about half the layout of the wind farm that contains 100 wind turbines, 66 m in diameter, rated at 1.5 MW.

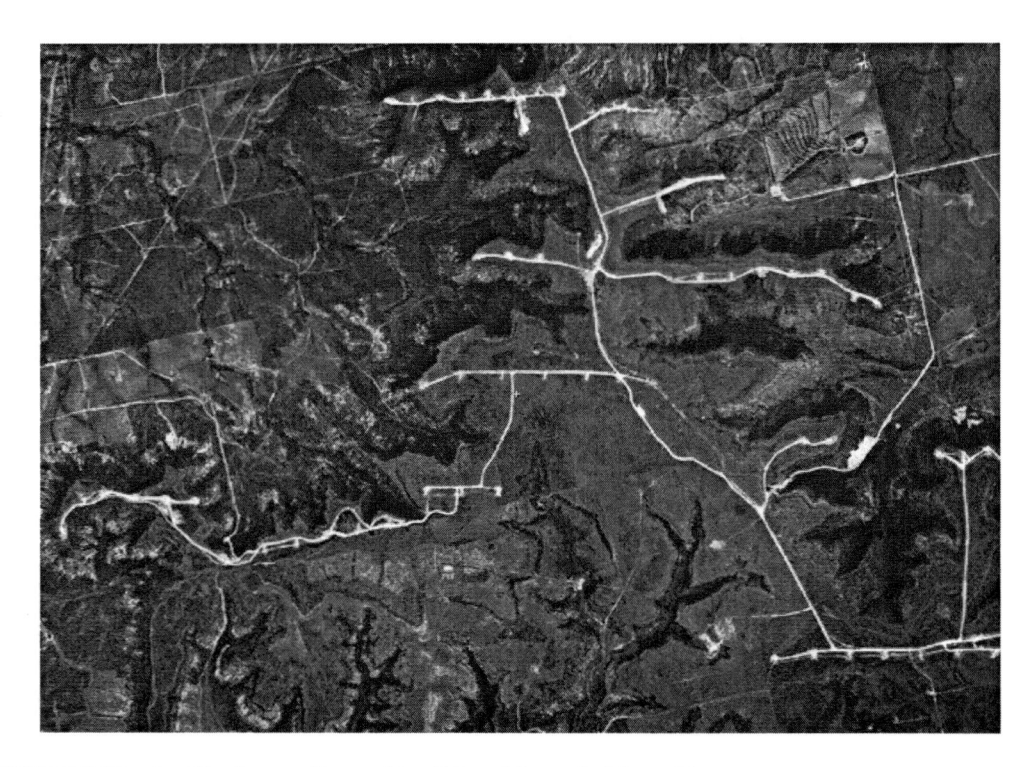

FIGURE 9.16 Satellite image of west side of Trent Mesa wind farm, Texas.

Economic and institutional issues also affect micrositing. An example is the Waubra wind farm project (192 MW) in Australia [35] that involves environmental, cultural heritage, and environmental management issues. Since installation, many residents have expressed opposition, claiming health effects caused by wind turbines. One vocal landowner was bought out by the wind farm.

9.8 OCEAN WINDS

Ocean wind observations (see Section 4.4) provide complementary sources of information for siting of offshore wind farms. The advantages of ocean wind maps are:

Some satellite wind maps are public domain.
All offer global coverage allowing observation of large areas without large numbers of meteorological towers.
All are accessible in archives spanning several years.
Accuracy is sufficient for wind resource screening.
They quantify spatial variations.
They are available at resolutions of 400 m, 1.6 m, and 0.25 degree.
Software has been developed for their use.

The major problems with ocean winds are:

Data are for 10 m height and values of wind shear are not known.
Standard deviations are around 1.2 to 1.5 m/s on mean wind speed.
Data are not available or not as reliable within 25 km of shore.

Ocean winds were used for wind resource estimation for Denmark [36]. Weibull parameters were calculated from the wind speed data to determine a wind speed distribution from which wind energy production could be estimated.

The average wind speed for Padre Island, a barrier island off Corpus Christi, Texas, is 5.1 m/sec at 10 m height—the same value as ocean winds 25 km from the coast. Data from 10 to 40 m height indicated an annual average shear exponent of 0.19. A shear exponent of 0.15 was noted for a site 15 km off Cape Cod, Massachusetts [37]. Also, ocean winds, terrain, and predominant wind direction will indicate regions of wind potential for islands and near shores. For example, ocean winds indicate excellent wind resources for the islands of Aruba, Bonaire, and Curaçao off the northern coast of Venezuela.

9.9 SUMMARY

GISs provide very flexible and powerful tools for terrain analysis relevant to wind energy prospecting. They can help reclassify existing wind maps and identify areas showing potential as possible wind farm sites. In addition, GIS can be used to quantify wind power potential and, in conjunction with numerical models, estimate annual energy production.

After a location is selected, GIS and topographical maps can be used for micrositing. Wind turbines should be located within a wind plant area to maximize annual energy production. However, the normal 90-m resolution may not be detailed enough for micrositing on complex terrain. PNL used a technique of spline interpolation to develop a finer grid from the 90-m data. Of course, if the DEM data at 10-m resolution are available, the interpolation is not needed.

A number of numerical models for micrositing are available and most run on personal computers. More powerful programs for weather forecasting and micrositing that run on large computers or clusters of PCs, are also available. In general, these must be purchased.

LINKS

3TIER software, models. www.3tiergroup.com/en/
EMD, WindPro. www.emd.dk/WindPRO/Frontpage Federal Wind Siting Information Center, www.windpoweringamerica.gov/siting.asp
D.M. Heimiller and S.R. Haymes. 2001. *Geographic information systems in support of wind energy activities at NREL*. REL/CP-500-29164. www.osti.gov/bridge
MesoMap. software, models. www.awstruewind.com
National Renewable Energy Laboratory. GIS publications and representations, information about creation and validation of NREL maps, regional and modeling data. www.nrel.gov/analysis/workshops/pdfs/brady_gis_workshop.pdf
Northwest mapping project. www.windmaps.org
ReSoft software, models. www.resoft.co.uk/English/index.htm
RETscreen, free software, decision-making tools. www.retscreen.net
Trent Mesa Wind Project. www.trentmesa.com/default.htm
TRC, CAMET, and MM5 software, models. www.src.com/windenergy/windenergy_main.htm
WAsP, software, models. www.wasp.dk
WindFarmer, software, models. www.garradhassan.com/products/ghwindfarmer/
Wind Logics, software, models. www.windlogics.com
Wind Resource Assessment Handbook. www.nrel.gov/docs/legosti/fy97/22223.pdf

REFERENCES

1. Wind Powering America. www.windpoweringamerica.gov/windmaps/residential_scale.asp
2. M.N. Schwartz and D.L. Elliott. 1995. Mexico wind resource assessment project. In *Proceedings of Windpower Conference*, p. 57.
3. H. L. Wegley et al. 1980. *Siting Handbook for Small Wind Energy Conversion Systems*. Report PNL-2521. Washington: U.S. Department of Energy.
4. P. Gipe. www.wind-works.org/index.html
5. P. Gipe. 1999. *Wind Energy Basics: A Guide to Small and Micro Wind Systems*. Post Mills, VT: Chelsea Green.
6. P. Gipe. 1993. *Wind Power for Home and Business*. Post Mills VT: Chelsea Green.
7. American Wind Energy Association. Distributed wind. www.awea.org/smallwind
8. Canadian Wind Energy Association. Small wind energy. www.smallwindenergy.ca/wind-energy/smallwind_e.php
9. U.S. Department of Energy. *Small Wind Electric Systems: A U.S. Consumer's Guide*. www.nrel.gov/docs/fy07osti/42005.pdf
10. RenewableUK. Small and medium scale wind. www.renewableuk.com.
11. Nobal Wind Map. www.rensmart.com/Weather/BERR
12. A.G. Dutton, J.A. Halliday, and M.J. Blanch. 2005. The feasibility of building-mounted/integrated wind turbines (BUWTs): achieving their potential for carbon emission reductions. www.eru.rl.ac.uk/pdfs/BUWT_final_v004_full.pdf
13. Wind Energy Integration in the Urban Environment. www.urbanwind.org
14. Aerotecture International. www.aerotecture.com
15. Urban Wind. Kirklees Council case study. www.urban-wind.org/admin/FCKeditor/import/File/Case_Study_UK1.pdf
16. Aerovironment Energy Technology Center: Architectural wind. www.avinc.com/Energy_Lab_Details.asp?Prodid = 52
17. Bahrain World Trade Center. www.bahrainwtc.com/index.htm
18. Public Service of Vermont/ Siting a wind turbine on your property: putting two good things together. http://publicservice.vermont.gov/sites/psd/files/Topics/Renewable_Energy/Resources/Wind/psb_wind_siting_handbook.pdf
19. R.L. Simon and R.H. Gates. 1991. Long-term interannual wind resource variations in California. In *Proceedings of Windpower Conference*, p. 236.
20. U.S. Geological Survey. www.USGS.gov; Topozone. www.topozone.com
21. R.L. Simon and R.H. Gates. 1992. Two examples of successful wind energy resource assessment. In *Proceedings of Windpower Conference*, p. 75.

22. M. Swartz and D. Elliott. 2006. Wind shear characteristics at Central Plains tall towers. In *Proceedings of Windpower Conference*. CD.
23. AWS Truewind. windNavigator. www.awstruewind.com
24. New York State. Small windExplorer. http://nyswe.awstruepower.com
25. 3Tier. FirstLook. www.3tiergroup.com
26. M.C. Brower et al. 1993. *Powering the Midwest: renewable electricity for the economy and the environment*. Report for Union of Concerned Scientists.
27. L. Shitao. 1994. Wind resource screening in the Texas Panhandle. Master's thesis, West Texas A&M University.
28. L. Shitao, J. McCarty, and V. Nelson. 1994. *Wind resource screening in the Texas Panhandle*. Report 94-1. Alternative Energy Institute. West Texas A&M University.
29. V. Nelson. 1995. Wind energy. In *Texas Renewable Energy Resource Assessment: Survey, Overview, and Recommendations*. Report for Texas Sustainable Energy Development Council. www.infinitepower.org/resintro.htm
30. C.M. Yu. 2003. Wind resource screening for Texas. Master's thesis, West Texas A&M University.
31. Alternative Energy Institute. 2004 Texas Wind Class Map. www.windenergy.org
32. USDOE, Energy Efficiency and Renewable Energy. Wind power today. www1.eere.energy.gov/windandhydro/
33. Wind Powering America. Handbook. www.eere.energy.gov/windandhydro/windpoweringamerica/pdfs/wpa/34600_wind_handbookpdf
34. J. Rohatgi and V. Nelson. 1994. *Wind characteristics, an analysis for the generation of wind power*. Alternative Energy Institute, West Texas A&M University.
35. Waubra Wind Farm. www.acciona.com.au/business-divisions/energy/operational-projects/waubra-wind-farm
36. C. Jasager et al. Wind resources and wind farm wake effects offshore observed from satellite. Risoe National Laboratory, Wind Energy Department. www.risoe.dk/rispubl/art/2007_79_paper.pdf
37. M. Swartz, D. Elliott, and G. Scott. 2007. Coastal and marine tall-tower data analysis. In *Proceedings of Windpower Conference*. CD.

PROBLEMS

1. A building is 20 m long, 15 m wide, and 15 m tall. You want to install a 10-kW wind turbine. How tall a tower will you need and how far away from the building should you place it?
2. Several trees 20 to 30 m tall are near a house. You want to install a 10-kW wind turbine. What is the minimum height of the tower? What is the approximate cost of the tower?
3. Refer to Figure 9.3. The building is 15 m tall. What is the power reduction at 15 m height at a distance of 60 m downwind? At 150 m downwind? Would it be cheaper to use a taller tower or to move the tower farther away from the building? Show all cost estimates.
4. Is there a small wind turbine in your region? If yes, what are the visual impacts from the neighbor's view and from the public view? Use Tables 9.1 and 9.2 to estimate scores.
5. Using Equation (9.1), calculate the corrected power for a class 3 wind area if the terrain exposure is 80 m and area is grassland. Use the bottom and middle values for class 3.
6. Estimate the annual energy production for a 50MW wind plant where the average wind power potential is 500 W/m^2 at 50 m height. Select the size of turbine from commercial turbines available today.
7. Do Problem 6. The land is now high priced. Select close spacing and estimate array losses.
8. What land area must you lease for a 50-MW wind farm? Select the size of turbine from commercial turbines available today and calculate spacing. Remember, spacing your turbines too closely will cause array losses. How many megawatts can you install per square kilometer?
9. Array spacing is 4D × 8D, for a 3-MW wind turbine 90 m in diameter. How many can be placed in a square kilometer?

10. The row spacing for 3-MW turbines 90 m in diameter is 2D. How many can be placed per linear kilometer on a ridge?
11. Assume you have complex terrain. What size of land area must you lease for a 50-MW wind farm? Select the size of turbine from commercial turbines available today and calculate the spacing. How many megawatts can you install per square kilometer?
12. In your opinion, what are some advantages and disadvantages of using vector- or raster-based GIS to determine wind energy potential?
13. Check two of the links on numerical models listed in the Links section. See whether they contain examples of wind maps. List website chosen, geographical region of wind map, and map resolution.
14. For the White Deer wind farm (Figure 9.11), what is the land area allocated for each turbine? How many turbines can be placed in a square kilometer?
15. For the White Deer wind farm (Figure 9.11), if the roads are 7 m wide, estimate the amount of land taken out of production for the wind turbines within the square mile shown in the figure. Do not forget the spaces between turbines.
16. Go to Flash Earth (www.flashearth.com) and search for White Deer, Texas (latitude, N 35 degrees, 27 minutes; longitude, W 101 degrees, 10 minutes). The wind farm is just northwest of the town. Zoom in to see the layout of the wind farm. Estimate the number of wind turbines per square mile for the farm. Remember, not all the land within the farm will have wind turbines on it.
17. Go to Google Earth and search for the wind farms in San Gorgonio Pass, California, just northwest of Palm Springs. Estimate the spacing for one of the densely packed wind farms.
18. How many meteorological stations, at what height, and over what period are needed to determine the wind potential for a 50-MW or larger wind farm? In general, terrain will not be completely flat. Also remember, wind turbines are getting larger and thus hub heights are larger. For your selection of number, height, instrumentation, and time period, estimate the costs for obtaining the data.
19. Go to www.remss.com and look at QSCAT data for area off Cape Cod during September 2007. Choose region "Atlantic, Tropical, North." What is the average wind speed and from what direction?
20. In a preliminary data collection for a wind farm, for how long should data be collected if: (a) no regional data are available; (b) good regional data are available; and (c) other wind farms are in the area.
21. Find a quadrangle map that shows Mesa Redonda in Quay County, New Mexico (www/newmexico.org/map). What is the elevation of the mesa? You can see all of the mesa in a 1:200,000 view. You will need a 1:50,000 view to read elevations.
22. What is the general rule for calculating megawatts per square kilometer (MW/km^2) in plains and rolling hills? What is the rule for calculating megawatts per kilometer for ridges and narrow mesas?
23. From Table 9.3, estimate megawatts per square kilometer.
24. From Table 9.3, using the general rule for square kilometers, what is the maximum megawattage of wind that could be installed? What is the maximum capturable power?
25. What is the annual wind speed at 100 m height on Mesa Redonda, south of Tucumcari in eastern New Mexico?

10 Applications and Wind Industry

The main applications of wind power are the generation of electricity and water pumping (Table 10.1). Except for the installed capacity of wind farms, the other statistics are best estimates, as data are difficult to acquire. Applications for generation of electricity are classified as utility-scale wind farms; small wind turbines (both remote and stand-alone systems); distributed; community; wind–diesel; village power (generally hybrid systems); and telecommunications (high-reliability hybrid systems) and street lighting. Many village power systems use photovoltaic panels with one- to three-day battery storage; some are wind hybrid systems and use only wind power. In some cases, village power includes diesel or gas back-up systems. Stand-alone systems generally use batteries for storage.

In wind-assist and stand-alone systems, two power sources work in parallel to produce power on demand. All wind turbines connected to a utility grid are wind-assist systems. Wind turbine sizes range from the utility-scale megawatt turbines for wind farms to small systems (≤100 kW) also connected to grids and 20- to 300-W remote units for sailboats and households, primarily in the developing world. Small systems are sometimes called micro wind turbines. Some vendors claim that micro and small wind turbines will produce electricity at less cost than utility-scale wind turbines but a user must connect a large number of them together.

10.1 UTILITY SCALE

The 282 GW capacity (Figure 1.15.b) installed by the end of 2012 can produce an estimated 620 TWh by 2013 (using a capacity factor of 25%). The major installations were in Europe (36%), China, and the U.S. and estimated energy production from major installations is 220 TWh, 150 TWh, and 130 TWh, respectively. World growth rate reached 28% per year from 1995 through 2012 (see Figure 1.15), but at some point shifted to linear increments, for example, Europe installations have produced around 10 GW annually since 2006.

Wikipedia has lists of onshore and offshore wind farms and The Wind Power [1] maintains a database of wind farms, wind turbines, manufacturers, developers, and operators. Lists of wind farms plus graphs of capacity per year and cumulative capacity are generally available from national wind energy associations and other groups, some of which provide more detailed information such as energy production and capacity factor by year [2]. Also check regional associations such as the Latin American Wind Energy Association for information on installed capacity. Interactive maps showing locations and information about wind farms based on Google Earth are now available for a number of websites.

In Europe, the 1995 goal of 4 GW of wind by 2000 was surpassed greatly and a 2010 goal was set at 60 GW. In 2003, the goal was raised to 75 GW. By 2010 both goals were exceeded and 86 GW had been installed. The 106 GW installed by the end of 2012 produced around 230 TWh in a normal wind year and this represented over 6% of the electrical demand. Now the European goal is 20% of electricity generated from renewables by 2020, of which 12 to 14% would be from wind. Of course, predictions are always risky, and the predicted megawatts change as projects and legislation are implemented and changed. In Denmark, wind turbines supplied 28% of the electric consumption in 2011.

TABLE 10.1
Comparison of Wind Industry Estimates for 1995, 2002, and 2007

	1995	2002	2007	2012
Utility scale, number	22,000	50,000	100,000	180,000
Installed capacity, GW	4.8	31	94	286
Production, TWh/yr	0.005	10	30	752
Small systems, number	150,000	370,000	500,000	840,000
Installed capacity, MW	15	55	200-300	850
Wind diesel, number			200	250?
Village power, number	10–30	150?	1800	2000?
Telecommunication	20–50	150	200?	3,500?
Farm windmills, number[a]	300,000	305,000	310,000	310,000
Production/yr	3,000	3,000	3,000	3,000

[a] Farm windmills are being replaced by electric and PV pumps. Production primarily replaces 30- to 40-year-old windmills.

Offshore wind farms have been installed in Europe [3], with a capacity of over 5 GW by the end of 2012. Early examples are Horns Rev at 160 MW [4] and Nysted at 158 MW in Denmark. The U.K. now has over half the installed offshore capacity in the world and also the largest wind farm, London Array, 630 MW. China has 261 MW offshore and plans for thirty-eight projects with 16.5 GW in the early development stage. Offshore and Great Lakes wind farms are being considered in the U.S., despite substantial opposition to a wind farm off Cape Cod, Massachusetts.

In the U.S., the Alta wind farm has a capacity of 1,320 MW, Shepard Flats in Oregon has a capacity of 845 MW, and four wind farms in Texas range in size from 523 to 736 MW. John Deere installed clusters of 10 MW wind farms (10 MW size is subject to less regulation). The clusters house enough wind turbines to obtain economies of scale in installation. The American Wind Energy Association has a database of utility-scale projects (member accessed) and publishes quarterly reports (public) of projects commissioned.

States and other entities may also compile information on wind projects, for example, the Kansas Energy Information Network [5] has a list of large wind projects in Kansas, Missouri, Nebraska, Oklahoma, and Wisconsin. The Alternative Energy Institute at West Texas A&M University maintains a list of projects for Texas, New Mexico, and Oklahoma.

By 2011, about 20,000 wind turbines with a capacity around 45 GW had been installed worldwide. The average size of a wind turbine is now around 2 MW. The Global Wind Energy Council lists installed capacity in 2011 by region (Table 10.2) and by countries within regions [6]. The growth of wind power in China was phenomenal with the installation of 70.5 GW between 2008 and 2012 [7]. China's goal of 5 GW by 2010 was reached three years ahead of schedule.

Three driving forces for installation of wind farms are economics, policy and incentives at national and state levels, and the negative public perception of nuclear power (especially since the disaster at the Fukishima Daiichi power plants in Japan). Some countries have voted to shut down their nuclear plants. Green power and reduction of pollution and emissions also assist in expanding the wind energy market.

The top fifteen manufacturers based on total installed capacity (Table 10.3) dominate the world market and European manufacturers still maintain major shares. Vestas is still the leader in terms of production (Table 10.4), and the top fifteen manufacturers producing 3 to 4 GW per year accounted for 88.4% of the market in 2011.

The domestic wind turbine industry in China accounted for 93% of the country's 2011 market. Seven Chinese companies are now among the top fifteen suppliers in the world, due to the huge size

TABLE 10.2
World Installed Wind capacity (MW) by Region for 2006, 2010, and 2011

	2007	2010	New 2011	2011
Africa and Middle East	538	1,065	31	1,093
Asia	10,091	64,106	21,298	82,398
Europe	57,136	86,106	10,281	96,616
Latin America and Caribbean	537	1,997	1,206	3,203
North America	18,664	44,306	8,077	52,148
Pacific	1,158	2,516	642	2,858
World total	94,123	197,637	41,236	238,351

Source: Global Wind Energy Council.

TABLE 10.3
Estimation of Global Cumulative Numbers, Cumulative Capacities, and 2011 Production of Large Wind Turbine Manufacturers[a]

2011 Rank	Company	2007 Rank	By end 2011 Total Units	By end 2011 Total GW	2011 Production MW
1	Vestas	1	47,335	51.7	5217
2	Enercon	4	20,000	28.0	3318
3	GE Wind	3	18,000	28.0	3542
4	Gamesa	2	15,500	24.1	3308
5	Suzlon	7	13,329	20.5	3116
6	Siemens	5	8,600	14.7	2591
7	Sinovel	15	8,479	13.0	2939
8	Goldwind	8	10,627	12.7	3790
9	Acciona	9	7,759	8.5	3844
10	Nordex	6	5,014	7.8	
11	Repower	12	4,239	7.8	
12	Dongfang Turbine		4,590	6.9	946
13	United Power		3,499	5.3	2,847
14	Mingyang		2,063	3.1	1,778
15	Alstom (Ecotécnia)	11	2,100	2.7	

[a] Most data obtained from manufacturers' websites.

of the Chinese market. The U.S. has only one manufacturer (GE Wind) in the top fifteen. Suzlon of India was the other non-European and non-China manufacturer.

Notice with the large megawatt units that the number of units installed is around half the number of megawatts installed. One problem with the data presented is the absence of good information on repowering of wind farms and decommissioning of old turbines. The Global Wind Energy Council market projection from 2007 was for 240 GW of wind power by 2012, with many new installations in Asia. That number was essentially reached in 2011.

In 2012 installed capacity in the U.S. was 13.1 GW and in China it was 14 GW. In preliminary results from BTM Consult, GE Wind displaced Vestas as the top producer in 2012 due to the rush to complete projects before the production tax credit expired in the U.S.

TABLE 10.4
Top 15 Wind Turbine Producers in 2011

	Accumulated MW 2010	Supplied MW 2011	Share % 2011%	Accumulated MW 2011	Share % Accumulated
Vestas (DK)	45,547	5,213	12.9	50,760	20.6
Goldwind (PRC)	9,055	3,789	9.4	12,844	5.2
GE Wind (US)	26,871	3,542	8.8	30,413	12.3
Gamesa (S)	21,812	3,309	8.2	25,121	10.2
Enercon (GE)	22,644	3,188	7.9	25,832	10.5
Suzlon Group (IND)	17,301	3,104	7.7	20,405	8.3
Sinovel (PRC)	10,044	2,945	7.3	12,989	5.3
United Power (PRC)	2,435	2,859	7.1	5,294	2.1
Siemens (DK)	13,538	2,540	6.3	16,078	6.5
Mingyang (PRC)	1,799	1,178	2.9	2,977	1.2
Nordex (GE)	6,994	970	2.4	7,964	3.2
Dongfang (PRC)	6,389	945	2.3	7,334	3.0
Hara Xemc (PRC)	1,081	718	1.8	1,799	0.7
Sewind (PRC)	1,080	708	1.8	1,788	0.7
Enercon (IND)	504	666	1.7	1,170	0.5
Others	18,834	4,686	11.6	23,520	9.5
Total	205,928	40,360	100	246,288	100

Source: BTM Consult. March 2012.

DK = Denmark. PRC = Peoples' Republic of China. US = United States. S = Spain. GE = Germany. IND = India. Suzlon Energy (IND) and REpower (GE) are listed as Suzlon Group for the second year by BTM.

Another aspect of utility scale relates to installations of wind turbines by cooperatives and individual farmers that then sell to the grid. This overlap is also known as community wind. Large businesses have also installed utility scale projects, some are independent and others are distributed wind producers.

10.2 SMALL WIND TURBINES

Small wind turbines overlap with village power systems since most wind turbines for village power generate less than 100 kW, as do some of the distributed and wind–diesel systems, so the numbers reported for production of small wind turbines will include all areas. In the U.S. and Europe, much of the small wind turbine capacity is grid connected. In China and other developing countries, most are stand-alone systems for households and generate 50 to 300 W.

A very rough estimate of small wind turbines (100 kW or less) is around 940,000 worldwide with a capacity of 860 MW (Table 10.5)—very impressive numbers that show the impact of small wind. To reflect current numbers, reduce the number of units by 120,000 and reduce capacity by 25 MW to compensate for replacements, upgrades to larger turbine, lifetimes, and operational failures (Table 10.1). The rough estimates are due to questionable accuracy of production figures for China and the difficulties of obtaining data from small manufacturers in all parts of the world. All reported data indicate that average sizes of units are increasing.

Another unknown in China is how much of the present production represents replacement of old wind turbines and how much results from upgrades of 50- to 100-W units 200- to 500-W production. The China estimate used a reduction of 30% on reported numbers (2002 through 2011) and then a reduction of 100,000 due to replacements of old turbines at the ends of their life cycles.

TABLE 10.5
Estimate of Small Wind Turbine Production as of 2012 by Region[a]

	Number of Major Companies	Cumulative Number Produced	Cumulative MW	2011 Average Size (kW)
China	34	665,000	482	0.78
US	4	181,000	248	3.0
Europe	8?	80,000	115	2.6
Other	4?	15,000	15	
Total		941,000	860	

[a] Some units are exported.

FIGURE 10.1 Small wind turbine (100 W) of remote household in Inner Mongolia, China. Note rope on tail for manual control even though turbine has hinged tail for automatic furling.

Without that reduction, 775,000 units were reportedly produced in China through 2012. Even in 2011, the largest number produced was for 300-W wind turbines and the average size was 0.76 KW.

Among approximately 330 manufacturers [8], about fifty are major producers and thirty-four of those are in China. The largest production was in China, with 116,000 units in 2011; 57,000 were exported [9], primarily to other Asian countries. In the past, most of the Chinese production consisted of 50- to 100-W wind turbines for remote households (Figure 10.1). Small wind turbines provide

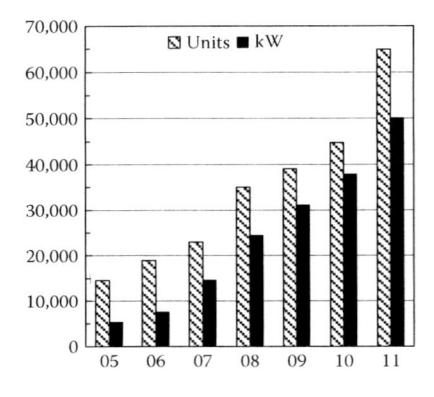

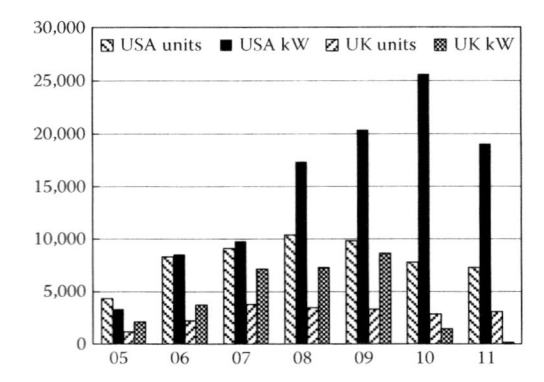

FIGURE 10.2 Cumulative installed numbers and capacity (kW) of small wind turbines, 2005 through 2011. Left: China installations. Right: U.S. and U.K. installations.

enough electricity for a few lights, a radio, and a small black-and-white TV. Marlec and Ampair in the United Kingdom and Southwest Windpower (in 2013, not in business) in the United States have produced large numbers of micro and small wind turbines, and units for sailboats constitute a big market.

The trend for the installations of small wind turbines in three major countries (China, U.S., and U.K.) was the same from 2005 through 2008: increased numbers and increased size (Figure 10.2). In the U.S. in 2011, 7,300 units, capacity 19 MW, and average size, 5.8 kW (on-grid) and 0.4 kW (off-grid) were installed, with 96% of the units produced in the U.S. [10]. A significant decline in installed capacity from 25 MW occurred in 2010, probably due to fewer installations of the 50- to 100-kW turbines. The export market for U.S. manufacturers in 2011 was 6,000 units (18-MW capacity, 3.5-kW average size).

In the U.K. in 2011, 2,335 units were installed with a capacity of 15.8 MW, average size 5.1 kW; 2,986 units were exported. The U.K. online market report [11] includes wind turbines from 100 to 500 kW. From 2004 through 2011, 2,549 units were mounted on buildings (10% of the number installed). Kestrel, South Africa, sold over 700 wind turbines from 2010 to 2012, and two thirds were exported to other countries

The National Renewable Energy Laboratory (NREL) of the National Wind Technology Center (NWTC) maintains a development and testing program for small wind units [12]. The American Wind Energy Association has a distributed energy (formerly small wind) section that includes annual reports from 2007 through 2011. Its 2002 U.S. map is probably outdated. The map estimated that small wind could provide 3% of U.S. electrical demand by 2020.

The American Wind Energy Association promulgated a program for the independent certification of small wind turbines, and a working group was established in 2006. In 2008, the Small Wind Certification Council (SWCC) was established [14] to work on problems related to the problems for adoption of small wind turbines:

- Non-standardized performance specifications
- Optimistic and inconsistent claims by suppliers
- Lack of consumer-friendly tools to compare small wind turbines and accurately estimate energy performance
- Need for greater assurance of safety, functionality, and durability for consumers and agencies providing financial incentives
- Inadequate field testing (fewer than half the small models on the market were tested)

The SWCC is an independent body that certifies small wind turbines that meet or exceed the performance and durability requirements of the American Wind Energy Association (AWEA) standard.

Applications and Wind Industry   225

Small Wind Certification Council
Certified Small Wind Turbine

Manufacturer/Model

Bergey Windpower Company
Excel 10 (240 VAC, 1-phase, 60 Hz)

(SWCC Certified logo)
CERTIFIED
SMALL WIND TURBINE
SWCC-10-12

Rated Annual Energy
Estimated annual energy production assuming an annual average wind speed of 5 m/s (11.2 mph), a Rayleigh wind speed distribution, sea-level air density and 100% availability. Actual production will vary depending on site conditions.

13,800
kWh/year

Rated Sound Level
The sound level that will not be exceeded 95% of the time, assuming an annual average wind speed of 5 m/s (11.2 mph), a Rayleigh wind speed distribution, sea-level air density, 100% availability and an observer location 60 m (~ 200 ft) from the rotor center.

42.9
dB(A)

Rated Power
The wind turbine power output at 11 m/s (24.6 mph) at standard sea-level conditions.

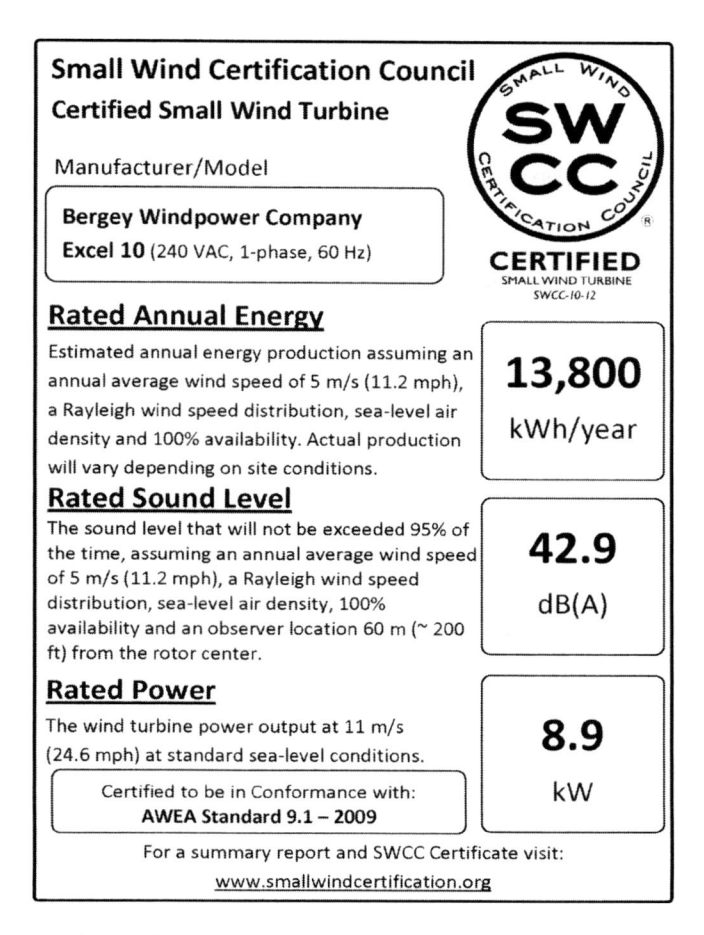

8.9
kW

Certified to be in Conformance with:
AWEA Standard 9.1 – 2009

For a summary report and SWCC Certificate visit:
www.smallwindcertification.org

FIGURE 10.3 Certification label for Bergey Windpower's Excel 10.

The Small Wind Turbine Performance and Safety Standard of AWEA is a common North American framework. SWCC issues labels (Figure 10.3) for rated annual energy output, rated power, and rated sound level, and also confirms that a turbine meets durability and safety requirements.

SWCC also publishes power curves, annual energy performance curves, and measured sound pressure levels. The rated power is at 11 m/sec and the rated annual energy is calculated from an annual average wind speed of 5 m/sec (11.2 mph), a Rayleigh distribution (see Section 3.11), and the wind turbine power curve.

SWCC also grants time-limited conditional temporary certifications to small wind turbines that meet the requirements of the IEC 61400 series of standards or the British Wind Energy Association (BWEA) standard. The International Electrotechnical Commission developed an international standard, IEC 61400l-2. Part 2 covers design requirements for small wind turbines. The BWEA Small Wind Turbine Performance and Safety Standard currently supports the microgeneration certification scheme for manufacturers of small wind turbines that wish to sell in the British market.

Field tests can be conducted at independent testing entities or by manufacturers, but self-testing requirements are more restrictive. Towers and foundations are not part of the certified system. Accredited testing organizations include NREL and a few other laboratories in Canada, the U.K., and Germany. Non-accredited organizations undergo SWCC evaluation and approval and include the NREL regional test centers (Kansas, New York, Texas, and Utah). In 2010, NREL provided funding for the centers [15] for certifying two turbines at each site that are or would be sold in the U.S. market.

A fairly new market for small wind turbines is for street lighting and telecommunications use. While a transmission line may be nearby, the cost of a transformer, connection, meter, and power may exceed the cost of electricity from a combination photovoltaic (PV)–wind system (Figure 10.4 and Figure 10.5). PV–wind system sizes are around 50 W for the PV and 100 W for the wind turbine.

FIGURE 10.4 Wind and photovoltaic energies power streetlights and flashing red lights at stop signs on McCormick Road at Interstate Highway 27 between Canyon and Amarillo, Texas.

FIGURE 10.5 Wind–photovoltaic streetlights in China. Wind = 300 W. PV = 120 W. (Photo courtesy of Ryan Rendleman, Verdegy.)

FIGURE 10.6 Two 1.5-kW wind turbines of telecommunications system in the Philippines. (Photo courtesy of Charlie Dou.)

It is estimated that there are more than 1,000 telecommunications stations having small wind turbines as parts of their power supplies (Figure 10.6). In China alone, there were around 2,500 wind turbines installed for telecommunication sites from 2009–2012. The number of telecom sites will be fewer as sites could have more than one wind turbine, for example one site had 16, 1 kW units. This a growing market due to the increased use of cellular phones, especially in remote areas of the world. Check the websites of different manufacturers for information about these two applications.

10.3 DISTRIBUTED SYSTEMS

Distributed systems are installations of wind turbines on the retail side of the electric meter for farms, ranches, agribusinesses, and small industries. They range from one or more small units to mid-size (100 to 1,000 kW) to megawatts for large industries. Two 10-kW units were installed at a liquor store near Amarillo, Texas, and in Lubbock, Texas, the American Wind Power Center and Museum installed a 660-kW unit and a cottonseed oil plant installed ten 1-MW units. There is an overlap among small wind turbines for residential use connected to a grid, large wind turbines (over 100 kW) attached as distributed systems (inside the meter), and community wind facilities ranging from small wind turbines to megawatt size turbines. Community wind is covered in the next section.

The U.S. market for farm, industrial, business, and community installations was estimated at 3,900 MW by 2020 [16]. Distributed wind systems will exert impacts on smaller utilities and electric cooperatives [17]. The international market is difficult to measure because much of the market consists of village power and remote systems; however, for farm, industrial, and business uses, the market was estimated at 400 MW for 2010 and 600 MW by 2020.

The issues related to distributed wind turbines for farmers, ranchers, and agribusinesses are somewhat similar to the considerations surrounding farm equipment purchases. The barriers and possible incentives for distributed wind applications are:

1. Cost (insufficient products to achieve economies of scale).
   a. Favorable life cycle costs will not sell wind turbines.
   b. Payback has to be four to six years.
   c. Wind turbines must compete almost directly with electricity from utilities that costs $0.10 to 0.15/kWh.
2. Lack of infrastructure.
   a. Enough units have to be installed to provide sales and maintenance opportunities for local businesses. Within a 250-km radius, a dealer would need $1,000,000/year in sales. At $50,000/unit, he would have to sell twenty units per year.
   b. To provide operations and maintenance (O&M) services, a business would need a customer base of about 300 units installed within the 250-km radius.
   c. Over time, distributed wind equipment business would be similar to the farm implement business, for example, a large tractor costs over $200,000.
3. Insufficient selection of wind turbine sizes from 50 to 500 kW. In 2008, Fuhrländer suspended production of its 30- to 600-kW units due to shortages of components and the great demand for utility-scale turbines. Wind Eagle sells 30- and 50-kW units, depending on wind regime. In a sense, wind turbines should have modular components. The trend toward larger sizes will continue. Suggested sizes for rural grid-connect systems are:
   Residential: 10 kW
   Farm and ranch residential: 50 kW
   Agribusiness: 100 to 250 kW
   Large agribusiness: 500 to 1,000 kW
4. Need for total package for agribusinesses.
   a. Wind turbine, electrical energy.
   b. Demand side management.
   c. Service.
5. Incentives.
   a. Ability to use production tax credit or have feed-in tariff.
   b. Irrigation market requires wind class 3 and above and net energy billing on annual basis for units up to 500 kW. The introduction of net energy billing of 50 kW (residential size wind turbines) in Texas resulted in essentially zero sales.
   c. Benefits for nitrogen oxide (NO_X) and sulfur oxide (SO_X) emission reductions. When carbon trading arrives, distributed wind turbines should be included.
   d. Installation of distributed wind turbines on rural electric cooperative grids.

The Rural Energy for America Program (REAP) provides financial assistance to agricultural producers and rural small businesses. Under its program, 325 wind projects have been funded and in October 2012, fifty-three wind projects were approved. Grants can be up to 25% of total eligible project costs and are limited to $500,000 for renewable energy systems. All projects must be located in rural areas, must be technically feasible, and must be owned by the applicants. These projects are primarily distributed wind and range in size from 10 kW to megawatt wind turbines. Some are refurbished wind turbines from repowering of wind farms, primarily from California.

A case study for a distributed wind and solar system is Cross Island Farms, Wellesley Island, New York. The system consists of a 10-kW wind turbine, a 5.5-kW PV unit, and a 17-kW propane generator [18]. The costs were $73,000 for the wind turbine, $40,000 for PV equipment, and $13,000 for the propane generator. The first developer received $9,000 for wind resource assessment and siting and concluded that the economics indicated an inadequate return. A second developer, more experienced in wind operations, recommended a taller tower [25 m (80 ft) to 37 m (120 ft)] and the payback was estimated at five years for the wind turbine and ten years for the PV.

One problem of distributed and/or community wind is the higher surface roughness leading to lower annual capacity factors if a wind turbine is located in a city. For example, the capacity factors for the 660-kW unit at the American Wind Power Center and Museum ranged from 13 to 18%, even though wind turbine was located in an exposed area. Expected capacity factors were 25 to 30%. A taller tower would have improved performance, but zoning laws limited tower height.

10.4 COMMUNITY WIND

The definition of community wind varies and again there are overlaps in reported numbers of turbines and capacities for utility scale, small wind, and distributed wind. The AWEA defines community wind as projects that incorporate local financial participation and control. Types of community wind projects are:

Nonprofit municipal electric utilities
Rural electric cooperatives
Owners of wind generation facilities
Purchasers of wind power from local residents
Local owners of wind turbines who sell power to for-profit utilities
Schools or colleges using power from their wind turbines
Large farm operations using power from their wind turbines
Other businesses using power from their wind turbines

The last two entities on the list are also distributed wind systems. Another definition of community wind applies to projects using turbines over 100 kW and completely owned by villages, towns, cities, commercial customers, and farmers (and excluding publicly owned or municipal utilities). Based on this definition, the three 100-kW wind turbines at North Texas University and the eight 60-kW units at three school districts near Lubbock, Texas, would be classified as distributed wind, not community wind.

Community wind projects leverage local distribution grids for the economic and environmental benefit of the local community and provide investment opportunities by allowing community members to become financial partners. As with larger wind farms, community wind brings rural economic development with local participation.

10.4.1 UNITED STATES

The AWEA gives some examples of community wind: 1- to 100-kW net metered home and farm-based systems, mid-size single turbine projects at schools and businesses, village wind–diesel projects, multi-megawatt wind farms owned by cooperatives and municipalities, and wind farms generating tens of megawatts via independent power producer arrangements. More local examples will appear as projects are brought online in the coming years and wind farms permit local farmers and/or investors to purchase shares. The association formed a Community Wind Working Group for developing policies to expand the development of community wind projects. Example policies are shown in the box 10.1.

AWEA maintains a database of U.S. community wind projects that lists installed numbers and capacities for large projects and large (>100 kW) wind turbines and also publishes such data in its quarterly market reports. For example, the 78.2-MW wind ranch of the Golden Spread Electric Cooperative in the Texas Panhandle housing 34 wind turbines (2.3 MW each) is listed

> **BOX 10.1 EXCERPTS FROM AMERICAN WIND ENERGY ASSOCIATION COMMUNITY'S WIND WORKING GROUP POLICIES**
>
> 1. Any wind project of 20 megawatts (MW) or smaller in nameplate capacity is a community wind project if it meets condition (a) and one or more of the following conditions from (b):
>   (a) Projects larger than 20 MW cannot be separated into smaller projects to meet this 20-MW project size limit. Specifically, more than one qualifying community wind project cannot be built within 5 miles of another qualifying project within a 12-month period and using the same interconnect.
>   (b) i. A local governing body (e.g., town, county) passes a resolution supporting the project;
>   ii. Members of the community are offered the opportunity to participate in an ownership interest in the project and are involved in the decision-making process during the project's development; or
>   iii. The project's local benefit is demonstrated in terms of retail power costs, benefits to the local grid (or the project is incorporated into a micro-grid) or resolving remote power issues.
> 2. A project larger than 20 MW up to 100 MW in nameplate capacity is a community wind project if local owners own at least 33.3 percent of the project.
>   (a) Local owner includes any:
>   Individual who resides in the same state as the project or within 250 miles of the project (and within the U.S.).
>   State department or agency, tribal council, school, town and other political subdivision located in the same state as the project.
>   Municipal, cooperative and similar publicly owned utility.
>   Corporation or other similar business entity of which at least 51% is owned by one or more individuals who reside in the same state as the project or within 250 miles of the project (and within the U.S.).
>   Not-for-profit corporations and similar nonprofit entities.
>   (b) A 33.3 percent ownership is measured at the "commercial operation date," a recognized term in the wind industry typically meaning the date at which the project is capable of and actually produces electricity. Local owners must own the project for some period of time thereafter.

in the community wind database. Others community operations such as Kotzubue and Kodiak, Alaska, are examples of wind–diesel projects.

Case studies are given for three projects by AWEA and photos and project summaries are given for some other projects [19]. The database indicates that approximately 626 MW of community wind projects were installed in the U.S. in 2010 and 2011. If we exclude 10-MW and larger projects, the capacity is 240 MW. This does not count a large number of smaller projects. Only three projects under 1 MW were listed. Many small-scale community wind projects for schools, public lighting, government buildings, and municipal services were not listed.

Windustry has information on community wind [20] that includes overviews, projects, and a toolbox that offers practical information for farmers and rural landowners looking to develop commercial-scale projects. The core content came from the *Community Wind Development Handbook* [21] developed on behalf of the Rural Minnesota Energy Board. *Community Wind*

101 [22] is a primer for policymakers and clean energy advocates that looks at economic benefits, examines obstacles facing community wind developers, and describes examples.

10.4.1.1 Minnesota

In 2005, the Minnesota legislature passed an energy bill that promoted community-based energy development (C-BED) to facilitate community wind projects without placing excessive burdens on utilities. The bill requires utilities to create a new tariff utilizing a net present value rate for electricity and provides the option of front-loading the rate in the first half of the contract's lifespan. The community wind projects average 10 to 20 MW and other states are adopting the mechanism to promote rural economic development. Dan Juhl [23] was an early proponent and by 2012 completed twenty-one projects (1,905 MW), most in Minnesota, and pursued another twenty-five projects under development (405) MW.

The Minwind arrangement consists of nine farmer-owned projects near Lucerne, Minnesota, and the case study [24] describes the ownership criteria. Early wind projects were farmer-owned, most under 2 MW, with one or two large turbines. The state provided a production incentive of $0.015/kWh for the first ten years. Sixty-six farmers raised 30% of the $3.6 million cost of four turbines (950 kW) for two projects. The remaining 70% was raised through local banks. Later projects were larger based on economics of scale, especially for the installation phase.

10.4.1.2 Schools, Colleges, and Universities

Eleven states are now participants in the Wind for Schools Program of Wind Powering America [25], with 124 installations at the end of 2012 (year 5 of program). The primary goals of the project are to (1) educate college juniors and seniors in wind energy applications; (2) engage communities in wind energy applications, benefits, and challenges; and (3) introduce students and teachers to wind energy. The general approach is to install small wind turbines at rural elementary and secondary schools and develop Wind Application Centers at higher education institutions. The standard system consists of a SkyStream (3.7 m diameter, 2.4 kW) wind turbine on a 21-m guyed or 19-m monopole power.

A number of Enertech wind turbines (13.5 m diameter, 40 or 50 kW) have been installed at schools, colleges, and universities. See Section 10.4 for detailed data on Enertech 44 performance. The original Enertech units are still being manufactured in Newton, Kansas. Later versions are built by Atlantic Orient, Entegrity, and Seaforth. At Natoma Rural High School in Kansas, an Entegrity unit produces 60,000 kWh per year. Units in the Texas Panhandle should produce around 100,000 kWh per year. Entegrity installed ten units in Texas and five in Kansas.

In Texas, stimulus money from the American Recovery and Reinvestment Act of 2009 provided funds for wind projects at three schools and two universities. A 50-kW unit was installed in 2012 by the Alternative Energy Institute at the feedlot of the Nance Ranch, West Texas A&M University. Performance data are available online [26]. Three 100-kW units are expected to produce around 8% of the electricity used at the Eagle Point grid that includes a residence hall and football stadium at the University of North Texas. The public can view performance online [27]. In the first year of operation, the turbines generated 283,000 kWh (12% capacity factor).

Northern Power [28] initiated a wind for schools program whereby the installation of its wind turbine also provides a web-based interface for students and teachers to view turbine performance, a public access website, and curriculum resources. The student view covers energy production, wind speed data, and historical trends. The public view lists current status (turbine power, wind speed, temperature), overviews (energy produced, environmental aspects, and cost savings), and carbon offset data. The company lists twelve schools and universities that utilize its units and also provides case studies including three school projects.

Other states utilize programs and/or policies for promoting wind turbines at schools. A U.S. DOE report [29] describes early examples. In Iowa [30], nine schools utilized installations ranging from 50 kW to 750 kW. Spirit Lake [31] installed a 250-kW turbine in 1993 and a 750-kW wind turbine in 2001 and they now supply 46% of the school's electricity. Some colleges with wind energy training programs for wind energy make large wind turbines available for their students. Examples are Mesalands Community College in New Mexico; Spirit Lake Community Schools in Iowa; and the Texas State Technical Institute in Sweetwater.

10.4.1.3 Electric Cooperatives

The Golden Spread Electric Cooperative [32] consists of sixteen member rural electric cooperatives serving a large area (182,000 km^2 or 69,700 sq. miles) across West Texas (92% of the area) and the Oklahoma Panhandle. Electricity was previously purchased from investor-owned utilities. The cooperatives now generate some of their power. The thirty-three wind turbines (3.2-MW) on an 84-MW wind farm commissioned in 2012 are expected to generate around 342 GWh per year.

The Fox Islands Electric Cooperative [33] installed three 1.5-MW wind turbines because its 2008 rates were about $0.28/kWh—two to four times the national average—and its undersea cable was becoming unreliable. In 2010, the wind project generated 12.1 GWh, which was in line with the projected 11.6 GWh per year.

The Basin Electric Power Cooperative [34] added 719 MW of wind energy through joint projects and purchase agreements. Crow Lake (162 MW) has 108 wind turbines of 1.5-MW size and 100 more (Prairie Winds SD 1) are owned by Basin Electric. One turbine was sold to the Mitchell Technical Institute, and seven are owned by a group of local community investors. Prairie Winds ND 1 is a 115-MW wind farm; other Basin Electric small projects use nine turbines around 10-MW size.

Other electric cooperatives pursue smaller projects and purchase wind power from independent producers or are considering wind power options. The Iowa Lake Electric Cooperative has two 10.5-MW projects involving seven 1.5-MW wind turbines. The Farmers' Electric Cooperative of Clovis, New Mexico, purchases 18 MW from a wind farm in Curry County.

10.4.1.4 Municipal and City Operations

The Berkshire Wind Power Project [35] is a 15-MW, 10-turbine wind farm atop Brodie Mountain in Hancock, Massachusetts. The project is owned and operated by the nonprofit Berkshire Wind Power Cooperative consisting of the Massachusetts Municipal Wholesale Electric Company and fourteen consumer-owned municipal utilities. The project is expected to have a capacity factor of 40% that would generate about 52 GWh per year.

Hull, Massachusetts [36] installed its first wind turbine of 40 kW in 1985. It operated until 1997 when damage from a 70-mph windstorm caused malfunctions of the tip brakes on the blades. In 2001, a 660-kW unit was installed and a 1.8-MW unit was added in 2006. Hull's website shows kilowatt hours generated, connect hours, and capacity factors for each turbine. As of February 2013, statistics for the two units are 16,684 and 23,921 MWh; 61 and 58% connect times; and capacity factors of 26.3 and 22.9%.

Lamar, Colorado installed four 1.5-MW wind turbines and Springfield, also in Colorado installed a single 1.5-MW wind turbine in 2004. Three turbines are owned by Lamar Light and Power and the other two are owned by Arkansas River Power Authority. The Lamar utility operates all five wind turbines from it power plant control room. This project is one of the case studies cited by Flowers [19].

The Jersey-Atlantic wind farm [37] utilizes five 1.5-MW wind turbines to supply electricity to a wastewater treatment facility and it sells excess power to the grid. The project cost in 2005 was $12.5 million, and the wind farm saved $3.17 million between January 2006 and May 2012. Web data includes monthly values for energy produced and revenue generated. Real time monitoring is available for the wind farm and solar array along with energy flow data.

Other municipalities and cities are considering wind power, and large municipal utilities have commitments to purchase power from wind farms as part of their goals for increasing renewables.

For example, Austin Energy has a wind portfolio of 851 MW from wind farms across Texas; its goal is 35% from renewable energy by 2020. CPS Energy of San Antonio, Texas, is the largest municipally owned utility (gas and electric) in the U.S. Renewable energy (wind, solar, and landfill gas) accounts for 9% and most of that comes from 1,059 MW of wind power from West Texas and coastal wind farms. CPS plans to achieve 1,500 MW of renewable energy or 20% of generation by 2020.

10.4.2 OTHER COUNTRIES

Wikipedia publishes articles on community wind energy in Europe, Australia, and Canada. Most community and cooperative wind projects are in Denmark and Germany. In Denmark, to encourage investment in wind power, families were offered tax exemptions for generating their own electricity within their own or adjoining commune. While this could have required a family to purchase a turbine, families often purchased shares in wind turbine cooperatives that then invested in community wind turbines.

At the end of 2005, individuals or cooperatives owned 83% of the 5,293 installed wind turbines. Privately owned wind turbines accounted for 77% of the capacity. The capacity percentage decreased somewhat with the installation of more offshore wind farms, but over 100 cooperatives still owned 75% of the wind turbines in 2011.

A famous example of a community project is the Middelgrundens offshore wind farm [38] in the harbor of Copenhagen, Denmark. The Middelgrundens Wind Turbine Cooperative owns ten of the twenty Bonus 2-MW wind turbines. Each share corresponds to 1/40,500 of the partnership (8,552 electric consumers, most of whom reside in Copenhagen). The weekly and monthly energy production figures and instantaneous values of power output, wind speed and direction, and other parameters can be followed online.

On Lolland Island, Denmark, wind power produced 50% excess energy. In 2007 the Lolland Hydrogen Community test project [39] was formed. Excess wind power is used to electrolyze water, producing hydrogen and oxygen. The hydrogen powers two cells (2 kW and 6.5 kW) for producing heat and power. The oxygen is used at the municipal water treatment plant to accelerate biological processes. The 4,200 residents of the island of Samsø own shares in twenty of the twenty-one wind turbines installed on two wind farms (one offshore) [40]. The turbines produce more energy per year than used on the island and the excess is exported to the mainland.

Germany had only two wind cooperatives in 2006. The number increased to 111 by 2011 [41]. Eighty-eight people hold shares in the community-owned Hilchenbach wind farm, where five wind turbines produce around 23.5 GWh per year. The city of Hilchenbach then purchased Rothaarwind and two thirds of the shares belong to local residents. The district of North Frisia has more than sixty wind farms producing about 700 MW, and 90% are community owned.

In The Netherlands, twenty-five wind cooperatives were formed from1986 to 1992, but eleven disbanded or merged with others. About half the onshore capacity is owned by local farmers, investors, and cooperatives. The largest project is a 122-MW wind farm (36 3.5-MW wind turbines) partnership of Nuon and sixty-three farmers that constitute the De Zuidlob wind turbine consortium.

In the U.K., the community planning website includes information about community wind [42]. The Baywind Energy Cooperative, the U.K.'s first community wind project is run by its 1,300 members who receive annual returns around 7%. Energy4All is a nonprofit company that supports community renewable energy projects. It set up seven community wind projects and detailed information about the twelve community wind projects in the U.K. is available online [43]. Falck Renewable Wind owns and operates five wind farms. It introduced community-owned turbines (2,500 investors in 2012) into its commercial wind farms.

Sweden has around 25 MW of community wind projects. No examples of community wind from other European countries were found on the web at this writing. Some web sites that promote community wind are maintained by developers of commercial wind farms.

The Hepburn Community Wind Park [44] in Australia restricts shareholders to members (1,900) of the cooperative and members receive dividends in proportion to investment. The project went on line in June 2011 and has generated 16.8 GWh since then (monthly production reports are available). The Denmark Community Wind Farm [45] utilizes two 800-kW wind turbines.

10.5 WIND–DIESEL GENERATION

For remote communities and rural industries, the diesel generator is a standard. Remote electric power is estimated at over 50 GW and generators range in size from 5 to 1,000+ kW. Canada utilizes more than 800 diesel generation sets with a combined installed rating of over 500 MW. In the State of Chubut, Argentina, village systems use diesel generators ranging from 75 kW in small villages to 1,250 kW for large ones. Because the systems are subsidized, it is difficult to determine the actual cost of electricity at state and national levels. Past costs ranged from $0.20 to $0.50/kWh but are probably higher because oil cost exceeds $100/bbl. Some remote communities in Canada are paying over $0.70/kWh but the cost is generally subsidized.

Diesel generators are inexpensive to install but expensive to operate and maintain. Major maintenance is needed every 2,000 to 20,000 hours, depending on generator size. Most small village systems generate electricity only in the evenings. More than 300 remote communities (about 200,000 people) in Canada use diesel-generated electricity. About ninety villages in coastal Alaska have the potential to displace displacing diesel fuel with wind. Section 8.6 discusses diesel generation in Kotzebue, Alaska. Australia, Argentina, northeast Brazil, Chile, China, Indonesia, Philippines, coastal sub-Sahara Africa, and other countries with populations in isolated villages and on islands have the potential to use wind–diesel systems. The techniques for designing, modeling, and simulating wind–diesel systems have improved as the industry has accumulated more on-site operational experience.

Wind–diesel systems were developed and tested at the Risoe National Laboratory, Denmark; The Netherlands Energy Research Center, Petten; the Wind Energy Institute of Canada (formerly Atlantic Wind Test Site), Prince Edward Island; the U.S. National Renewable Energy Laboratory; the U.K., and other locations. Primary work focused on developing wind–diesel systems for the retrofit market. This market would be for installing existing diesel generators in windy locations where wind power would exceed 50% of the installed capacity. A wind–biodiesel system was tested at the Agricultural Research Service operation at Bushland, Texas [46].

Wind–diesel [47,48] systems have potential because of the high costs of generating power from isolated diesel systems. By 1986, more than a megawatt of wind turbines were combined with existing diesel systems. Simulation models for wind–diesel systems are available. A problem common to all new applications is that early prototypes may not be operational over the long term. For example, thirteen projects in remote communities in Canada were no longer operational by 2011.

Wind turbines may be added to existing diesel power plants to save fuel and integrated wind–diesel or wind hybrid systems are now available to provide village power. Wind–diesel power systems can vary from simple designs in which simple wind turbines are connected directly to diesel grids (Figure 10.7) to very complex systems [49]. Wind–diesel power systems have peak demands of 100 kW to a megawatt, based on AC bus configurations, and storage is needed for high penetration. Among the problems of integrating a wind turbine and existing diesel generator are voltage and frequency control, frequent stops and starts of the diesel, utilization of surplus energy, and the use and operation of new technologies. These problems vary by the amount of penetration (Table 10.6). Wind turbines at low penetration can be added to existing diesel power for large communities with few problems because the wind segment is essentially a fuel saver. One solution for high wind penetration is the use of flywheels or battery storage [50,51].

Based on data for Alaska wind–diesel systems, a dump load or storage is needed to reach predicted energy production from the wind turbine even for low penetration. Otherwise the capacity factors will be lower than predicted. One company recommends wind turbine capacity at 130 to

Applications and Wind Industry

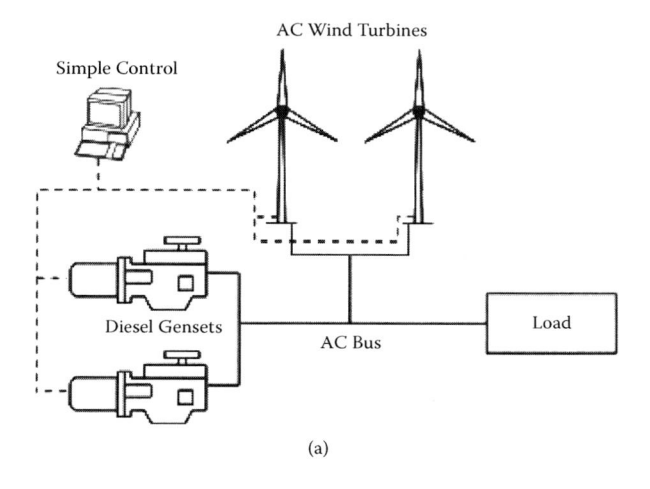

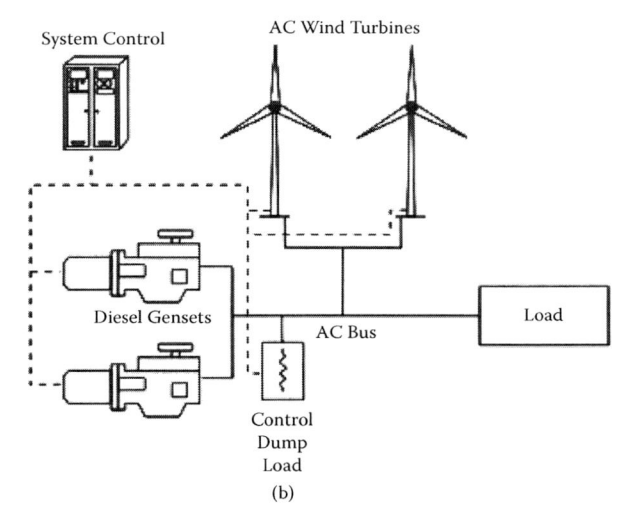

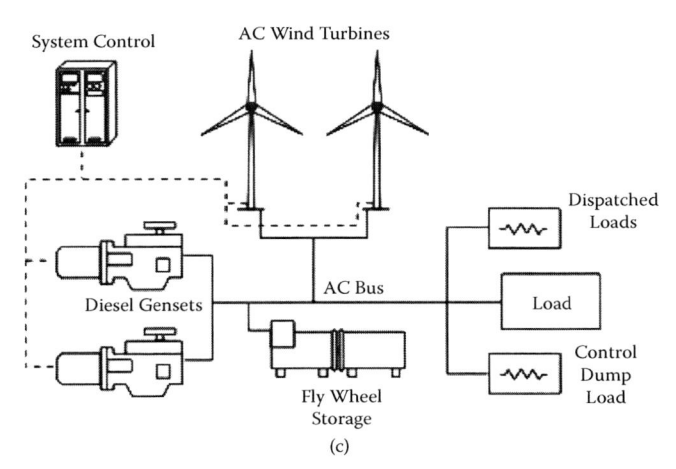

FIGURE 10.7 (a) Low penetration diesel without storage. Diesel governor and voltage controls maintain system power quality. (b) Medium penetration with system control and dump load for high winds and medium diesel power. (c) High penetration with flywheel storage.

TABLE 10.6
Penetration (Class and Percentage) for Wind–Diesel Systems

Class	Operating Characteristics	Peak Instantaneous	Annual Average
Low	Diesel runs full-time	< 50%	< 20%
	Wind power reduces net load on diesel		
	All wind energy goes to primary load		
	No supervisory control system		
Medium	Diesel runs full-time	50 to 100%	20 to 50%
	At high wind power levels, secondary loads dispatched to ensure sufficient diesel loading or wind generation is curtailed		
	Requires relatively simple control system		
High	Diesels may be shut down during high wind availability	100 to 400%	50 to 150%
	Auxiliary components required to regulate voltage and frequency		
	Requires sophisticated control systems		

150% of engine power to provide 70 to 80% wind penetration. If the surplus energy from the wind and heat from the diesel engines is utilized for central heating, ice storage, and electric cars, the wind penetration can reach 90%. Other options are adding desalination or hydrogen production packages to wind–diesel systems.

About 250 wind–diesel projects from prototypes to operating systems have been installed but the market is changing with the high cost of diesel fuel. Wikipedia has an incomplete list of forty-six wind–diesel installations that produce 50 MW of wind power. Reports on operational experiences from eleven wind–diesel facilities are available from the 2004 International Wind–Diesel Workshop [52].

The U.S. Air Force installed four 225-kW wind turbines connected to two 1,900-kW diesel generators (average load, 2.2 to 2.4 MW) for a low-penetration system on Ascension Island [53]. Average penetration was 14 to 24%. Tower height was limited to 30 m because of local crane capacity. In 2003, the addition of two large wind turbines (900 kW), controllable electric boilers, and a synchronous condenser increased the average penetration from 43 to 64% [54]. Fuel consumption was reduced significantly, with a savings of approximately $1 million per year. Wind penetration ratios exceeding 40% usually have stability problems but in this case reliable and stable power was delivered at 80% power penetration.

Cape Verde had eleven wind–diesel systems, with energy penetration of 14% and power penetration of 35% along with some problems [55]. Three of the systems were not working by 2005. Other systems are Alto Baguales-Coyhaique, Chile (2 MW wind) and wind–hydro–diesel systems (1.4 MW wind) on Flores and Graciosa Islands in the Portuguese Azores.

Data from the Alaska Energy Authority [56] show that from 2007 through 2012, 63.8 MW (144 turbines at 27 locations) were installed. Online data for wind turbines also cover manufacturers, rated power, commissioning dates, and costs. Installed costs ranged from $3,000 to over $20,000 per kilowatt. Diesel fuel cost was about $1.20/L ($5/gal) or $0.38/kWh assuming 3.4 kWh/L (13 kWh/gal) generation. The 48.5 MW installed in 2012 came primarily from eleven 1.5-MW wind turbines at Fire Island and twelve 2-MW turbines at Eva Creek. The Pillar Mountain hydro–wind–diesel system with a 1.5-MW wind turbine provides around 10% of the power. Online performance

Applications and Wind Industry

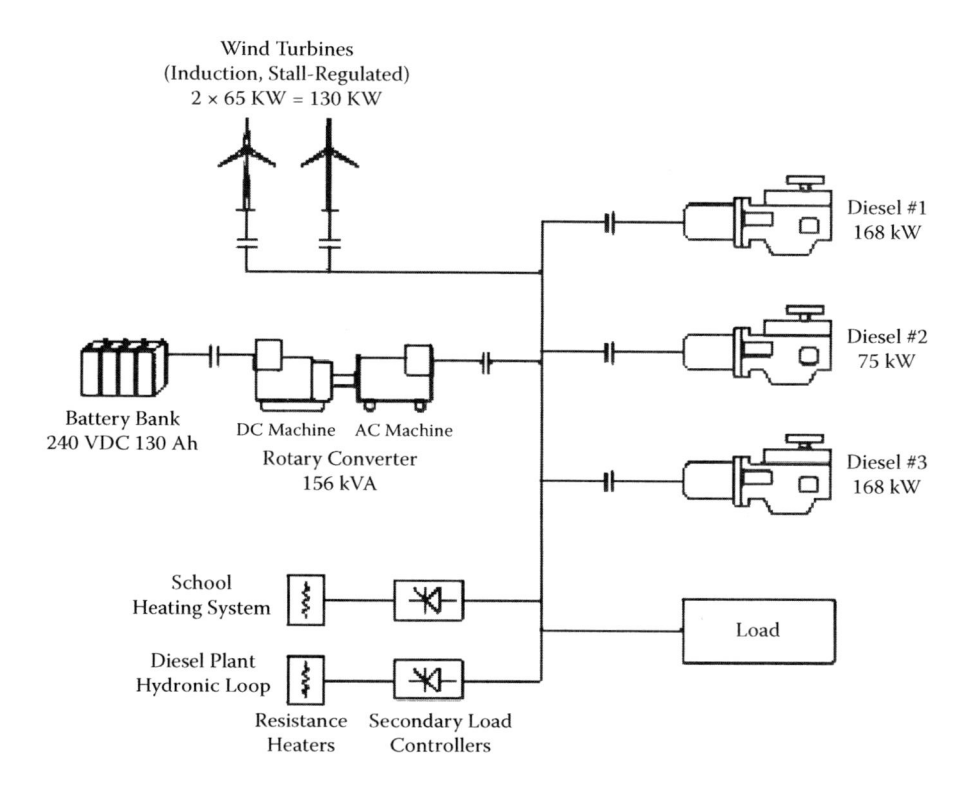

FIGURE 10.8 High penetration wind–diesel system with battery storage at Wales, Alaska.

data for wind–diesel systems shows power curves. Unalakleet expects to displace 90,000 gallons of diesel fuel per year from 600 kW of wind turbines.

Wales, Alaska's high-penetration system (Figure 10.8) utilized battery storage [57,58]. The system has not worked properly since 2006 because of high-penetration issues and it essentially needs to be reworked. The University of Alaska at Fairbanks wind–diesel applications center [59] produces a best practices guide and uses a wind hybrid simulator consisting of two 200-kW diesel generator sets, a 100-kW wind turbine, wind turbine simulator, grid simulator, battery bank, secondary load control, synchronous condenser, inverter, controls, and data acquisition equipment.

Three 250-kW wind turbines were added to the diesel system (four 1,200-kW generators) on King Island, between Tasmania and Australia. Wind provided 18% of the electrical demand. In 2003, another 1,700 kW of wind power and a 200-kW battery and inverter system were added [60] to produce around 50% of electrical demand. A large-flow vanadium redox battery reduces the variability of the wind energy. The battery went out of service in 2012 and Hydro Tasmania is evaluating rectification or replacement of the storage system.

Powercorp [61] published case studies of ten wind–diesel installations with 8.5 MW of wind power. To handle the instabilities of integrating wind power with diesel, a flywheel that can absorb or release 1 MW of power within 0.5 msec is part of the system. Power data for two wind turbines and percent electricity statistics are online for the Mawson Research Station [62] in Antarctica. Ross Island, Antarctica, installed three, 300-kW wind turbines in 2004 that provide 22% of the electricity and 11% fuel saving (463,000 liters per year). In the second phase during summers of 2010 and 2011, thirteen wind turbines (about 4 MW) were installed.

From 1992 through 2010, Vergnet [63] installed 7.7 MW of wind power, with turbines ranging in size from 12 to 275 kW. The company provided case studies for locations at Devil's Point, Vanuatu, (11 275-kW wind turbines) and Coral Bay, Australia (3 275-kW wind turbines with flywheel storage).

Danvest Energy [64] lists eight projects indicating a total of 4.6 MW of wind power and fuel savings of 50 to 85% per year. The site has downloads describing wind–diesel facilities, service and controls, desalination, dump load controls, and other topics. Ramea (population 700) on Northwest Island off the coast of Newfoundland, Canada, has a wind–diesel project [65] consisting of six remanufactured Windmatic turbines (total 390 kW).

Wind–diesel and wind hybrid systems are now available for village power, so the wind constitutes an integral part of the original design [66]. A number of wind turbine manufacturers offer wind–diesel or wind hybrid options. These range from simple, no-storage systems to complex integrated systems with battery storage and dump loads.

Wind–hydrogen systems are similar to wind–diesel devices if hydrogen is used to power a generator set. An added advantage is that fuel does not have to be transported to remote locations. Of course, the hydrogen can also be used as a fuel for heating and cooking. In 2009, wind–hydrogen capability [67] was added to the Ramea project [64]. The system consists of 300 kW wind and a 250-kW hydrogen genset. Hydrogen is produced by electrolysis of water. Seven projects using wind for hydrogen production including Ramea were discussed at a workshop [68]. In a pilot project on the island of Utsira in Norway [69], ten households receive power from two 600-kW wind turbines, a 5-kWh flywheel storage system, an electrolyzer with a peak load of 48 kW, and a 55-kW engine.

The Wind Energy Institute of Canada [70], maintains a wind–hydrogen village at its North Cape test site (formerly the Atlantic Wind test site). A 300-kW electrolyzer is powered independent of the grid by wind turbines. The 150-kW hydrogen genset supplements or replaces wind power when wind speeds are too low. A Wind Technology Center research project known as WIND2H2 [71] used two wind turbines (100 and 10 kW) and converted variable output DC to produce hydrogen via an electrolyzer.

Installations of wind–diesel systems and associated R&D have continued for several years. Many configurations were devised but little consensus and replication resulted. The technology is more mature than it was in 2007, but the number of installations of village power systems is insufficient despite the huge market potential. A lot of information from wind–diesel workshops is available on the Internet.

10.6 VILLAGE POWER

Around 1.4 billion people lack electric service because they are too far from transmission lines of conventional electric power plants. Extension of the grid is too expensive for most rural areas, and extended facilities exhibit poor cost recovery. Efforts are ongoing to bring renewable energy from wind, sun, mini and micro hydro, and biomass to these villages. Other components required to supply limited reliable energy include controllers, batteries, and conventional diesel and/or gas generators. In windy areas, wind is the lowest cost component of a renewable power supply.

Village power systems provide power from a central source or mini grid. Village hybrid power systems [72] can range from small micro grids (<100 kWh per day, about 15 kWpeak) to larger communities (tens of megawatt hours per day, hundreds of kWp). One or multiple wind turbines (10 to 100 kW) may be installed and software programs are available for modeling hybrid systems [73,74]. The main difference between village power and wind–diesel generation is system size although the categories overlap.

International conferences have focused on village power. The NREL maintained a database covering 146 projects (14 involving wind) from 1994 through 2000. The database has been archived and is not available online. The Global Village Energy Partnership provides information and publishes a newsletter [75]. Manufacturers' websites may show case studies; for example, Bergey Windpower case studies show different types of installations that include village power [76]. Renewable systems for villages include:

Fuel saver system (adding renewable energy system to existing diesel power plant (see Section 10.4)
Single or hybrid renewable energy source with battery storage (Figure 10.9)

Applications and Wind Industry

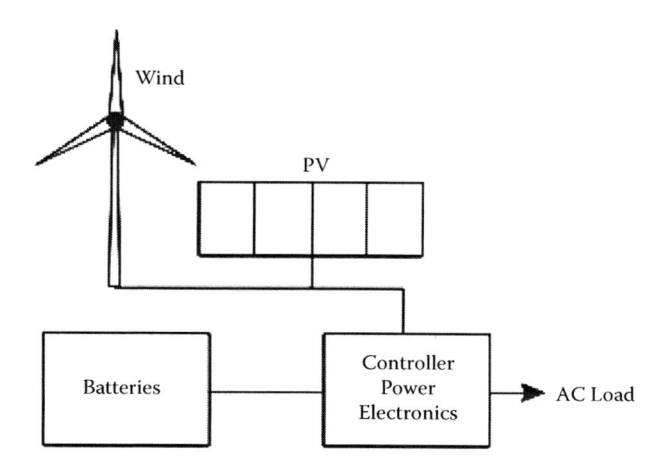

FIGURE 10.9 Hybrid wind–PV system with battery storage.

Single or hybrid renewable energy source with diesel or gas generator
Single or hybrid renewable energy source with diesel or gas generator and battery storage

The advantages of village power systems using renewable energy are:

Generation of AC or DC power for remote areas (AC is standard)
Available electricity for productive uses
Modular equipment
Reduced or no fuel costs
Lowest life cycle cost of electricity
Local ownership and operation

The disadvantages are:

High initial capital cost compared to diesel generators
Equipment complexity (sophisticated controllers, power conditioning, batteries)
High growth in demand (may require load management and limitation)
Little supplier support (few systems installed)
Need for infrastructure (trained staff, billing structure, administration)

Institutional issues are more important than technical issues, especially for demonstration projects. Many demonstration projects may become very political, and the social issues may dominate technical issues. Institutional issues are:

Pre-installation planning that includes local residence
Cost, subsidy, and maintenance requirements
Ownership
Operation and maintenance
Operator training
Financing possibilities: world (multilateral), national aid agencies (lateral), and nongovernment organizations; national, state, local, and private organizations
Tariff design, metering, ability and willingness to pay
Load growth
User education

TABLE 10.7
Configurations of Five Village Power Systems in Bulunkou Township, China[a]

	Bulunkou	Subashi N.	Subashi S.	Gaizi	Kahu Lake	Total
Wind	20 kW	20 kW	10 kW	20 kW	10 kW	80 kW
Solar	4 kW	4 kW	0	2 kW	0	10 kW
Diesel	30 kVA	30 kVA	15 kVA	30 kVA	15 kVA	120 kVA
Inverter	30 kVA	30 kVA	15 kVA	30 kVA	15 kVA	120 kVA
Battery bank	1000 Ah	1000 Ah	500 Ah	1000 Ah	500 Ah	4000 Ah
Mini grid length	1200 m	750 m	3600 m	590 m	350 m	6490 m
Number of posts	18	12	69	85	5	189

[a] Hybrid systems with batteries, 10-kW wind turbines.

Service quality
Economic development versus social services (schools, clinics)
Cultural response
Institutional cooperation: local, state, national, electric utilities, financing

10.6.1 CHINA

Among the estimated 2,000 renewable village power systems in the world, more than 1,250 are in China. About 100 renewable village power systems from 5 to 200 kW were installed in China by 2000, but more than 21,000 villages and 7 million households still lack electric service [77, chap. 2]. China now leads the world in installation of renewable village systems (100 include wind). One example is the electrification of five villages in Bulunkou Township (Table 10.7) in the Xinjiang Uigur Autonomous Region [77, chap. 5].

China started a Township Electrification Program in 2002 and its goal was to provide electricity to 1.3 million people in seven western provinces. The program called for 1,013 village power systems with a capacity over 18 MW: 292 small hydro, 689 PV, 57 PV–wind, and 6 wind systems (Table 10.8). Because the projects required some funding from the townships, not all the planned projects were installed.

Note the large difference in average sizes (780 kW versus 22 kW) of the mini hydro, PV, and PV–wind systems. In 2005, 66 PV–wind hybrid systems were visited to check on performance. A large hybrid village system (Figure 10.10) in Gansu Province had a projected load of about 235 kW. Mazhongshan Township is 158 km from the nearest utility grid. The village power system consists of 210 kW wind power and 90 kW PV, divided into three groups. A group consists of seven 10-kW wind turbines, a 30-kW PV cell, a battery bank of 240 V, 3,000 Ah, and a 100-kW DC–AC inverter. A system provides electricity to one part of the township.

10.6.2 CASE STUDY: WIND VILLAGE POWER SYSTEM

Huaerci [77] is a village in the mountainous area of eastern Xinjiang Province, China, with ninety households and 360 inhabitants. The primary economic activity is animal husbandry. The income per capita is well below the national poverty level. The distance to the nearest electricity grid is 110 km, and the roads are very poor. Lighting at night was provided by candles. Most families used two candles per night so the children could do their homework. The renewable resources are wind (annual mean wind speed, 8.3 m/sec) and solar (annual average, 3,100 hr).

The system configuration chosen consisted of a single 10-kW wind turbine, a 55-kWh battery bank, and a 7.5-kW DC–AC inverter. The system produces around 50 kWh per day. The project was financed by a government-subsidized five-year loan at 3% interest.

Applications and Wind Industry

TABLE 10.8
Renewable Village Power and Single Household Systems (SHS) of SDDX program in Western China Provinces

Province	PV–Wind	Capacity (kW)	Minihydro (kW)	Capacity (kW)	SHS	Capacity
Tibet	329	6763	72	16,470		
Qinghai	112	2715			6800	136
Xinjiang	110	1417	1	110	2886	144
Corps	49	961			4247	212.
Inner Mongolia	42	752			1525	610
Gansu	23	995	8	35,190		
Sichuan	46	1817	21	21,990		
Shaanxi	9	100	16	2,1985		
Chongqing			3	4840		
Yunnan			4	4460		
Jiangxi			2	1650		
Hunan	1	20	19	7070		
Total	721	15,540	146	113,765	15,458	1102

FIGURE 10.10 Fourteen of 21 wind turbines (10 kW each) at Mazongshan Township, Gansu Province, China. (Photo courtesy of Charlie Dou.)

The system provides 24-hr power for the ninety households, village offices, a school, and a TV transmitting station. All light bulbs are energy saving types, and since installation of the system, ten color TVs, thirty black-and-white TVs, and a CD player have been purchased. The peak residential load is about 5 kW, and energy consumption is around 300 kWh monthly with an additional 45 kWh per month for the institutional loads.

A village power management committee is composed of village officials, villagers, and the deputy director of the border control stations. A tariff of 1.2 yuan/kWh ($0.16/kWh) is charged to all customers. Most of the revenue will be used for maintenance costs. Cash flow should be sufficient to allow replacement of the battery bank, but the village power system is not fully commercialized. The system has a part-time operator. No productive loads are served to date due to limited system capacity. The lessons learned are:

Load analysis and prediction are important. Proper system configuration to match load is a critical factor for cost recovery.
Six renewable energy village power systems have been developed in Barkol County. They provide a great opportunity to develop a multiple project management entity and introduce a commercialized model to ensure sustainability.

Productive loads should have been established at the beginning.
A skillful technical operator should provide some services to users and encourage wise use of electricity.

The large initial investment for renewable energy village power is beyond the financial resources of the local residents and local government. Four villages in Barkol County have been powered by renewable energy since 1999. Each is powered by a wind turbine system, and another two large villages are powered with a 30-kW wind system.

10.7 WATER PUMPING

The water pumping and sailboat propulsion are the oldest and most consistent uses of wind power. The two common examples of mechanical water pumping are the historical Dutch windmill for pumping large volumes of water from a low lift and the farm windmill for pumping small volumes of water from a high lift [78–80].

For mechanical windmills and wind turbines, the important considerations are the wind power and method of transferring power. This means that the characteristics of a wind turbine (primarily the rotor) and the characteristics of the pump are combined into an operating system. The type of pump in many cases dictates the mode of operation of the rotor and how the rotational shaft power is transferred to pump power. Of course, the size of the system depends on the dynamic pumping head and the quantity of water to be pumped. Farm windmill efficiency depends on the load matching of the rotor to a positive displacement device, generally a reciprocating (piston) pump.

The American farm windmill (Figure 1.3) is still in widespread use around the world for pumping low volumes of water from wells or boreholes. About 80,000 operating windmills are estimated to operate in the Southern High Plains of the U.S. World production of farm windmills production is estimated at 3,000 per year, primarily from replacements for old units installed over sixty years ago. The American farm windmill is well designed for pumping small volumes of water for livestock and residential use and the design has not changed since the 1930s. The only change has been in materials used for bearings and the use of plastic pumps and drop pipes.

The American farm windmill is characterized by a high-solidity rotor (also called a wheel) consisting of fifteen to eighteen slightly curved blades (vanes); see Figure 10.11. The large number of blades provides the high starting torque needed to operate the piston pump. Most units have back gearing (reduction in speed) that transfers the rotating motion of the rotor to a reciprocating motion for pumping water. All wind turbines have characteristics that reduce efficiency and prevent capture of all the energy possible at high winds. On a farm windmill, the rotor axis and yaw axis are offset to rotate (yaw) the rotor out of the wind. This is called furling. At low wind speeds, the tail and the spring bring the rotor perpendicular to the wind.

The rotor has a peak power coefficient (C_P) of about 30% at a tip speed ratio of around 0.8. The efficiency for a reciprocating displacement pump is essentially constant at 80% over the operating range of wind speeds. The overall annual efficiency (wind to water pumped) is around 5 to 6% (see Section 8.5).

In the 1970s and 1980s, several research groups and manufacturers attempted to improve the performance of farm windmills [78, chap. 5] and reduce costs. Many of these projects were designed to pump water in developing countries. Designers believed that the performance could be increased by the following changes:

1. Reduce the solidity (number of blades or area of blades) to achieve higher rotor rpm.
2. Change pump characteristics by using variable stroke or variable volume to match rotor characteristics.
3. Develop a windmill for the low wind regions of the tropics.
4. Counterbalance the weight of the rods, pump, and water column.

Applications and Wind Industry   243

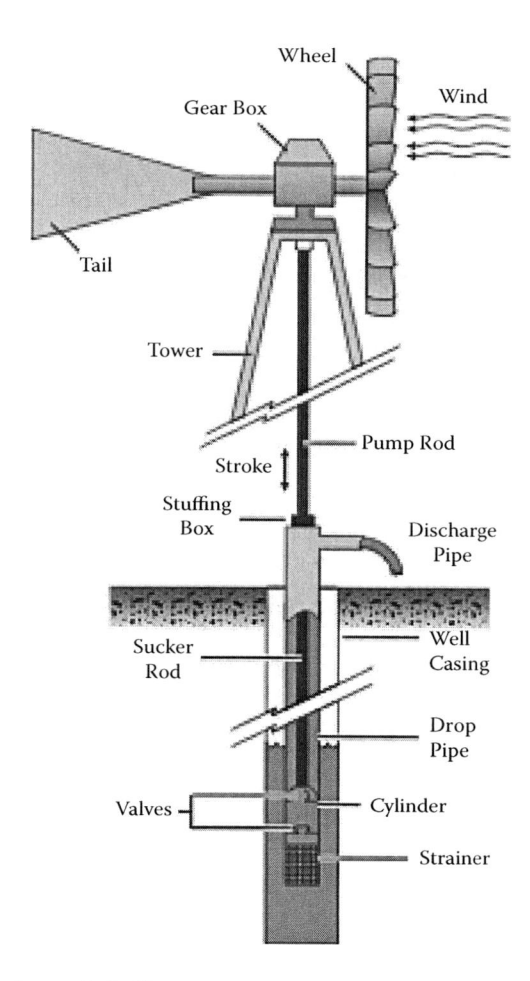

FIGURE 10.11 American farm windmill.

The Agricultural Research Service of the U.S. Department of Agriculture (USDA) and the Alternative Energy Institute at West Texas A&M University (WTAMU) tested some of these concepts in their cooperative program on wind energy for rural applications [81] but USDA withdrew from the program in 2012. A company in South Africa developed a windmill with a rotating helix pump [82].

Costs can be reduced by using local materials, buying from local light industry manufacturers, and considering new windmills designed for developing countries. One option is the use of a Savonius rotor for low-volume, shallow water. Another option is a wind turbine that drives an air compressor and airlift pump. However, the problem is still the same: the rotor must be connected to a constant-torque device.

10.7.1 Design of Wind Pumping System

The requirements for applications differ in that water for livestock and residential use requires a low volume storage tank. Villages require potable water and a low or high volume storage tank, depending on the size of the village. Irrigation requires large volumes of water, usually without a storage tank. The steps to consider in designing or sizing a water pumping system are:

1. Water demand: residence, livestock, village, irrigation
2. Water resource: surface, well; also available volume

3. Hydraulic power: volume times dynamic head
4. Wind resource
5. Comparison of other power sources
6. Design considerations

Other design considerations are economics, O&M costs, institutional issues, equipment life, and future demand requiring addition or expansion of the system. The average daily demand (m^3/day) is estimated for the month of high demand or the wind design month (month with lowest average wind speed). Demand must also consider growth during the design period, which should be at least ten years.

The water demand for livestock can be up to 90 L/day (Table 10.9). Evaporation from an open storage tank, especially in windy and dry areas, will require even more water. Also, animals travel only limited distances from a water source, so one water source is required for each 250 ha to harvest grassland. If the water supply and grassland are communal, then there is the distinct possibility that the growth in the size of the herds will result in overgrazing, especially near the water supply.

Domestic water needs depend on the number of people, usage, and type of service (Table 10.10). What is considered necessary in some countries or regions is considered a luxury in others. In addition, people will consume more water during hot, dry periods. Local water consumption is the best guide but remember that usage per person will probably increase if availability improves. Village water supplies must include clinics, stores, schools, and other institutions. Growth in demand will depend primarily on water availability, growth in size of herds or flocks, and population growth in villages. Growth in population should be estimated from local trends.

Water demand for irrigation (low or high volume) is based on local conditions, season, crops, and evapotranspiration. These data are generally available from regional or national government agricultural agencies.

TABLE 10.9
Livestock Water Requirements

Animal	Liter/Day
Cattle, beef	40 to 50
Cattle, dairy	60 to 75
Camels	40 to 90
Sheep and goats	8 to 10
Swine	10 to 20
Horses	40 to 50
Chickens (100)	8 to 15
Turkeys (100)	15 to 25
Evaporation	800 to 1200

TABLE 10.10
Typical Per-Person Water Consumption

Service	Liter/Day
Stand post	40
Yard tap	75
Home connection	100
U.S. farm residence	125

10.7.2 Large Systems

Large systems have been considered for pumping water for irrigation and villages. These can be classified as wind-assist and remote stand-alone systems. Wind-assist water pumping involves a wind turbine and another power source that work in parallel to provide power on demand. Wind-assist is essentially a fuel-saving mode of operation, since it does not require changes in irrigation applications.

Wind-assist systems are of the indirect and direct connection types. The advantages of the indirect connect are that the wind turbine does not need to be located at the well and electricity can be returned to the grid when the wind turbine produces more power than is needed by the load. A direct mechanical connection to a gear head has also been tested with a conventional electric or diesel power source [83,84]. The disadvantages of the mechanical connection are the need to place the wind turbine near the well and the use of the wind turbine only when water is needed.

The wind–electric water pumping system is a major improvement over the farm windmill in the areas of efficiency and volume of water. The annual efficiency is double that of a farm windmill. Because wind turbines are available in larger sizes (1 to 10 kW with 50-kW permanent magnet alternators), wind–electric systems can pump enough water for irrigation and villages.

A wind–electric system consists of a wind turbine generator connected directly to a standard three-phase induction motor driving a centrifugal or submersible turbine pump. The wind turbine output and the centrifugal pump constitute a good match because both have power proportional to rpm value cubed. Another advantage is that the wind turbine can be located some distance from the well or pump.

Two 10-kW wind–electric systems were installed in Naima, Morocco, in 1989 to supply water for villages and animals [85]. The spring is some distance from the villages. The first wind turbine pumps water from the collection tank to a large storage tank on top of a hill (Figure 10.12). There is gravity flow to two other storage tanks and a second wind turbine pumps water to another village. The wind–electric systems replaced inoperable diesel pumping systems. In 1997, two more 1.5-kW wind–electric systems were installed.

The Kestrel website shows examples of wind–electric water pumping. A 3-kW wind turbine is used to pump water for cattle from a well and then from one dam to another. The system delivers 2,500 L/hr and 1,300 L/hr. At another location, a 1-kW wind turbine pumps 1,800 L/hr from a 50 m depth. Southwest Windpower had a wind–solar hybrid electric water pumping system that used helical or centrifugal pumps.

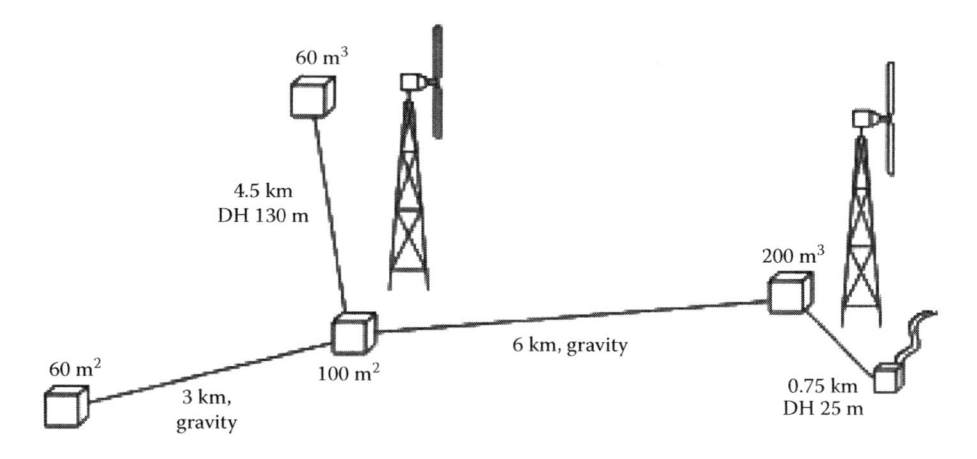

FIGURE 10.12 Layout of water supply for three villages at Naima, Morocco. DH = dynamic head.

10.8 WIND INDUSTRY

After the oil crisis in 1973, the first step was the development of small wind turbines (<100 kW). Most companies in the U.S. began by importing wind turbines, finding abandoned units to refurbish and sell, and then designing and building systems similar to the wind chargers of the 1930s and 1940s (direct current, 0.1 to 4 kW, up to 5 m diameter). A number of home builders turned to the Savonius types because of their simplicity and ease of construction. Electricity consumption also increased over the small demand of the 1930s. Larger wind turbines were needed because 5-m diameter rotors could not meet the demands of farmers and ranchers. In addition, many more uses of electricity required larger wind turbines.

Since the electric distribution system covered the entire U.S., there was a market for wind turbines that were fully compatible with utility systems: 120, 240, or 480 V, alternating current (AC). Inverters with solid-state electronics were now available to connect direct current (DC) units and alternators to utility lines. Enertech and Carter were early proponents of induction generators that could be connected directly to utility grids.

The second step was the influx of federal funding for research through the Energy Research and Development Agency (ERDA) and later the Department of Energy (DOE). Federal support for wind energy began with $300,000 in 1973, and increased by 1980 to $67 million. The federal program for development of wind turbines was geared to large units to connect to utility grids (Figure 10.13). These units were to produce power at $0.02 to 0.04/kWh. The program was managed by NASA–Lewis [86] starting with the MOD-0 (100 kW) and MOD-0A (200 kW) and progressing

FIGURE 10.13 Top left: MOD-5B in Oahu, Hawaii. Top right: WTS in Medicine Bow, Wyoming. Top of MOD-2 is visible at lower right. Bottom: Westinghouse 600 at Oahu, Hawaii. MOD-5B Is visible at right.

Applications and Wind Industry

FIGURE 10.14 MOD-2 wind turbines at Goodnoe Hills, Washington, near the Columbia River. Turbines were placed in a triangle to aid research on wake interference. (Photo courtesy of NASA–Lewis.)

to megawatt-sized wind turbines. Five of the MOD-2s (Figure 10.14) were built, and the original design of the MOD-5 was reduced from 7.2 to 3.2 MW. All these units had two blades.

During the 1980s, other large wind turbines were developed and installed in the U.S. and Europe (Table 10.11). The Hamilton Standard WTS-4, Wind Turbine Generator, Bendix-Schachle, and Alcoa units were developed mainly through private funds. However, a group of wind enthusiasts convinced federal officials to support a program for small wind energy conversion systems (SWECS). The SWECS prototype program awarded contracts in 1978 and 1979 (Table 10.12). By 1980, more than fifty companies produced wind energy conversion systems (1 to 100 kW) in the U.S. However, the installed capacity of SWECS was only around 3 MW from 1,700 units [87].

The third step was the passage of the National Energy Act of 1978. The section entitled Public Utility Regulatory Policy Act (PURPA) provided for connection of renewable power sources to electric grids without penalty and for payments to producers for electricity sold to utility companies. The value of that electricity was determined by the avoided cost concept implemented by the states.

10.8.1 1980 THROUGH 1990

The five years from 1980 to 1985 are considered the nascent phase of wind industry development. The boom of wind farms in California drove the exponential growth of the industry from 3 to 900 MW. The California wind market expanded from tax shelters (solar and investment tax credits) along with avoided costs and standard contracts set by the California Energy Commission. As with many new industries, a lot of manufacturers appeared. Only small wind turbines (<100 kW) were available commercially and they had reliability problems.

From 1980 to 1990, four features characterized the wind industry that evolved from the installations in California: (1) rapid growth; (2) development of intermediate-sized wind turbines (100 to 600 kW) without government funding; (3) the withdrawal from the market of U.S. aerospace companies, even those that received government funding for design and development; and (4) strong foreign competition, primarily from Europe. Foreign manufacturers, with Denmark leading the way, became important factors. Vertical axis wind turbines from Flowind and VAWTPower were installed on California wind farms although most U.S. installations were horizontal axis types.

The five years from 1986 to 1990 involved consolidation and shakeout within the industry. The tax credits ended in 1985 although earlier contracts meant wind turbines were still being installed in California, but not at the earlier fast pace. The U.S. had fewer than ten manufacturers in 1990; the only major producer was U.S. Windpower.

U.S. federal R&D support for wind energy fell to a low of $8 million in 1988. However, the Europeans increased their support for wind energy. Japanese companies, especially Mitsubishi,

TABLE 10.11
Large (≥500 kW) Wind Turbine Installations, 1975 through 1990

	Number	Diameter (m)	Rated kW	Year	Country
MOD-1	1	61	2000	79	U.S.
MOD-2	5	91	2500	82	U.S.
MOD-5B	1	88	3200	86	U.S.
WWG-0600	15	43	600	85	U.S.
Mehrkam		4	2000	80	U.S.
WTS-4	2	78	4000	80	U.S.
Schachle-Bendix		25	3000	80	U.S.
Alcoa		56 × 25	500		U.S.
VAWT 34 m test bed		34 × 42	500	89	U.S.
HMZ		33	500	89	Belgium
DAF-Indal		24 × 37	500	77	Canada
Eolé		64 × 94	4000	87	Canada
Nibe A		40	630	79	Denmark
Nibe B		40	630	80	Denmark
Tiareborg		60	2000	88	Denmark
Tvind		54	2000	78	Denmark
Windane		40	750	87	Denmark
M.A.N.		60	1200	89	Germany
Monopteros		48	650	89	Germany
Stork-FDO		45	1000	85	Netherlands
Windmaster		33	500	89	Netherlands
Newinco		34	500	89	Netherlands
Anisel. M.A.N.		60	1200	89	Spain
Nausdden		75	2000	82	Sweden
WTS-3		78	3000	82	Sweden
WTS-75		75	2000	83	Sweden
Howden		45	750	89	U.K.
Howden		55	1000	89	U.K.
WEG LS1		60	3000	88	U.K.

entered the world market and were determined to be major manufacturers. Many of the earlier large megawatt units were prototypes developed with government funding, but by the end of the decade, development was driven by the market as wind turbines increased in size above 100 kW.

Three hundred fifty million dollars, over half of the federal funding for wind energy from 1973 to 1990, was spent on the development of large wind turbines. This program was largely a failure because it proceeded to the next stage without fully developing the wind turbines from the previous stage. Design of wind turbines was much more difficult than aerospace engineers anticipated, and the aerospace industry was more interested in cost-plus government contracts than in developing commercial products. All the DOE prototypes were dismantled because of failures or excessive O&M costs.

10.8.2 1990 THROUGH 2000

World energy production in 1995 was estimated at 5 TWh/year from over 22,000 wind turbines with an installed capacity of around 4 GW. The American Wind Energy Association set a very optimistic goal

TABLE 10.12
Small Wind Energy Conversion System Prototype Development Program Funded by U.S. Department of Energy

Contractor	Type	Size (m)
1 kW at 8.9 m/sec: high reliability, remote, $1.95 million		
Aerospace/Pinson	Giromill	4.6 × 5.5
Enertech	HAWT	4.9
North Wind Power	HAWT	4.9
4 kW at 7.2 m/sec, $1.425 million		
North Wind Power	HAWT	9.1
Structural Composites	Dropped out at design stage	
TUMAC	Darrieus	9.1 × 11.5
8 kW at 8.9 m/sec, $2.26 million		
Alcoa	Dropped out at design stage	
Grumman	HAWT	10
United Technologies	HAWT	9.8
Windworks	HAWT	10
15 kW at 8.9 m/s, $3.23 million		
Enertech	HAWT	13.6
United Technologies	HAWT	14
40 kW at 8.9 m/s, $4.45 million		
Kaman	HAWT	19.5
McDonnell-Douglas	Giromill	18 × 12.8

for the U.S. of 10 GW by 2000. This was not achieved, although wind energy activity in states other than California increased due to the new production tax credit (PTC) incentive. The PTC was $0.015/kWh for ten years, with an inflation factor for wind farms installed in later years. The PTC has been extended a number of times. (now through 2013).

The Sandia National Laboratory managed the DOE program for VAWTs. A 34-m VAWT test bed, 500 kW, was tested at the USDA site at Bushland, Texas, from 1988 to 1998 (Figure 10.15). The DOE program, managed by the National Wind Technology Center, changed its focus to assistance and R&D to allow U.S. industries to meet foreign competition through the Advanced Wind Turbine Program [88–90].

Another measure in the U.S. was the EPRI/DOE Wind Turbine Performance Verification Program that was to provide a bridge from utility-grade development programs to commercial purchases. The 1995 goal of the Advanced Wind Turbine Program was to develop wind turbines for class 3 wind regimes (5 to 5.5 m/sec average at 10 m height) to produce electricity at $0.03 to 0.04/kWh, with O&M costs of $0.005/kWh. Another DOE R&D project goal was to achieve wind energy cost to $0.025/kWh or less at sites with 6.7 m/sec winds by 2002.

Government regulations and incentives in Europe, especially in Germany, resulted in rapid expansion of the industry and installation of wind turbines. More consolidation occurred and some manufacturing shifted from Denmark. Vergnet in France and Wind Energy Solutions (formerly Lagerwey) in Netherlands are selling commercial machines, and prototypes are still being tested. The manufacturers of two-blade, light-weight systems (Carter, for example) went out of business. Vergnet in France and Wind Energy Solutions (formerly Lagerwey) in The Netherlands are selling

FIGURE 10.15 Vertical axis wind turbine, 34-m test bed at USDA-ARS research center at Bushland, Texas. The VAWT is rated at 500 kW, peak power 625 kW. (Photo courtesy of USDA-ARS.)

commercial machines, and prototypes are still being tested. No VAWTs were produced for the wind farm market. This period was characterized by:

1. Continued rapid growth of the wind industry. Sizes of wind turbines increased from 200 kW to megawatts. Countries outside the U.S. and Europe installed wind farms; India installed 1,220 MW by the end of 2000.
2. European manufacturers dominated the market for large wind turbines.
3. Offshore wind farms were installed in Europe.
4. Large wind turbines with no gearboxes were developed.
5. The major U.S. manufacturer of large wind turbines (Kenetech) went out of business. The remaining manufacturer (Zond) was purchased by Enron and renamed Enron Wind.

The Utility Wind Integration Group in the U.S. published a number of brochures [91] on all aspects of the wind industry. This information is primarily for planners in utilities and policy makers in state governments.

10.8.3 2000 THROUGH 2010

Wind turbines were considered in planning processes for new electric plants in many countries. Europe continued to lead in installed capacity, development of multimegawatt turbines, and manufacturing. Global producers of equipment for electric power plants, General Electric (Enron), United States, and Siemens (Bonus), Germany, became manufacturers by purchasing wind turbine producers. Vestas absorbed NEG-Micon and remains the leading manufacturer of large wind turbines. The large three-blade, full-span, variable-pitch, upwind machines dominated the market.

The most economical sizes for wind turbines have not been determined. The trend for the wind farm market has been larger machines. However, in many locations, especially on islands and at other remote areas, the infrastructures (e.g., lack of cranes) limit the installation of megawatt wind turbines. Larger wind turbines are now being developed for offshore use.

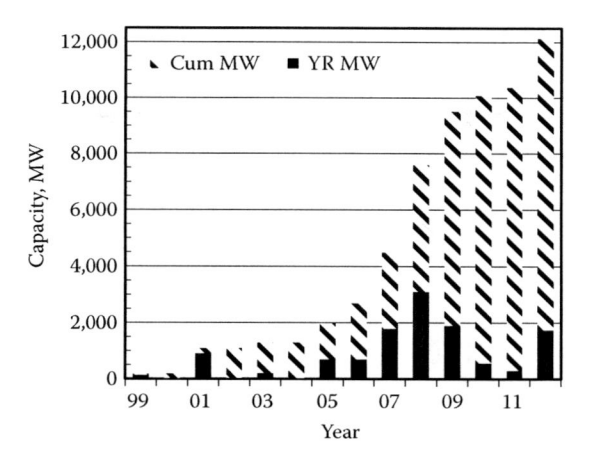

FIGURE 10.16 Wind farm capacity in Texas. Bars represent capacity installed for year. Stripes represent cumulative capacity.

The U.S. production tax credit and the Renewable Portfolio Standard in Texas led to a resurgence of wind farm installations. In 2006, Texas (Figure 10.16) surpassed California in installed capacity, and a number of other states have passed renewable portfolio standards. The late passage of the production tax credit meant that few wind turbines were installed that year (Figure 10.16). Since no wind turbines from the DOE program became major players in the commercial market, the research thrust by 2006 changed to assist manufacturers through a program called the Low Speed Wind Technology Project. The cost-shared R&D projects included a 2.5-MW wind turbine of Clipper Windpower; a modular power electronics package that can be scaled for small to megawatt wind turbines; a 1.5-MW direct-drive generator from Northern Power Systems; and a prototype multi-megawatt low wind speed turbine from GE.

Another emerging market is village electrification in developing countries. It is cheaper to have village power than extend transmission lines. The primary source of energy will be renewable: solar, wind, hydro, and biomass. Many of these sources will require hybrid systems with batteries and diesel for firm power. Large lending institutions like the World Bank and national government aid programs now understand they have alternatives to large-scale projects for producing power. Also, more U.S. companies make small wind turbines and China expanded its production.

Landowners now harvest wind much as they harvest the sun to produce crops. Since financing large wind power plants requires millions of dollars, landowners usually lease the land and receive royalties from the energy produced by wind turbines, similar to the practice in the oil and gas industry. The difference is that wind resources can be determined before large amounts of money are invested. Most importantly, the resource will not be depleted.

10.8.4 2010 Onward

Although information is only available through 2012, a major change has been the rapid growth of wind farms in China to 76.4 GW—an astounding addition of 50.6 GW in three years. Over the same period, Europe installed 26.7 GW and the U.S. added 25 GW [92]. The number of wind farms increased markedly in other parts of the world—more than double the number since the start of 2010. Vestas continued to lead in annual production until GE surpassed Vestas production in 2012.

Due to the huge market in China, five Chinese manufacturers ranked among the top fifteen in terms of cumulative production (Table 10.3) and seven were in the top fifteen based on production in 2011 (Table 10.4). More megawatt wind turbines with direct drive and permanent magnet generators have been installed. Sizes and hub heights continue to grow.

FIGURE 10.17 Mold for Siemens 75-m blade. (Photo courtesy of Siemens.)

The U.S. NREL research program for support of wind energy now focuses on offshore wind, mid-size and small wind turbine development, and gearbox reliability [93]. Sandia is focusing on increasing the viability of wind technology by improving performance and reliability, reducing cost, and enhancing blade design and manufacturing techniques [94]. Presentations from the 2012 Wind Turbine Blade Workshop are available online.

The Internet makes a lot of information available about the development and commercial production of wind turbines of 5 to 10 MW. In 2012, Vestas introduced a 80-m blade for the V164 8.0-MW offshore unit (prototype in 2014), Siemens makes a 75-m blade (Figure 10.17) for its 6-MW offshore unit, and Alstom uses an LM Wind Power 73.5-m blade on its 6-MW offshore unit. Some analysts predict wind turbines up to 20 MW by the end of the decade, primarily for the offshore market.

Community wind will increase as local investors and entities install small wind farms and/or smaller units in areas of less demand. Distributed wind will also continue to grow, primarily due to incentives by nations and states.

There is a resurgence of small wind manufacturers (more than 300 worldwide); unit production is still dominated by China followed by the U.S. and Europe in terms of capacity (Figure 10.2). Although no large VAWTs are used in commercial facilities, many small VAWTs are on the market.

Two markets that did not expand as expected from 2008 to 2012 were wind–diesel and village power. Despite the large number of wind–diesel systems installed in Alaska in 2012, the number in the rest of the world did not expand much after 2008. The primary problems of village power are the nascent aspect, financing, and lack of infrastructure. Without massive government support, that market will consist mainly of prototypes and demonstration projects.

10.9 STORAGE

Much of the material in this section was taken from *Introduction to Renewable Energy* [95, chap. 13]. Energy on demand comes from stored energy. The most common storage methods are damming, batteries, and biomass. Fossil fuels store solar energy from past geological ages. If a way could be found to store energy cheaply [96], we would have no need to construct new electrical power plants for some time. Storage is a billion-dollar idea and anyone who devises an economical method to store power will be richer than Bill Gates.

Economic storage would allow energy in existing power plants to be retained during periods of low demand [97, 98] and wind farms could provide some firm power. In general, the efficiency of storage systems is around 60 to 70%. Energy cannot be created or destroyed. It can only be transformed from one form to another. In reality, energy can be stored only as kinetic energy or potential energy. Kinetic energy storage can be achieved by flywheels or as heat (thermal storage). Dump loads (resistance heating of water) for storing heat are generally used in wind–diesel systems in cold regions. Compressed air is a mixture of mechanical and thermal change. Super (high rpm) flywheels made of composite materials to ensure strength have been used in wind–diesel systems, uninterrupted power supplies, and as prototypes on buses.

Potential energy arises from interactions such as gravity, electromagnetic forces, and weak and strong nuclear reactions. We will consider gravitational and electromagnetic concepts for storage systems. Most potential gravitational energy comes from water sources (dams, tidal basins, and pumped storage). Electromagnetic interactions include chemical reactions, phase changes, magnetic systems such as superconductor magnetic energy storage (SMES), electric devices (capacitors), and mechanical methods (springs). Chemical storage is by batteries, photosynthesis, and production of methane, hydrogen, fertilizers, and other compounds. Gas storage requires high pressure, conversion to liquid state, or inclusion in a chemical compound, for example, storing hydrogen in metal hydrides. Of course, solar energy is stored in chemical compounds such as foods and fibers. Solar energy is also converted into sugars, starches, and cellulose that are liquid precursors for biofuels.

The main components to consider in designing a storage system are (1) energy density, (2) efficiency and rates of charge and discharge, (3) lifetime (number of cycles), and (4) economics. Different storage technologies (Table 10.13) can accommodate power and/or energy. Energy density, size and weight are important factors for some applications (Figure 10.18). Liquid fuels have large energy density while hydrogen gas has low energy density.

Storage efficiencies generally range from 50 to 80%, and lifetimes vary widely from 100 years for dams to 5 to 10 years for lead acid batteries in a PV system and less than an hour for non-rechargeable batteries (Figure 10.19). The maximum rate and best rate of charging and discharging storage relate to the type and use of storage (Figure 10.20). The inclusion of energy storage in an application and the type of storage are driven by economics (Figures 10.21 and 10.22) and specific power and energy.

TABLE 10.13
Relative Ratings of Storage Technologies

Type	Advantages	Disadvantages
Pumped hydro	High capacity	Site requirements
Compressed air	High capacity	Site requirements
Flywheel	High power	Low energy density
Magnetic superconductor	High power	Low energy density
Capacitor	Many cycles, high efficiency	Low energy density
Batteries		
Flow	High capacity	Low energy density
Metal–air	High energy density	Difficult charging
Lead–acid		Limited life with deep discharge
Li ion, Ni-Cd, NaS	High power and energy density	
Hydrogen		Low energy density

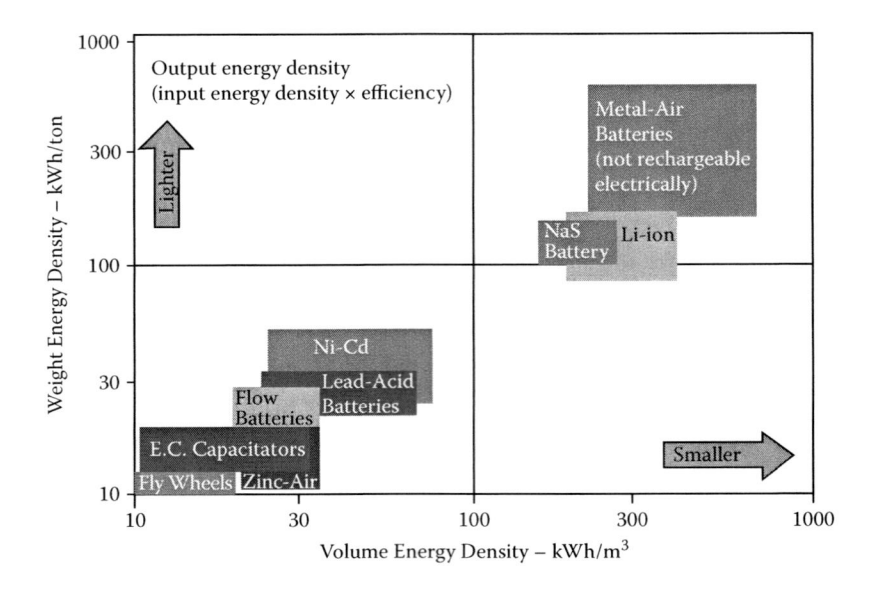

FIGURE 10.18 Energy density by weight and volume for electric storage. (*Source:* Electricity Storage Association. With permission.)

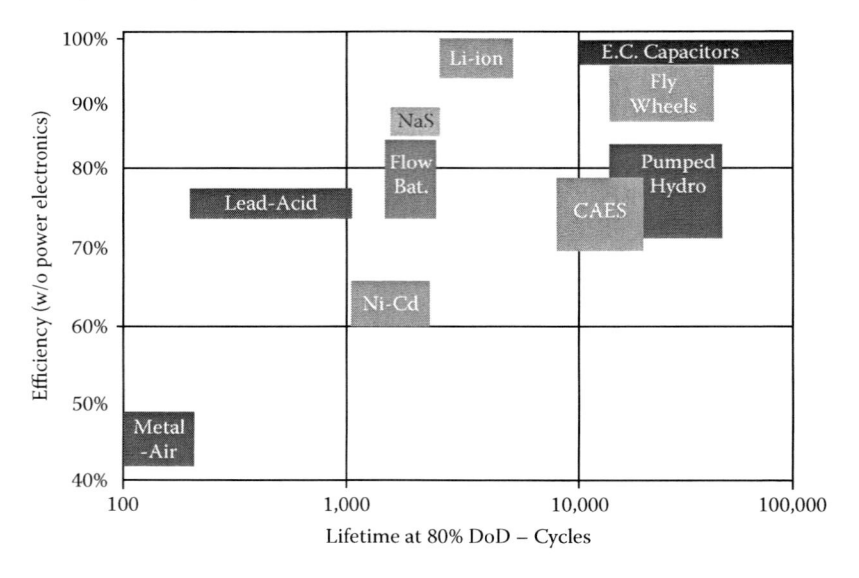

FIGURE 10.19 Efficiency and lifetime data for various electric storage systems. CAES efficiency is for storage only. (*Source:* Electricity Storage Association. With permission.)

An electric car whose batteries take three hours to charge could not be used on a cross-country trip even if charging stations were available. This demonstrates how versatile liquid fuels are for transportation. Also the requirements for storing high power for a short time differ from the requirements for storing energy for a few days. For utilities, large storage systems are limited to pumped storage and compressed air energy systems (CAES). However, battery systems for power shaving, conditioning, and reducing the variability from renewable energy sources have been installed on wind and solar systems. Pumped hydro and compressed air have long lives over many cycles.

Batteries are the most common storage devices for remote village power systems and stand-alone systems. A comprehensive report on energy storage solutions and applications for island

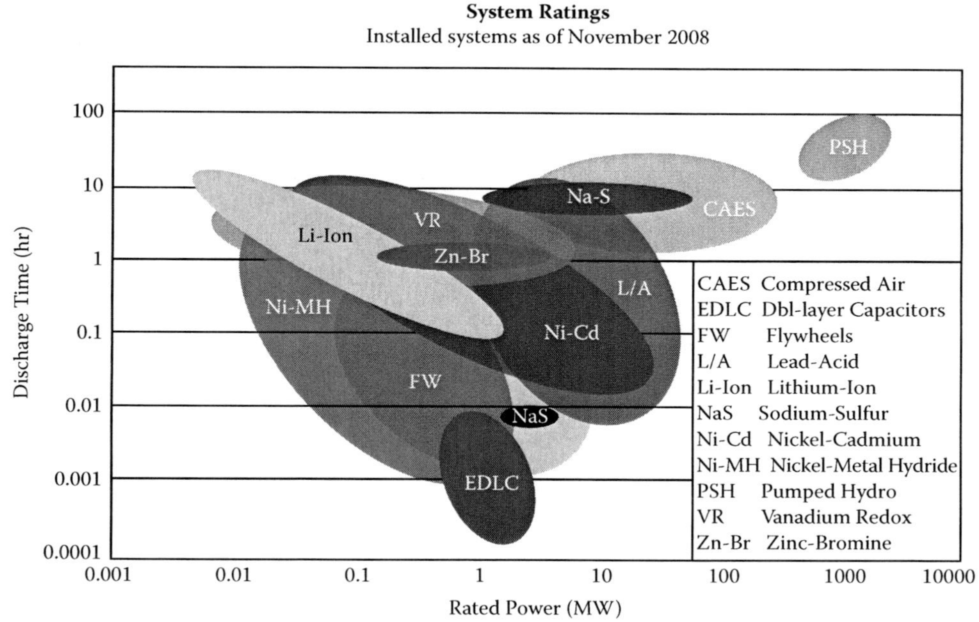

Applications and Wind Industry

FIGURE 10.20 Discharge times and rated powers of installed electric storage systems. (*Source:* Electricity Storage Association. With permission.)

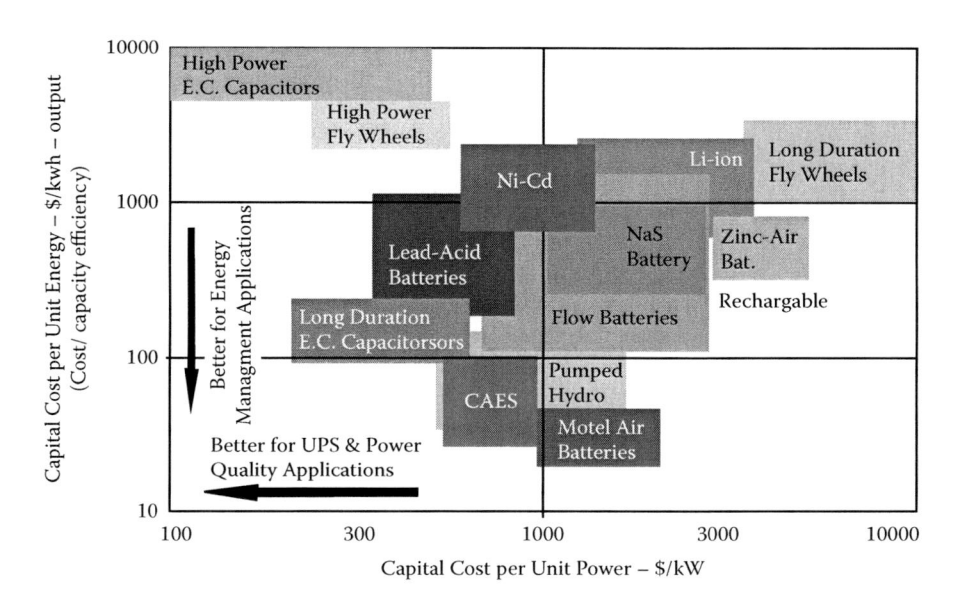

FIGURE 10.21 Capital costs for electric storage systems. (*Source:* Electricity Storage Association. With permission.)

communities [99] discusses case studies and various scenarios and strategies. It also contains estimates of size ranges and costs; capital cost (dollars per kilowatt) levelized cost of storage (dollars per kilowatt hour), and annual operating costs (dollars per kilowatt per year). The levelized cost for battery storage ranged from a low of $0.05 to $0.15 for sodium–sulfur (NaS) batteries to $0.30 to 0.45/kWh for lithium ion batteries.

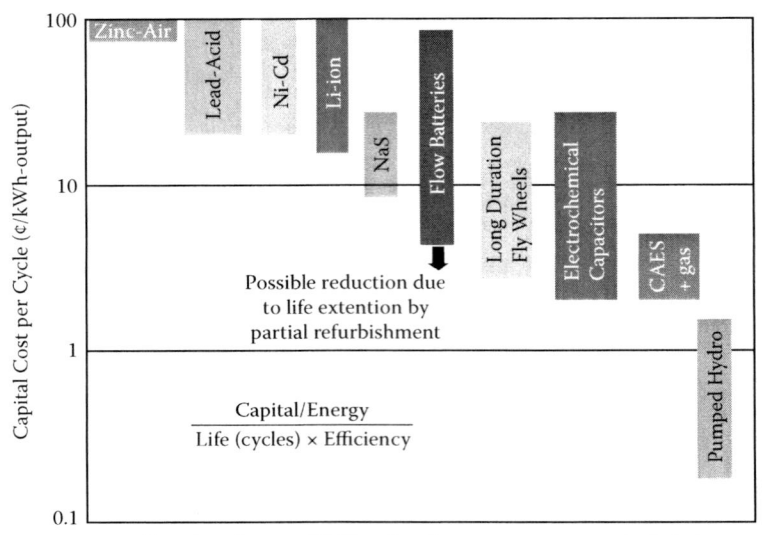

FIGURE 10.22 Carry charges for electric storage systems. (*Source:* Electricity Storage Association. With permission.)

The electricity produced by a generator cannot be stored. Energy in minus energy losses is the demand. Generation supplies that amount of demand, which varies by time of day and season. In addition, utility systems must meet peak demands and have spinning reserves for unforeseen conditions. If demand exceeds capacity, users are taken off the grid. In some parts of the world, rolling blackouts occur or electricity is available only for certain time periods. Finally, extreme events may force shutdown of a total grid. A very good source for information about storage for electricity is the Electricity Storage Association (www.electricitystorage.org/ESA/home). Check out the technologies section.

10.9.1 Compressed Air Energy Storage

Compressed air energy storage (CAES) is a peaking power plant that consumes 40% less natural gas than a conventional gas turbine that uses about two thirds of the input fuel to compress air. In a CAES plant, air is compressed during off peak periods and then utilized during peak periods. The compressed air can be stored in underground mines or salt caverns that take one and a half to two years to create by dissolving the salt. CAES has relatively low efficiency with a cost over $1000/kW of storage.

10.9.2 Flywheels

Flywheels store energy due to rotational kinetic energy, which is proportional to the mass and the square of the rotational speed.

$$E = 0.5 \times I \times \omega^2, J \tag{10.1}$$

where I = moment of inertia (kg m^2) and ω = angular velocity (rad/sec). For a mass M and radius$\approx R$, the moment of inertia for a ring is I = M × R^2, and for a homogeneous disk, I = 0.5 × M × R^2.

Applications and Wind Industry

Increasing the rpm increases the energy density so high speed flywheels have rpm values in the tens of thousands. Low speed flywheels are made from steel and high-speed flywheels are made from carbon fiber and/or fiberglass. High-speed flywheels are housed in a low vacuum and use magnetic bearings to reduce or eliminate frictional losses. Advances in power electronics, magnetic bearings, and materials have resulted in direct current (DC) flywheels. In case of a material failure, the container has to retain the energy inside. Cycle efficiency is around 80%.

Installed cost of a flywheel depends on type and ranges from $150 to 400/kW. Another application for flywheels is for cranes at ship and rail yards, where they provide short-time high energy for lifting. Energy is returned to the flywheel as objects are lowered.

10.9.3 Batteries

Batteries are common all over the world. Lead–acid batteries are used for vehicles and batteries are used for small power and energy applications. Batteries convert chemical energy into electrical energy using electrodes immersed in a medium (liquid, gels, and even solid) that supports the transport of ions or electrolyte reactions at the two electrodes. Individual cells are placed in series for higher voltage and in parallel for higher current. Internal resistance and other factors cause losses during charging and discharging a battery.

The power is the product of the voltage and current, but batteries are generally specified by volts and storage capacity C_B which is related to stored energy. C_B is the amount of charge that a battery can deliver to a load. It is not an exact number because it depends on age of the battery (number of cycles), temperature, state of charge, and rate of discharge. If you discharge a lead–acid battery to essentially zero a few times, you have drastically reduced its lifetime. As a first approximation, the energy is

$$E = V \times C_B, \text{ J} \qquad 10.2$$

where V = voltage and C_B = battery capacity in amps per hour.

Example 10.1

A 12 V battery rated at 100 Ah could deliver 5 Ah for 20 hr. E = 12 × 20 Wh or 1.2 kWh. However at a faster discharge rate, the values would be lower: 85 Ah for 10 hr, 70 Ah for 5 hr.

Decreased temperature results in less battery capacity and for a lead–acid battery, storage capacity decreased around 1% for every 1-degree Celsius drop in temperature. Remember those very cold mornings when the battery just had enough juice to start your car? Explosions from short circuits, generation of hydrogen, and disposal of used batteries and toxic chemicals are problems.

10.9.3.1 Lead–Acid Batteries

Lead–acid batteries are low cost and widely used technologies for power quality, uninterrupted power supply (UPS), and some applications for spinning reserves. However, their use is limited for large amounts of energy storage for utilities, primarily due to their short cycle lives. As noted, rates of charge and discharge (Figure 10.19) affect capacity (fast rate, fewer volts) and depth of discharge affects life (Fig. 10.20) so the trade-off is between the cost of new batteries and lifetime. Even with that disadvantage, lead–acid batteries are the most common energy storage facilities for remote village power and stand-alone and stand-alone wind systems [100, chap. 11]. Examples of lead–acid systems are a 10-MW 4-hr system in Chino, California; a 20-MW, 40-min system in San Juan, Puerto Rico; and a 3.5-MW, 1-hr system in Vernon, California.

Some useful battery practices for small renewable systems are:

Do not add new batteries to old sets.
Avoid more than two (three at most) parallel strings.
Do not use different types of batteries in same set.
Keep cable lengths the same.
Keep all components clean and connections tight.
Follow manufacturer's recommendations for charging and equalization.
Do not wear jewelry when working on batteries.
Use insulated tools.
Wear protective clothing and eye protection when working on batteries.
Do not smoke or generate sparks around batteries.
Maintain a battery log.

In general, car batteries are not suitable for storage in renewable energy systems, although they are available and have been used for some small wind systems.

The 5-min ramp rate at the Hampton Wind Farm (1.3 MW), New South Sales, Australia, is smoothed with an advanced system composed of lead–acid batteries and capacitors. A battery system consisting of modules of power cells (12 V, 1 kWh, dry cell battery) was installed at the Notrees Wind Farm in Texas to mitigate the variability of the wind power. The storage system consists of twenty-four 1.5-MVA/1-MWh modules for total of 24 MWh. Cost of the system was $43.6 million. Two other systems based on the same technology were installed in Hawaii. A 15-MW/10-MWh battery system was installed at the 30-MW Kahuhu wind farm on Oahu, and a prototype 1.5-MW/1-MWh system was installed to control the ramp rates of 3 MW of wind turbines at a 30-MW wind farm on Maui. The systems provide responsive reserves and ramp control (rates to 1 MW/min) to reduce curtailment and increase energy delivered by 70%.

10.9.3.2 Lithium (Li) Ion Batteries

The main advantages of Li ion batteries compared to other advanced batteries are: (1) high energy density (300 to 400 kWh/m^3, 130 kWh/ton), (2) high efficiency (almost 100%), and (3) long cycle life (3,000 cycles at 80% depth of discharge [DOD]). Li ion batteries have captured 50% of the small portable market and hybrid and electric vehicles are the main drivers behind their increased use.

The Tehachapi wind energy storage demonstration project plans to evaluate the performance of an 8-MW/32-MWh Li ion battery system to improve grid performance and integration with the large wind farms in the area. The cost of the project is around $54 million.

10.9.3.3 Sodium–Sulfur Batteries

The performance factors for commercial NaS battery banks are (1) capacity of 25 to 250 kW per bank, (2) 87% efficiency, and (3) lifetime of 2,500 cycles at 100% DOD or 4,500 cycles at 80% DOD. The cost is around $2500/kW.

NaS batteries of over 270 MW at 6 hr storage have been demonstrated at numerous sites in Japan. The largest installation is a 34-MW/245-MWh unit for a wind farm in Northern Japan. U.S. utilities have deployed around 20 MW for peak shaving, backup power, wind farm, and other applications. Presidio, Texas, frequently experienced power outages due to a long connection to the main grid via a single, sixty-year-old transmission line. A 4-MW/32-MWh NaS battery that stores enough electricity for the whole town was installed. The cost of the battery and substation was estimated at $23 million.

Xcel energy installed 1-MW battery storage next to an 11-MW wind farm [101]. The twenty 50-kW modules store about 7.2 MWh—enough to power 500 homes for 7 hr. For basic generation storage, the battery was discharged during on-peak periods and charged during off-peak periods at a rate proportional to power output of the wind farm. The optimal ratio of storage to wind for this operation should be 200 to 400 kW storage per megawatt of installed wind. The other operations tested were economic dispatch, frequency regulation, wind smoothing (ramp control), and wind leveling.

10.9.3.4 Flow Batteries

A flow battery converts chemical energy to electricity. Electrolytes containing one or more dissolved electroactive species flow through an electrochemical cell. Additional electrolyte is stored externally, generally in tanks, and is usually pumped through the reactor, although gravity feed systems are also known. The power and energy ratings are independent of each other. Flow batteries can be recharged rapidly by replacing the electrolyte liquid while simultaneously recovering the spent material for re-energization.

Vanadium can exist in four different oxidation states. Vanadium redox batteries [102] use this characteristic in a battery that contains only one chemical electrolyte instead of two. Hydrogen (H^+) ions (protons) are exchanged between two electrolyte tanks through a permeable polymer membrane. The net efficiency of this battery can be as high as 85%. Two installations for wind power: a 275-kW output balancer at Tomari Wind Hills, Hokkaido, Japan, and a 200-kW/800-kWh output leveler at Huxley Hill Wind Farm, King Island, Tasmania; one more is planned.

The Kotzebue Electric Association in Alaska has 1.14 MW of wind for its wind–diesel system and will install two additional wind turbines for a total of 2.94 MW. The association also plans to install a 500-kW/3.7-MWh flow battery to capture excess wind power.

10.9.4 OTHER TYPES OF BATTERIES

Other types of batteries are metal–air, nickel–cadmium (Ni-Cd), zinc–bromide (Zn-Br), and even organic compounds. The metal–air batteries may be less expensive, but they are at the preproduction stage at which the consumed metal is mechanically replaced and processed separately. Recharge using electricity is under development, but metal–air batteries have lives of only a few hundred cycles and efficiency around 50%.

A Zn-Br battery has two different electrolytes that flow past carbon–plastic composite electrodes in two compartments separated by a microporous polyolefin membrane. Battery systems are available on transportable trailers (storage plus power electronics) with unit capacities of up to 1 MW (3 MWh for utility-scale applications). Some electric utilities are installing 5-kW/20-kWh systems for community energy storage.

General Electric produces a sodium halide battery for utility, telecom, and UPS applications. A Ni-Cd battery system (40 MW, 7 min) was installed in Fairbanks, Alaska. Small rechargeable Ni-Cd and Li ion batteries are available.

10.9.5 HYDROGEN FUEL CELLS

A lot of information cites hydrogen as the fuel of future and discusses hydrogen storage [103, chap. 11]. Although hydrogen has low environmental impact and can be produced by renewable energy sources, its major disadvantage is the low energy density per volume. Hydrogen has one third the energy content of methane. However, existing natural gas pipelines could carry about the same capacity because hydrogen has lower viscosity.

Fuel cells are much more efficient than internal combustion engines, but the infrastructure for producing fuel cell cars is at the nascent stage. If hydrogen is produced by wind energy systems (see Section 10.5) through electrolysis, transportation and storage become major factors. Hydrogen can be stored as a compressed gas or liquid (cooled to 20K in non-pressurized containers) or by an extraction process. Storage in materials may involve adsorption (activated carbon); chemical compounds; compounds that can be reversibly transformed into other substances of higher hydrogen content; and metal hydrides that change hydrogen content with temperature.

Metal hydrides may be used storage of hydrogen for vehicles powered by fuel cells. As a comparison, 100 kg of hydride can store around 500 MJ and 100 kg of gasoline provides 4,700 MJ of energy—a large difference. However the efficiency of a hydrogen fuel cell is around 60% and that of a gasoline engine is around 20%. Hydrogen would produce fewer emissions and is essentially nondepletable. The production of hydrogen from water using wind and sun would be problematic in arid and semiarid areas that already lack water. Some considerations for hydrogen storage systems are:

Ratio of mass of hydrogen to overall mass of storage and retrieval system
Ratio of mass of hydrogen to total volume of storage and retrieval system
Cycle efficiency
Retention (amount of hydrogen remaining over long period)

Of course, as with other storage systems, economics (installed cost, O&M, replacement costs) are paramount. Other considerations are safety, ease of use, and infrastructure for transportation.

10.10 COMMENTS

Beside the emerging market of distributed systems, two other factors will increase the market: green power and reduction of pollution. People can purchase green power at a premium, and a lot of that power is generated by wind turbines (Figure 10.23) or wind farms. A number of urban areas are in nonattainment for clean air, and one way to reduce pollution is by producing electricity from renewable energy.

The reduction of carbon dioxide emissions per the Kyoto Treaty is included in many nations' regulations. As trading in CO_2 becomes regulated in the U.S., electricity produced by wind turbines will become more valuable and will eventually become the most cost-competitive power source on the market.

FIGURE 10.23 Two Vestas V-47 660-kW units on ridge of Hueco Mountains east of El Paso, Texas. Electricity sold under green power program. Notice car at the base of turbine. (Photo courtesy of Cielo Wind Power.)

Some wind applications are of interest because of prototype testing and commercial availability, for example, hydrogen production, compressed air storage, pumped hydro, stand-alone electric–electric systems for making ice, and desalination plants. Wind hybrid systems are used for telecommunications and remote military installations. Many of these sites are accessible only by helicopter, so fuel costs are high.

An ultraviolet water purifier powered by renewable energy was tested by USDA-ARS and AEI [104]. A controller was developed and five configurations were tested—two with photovoltaic cells (100 W), two with wind (500 W), and one hybrid wind–photovoltaic system. The photovoltaic-only system was more efficient and cost-effective than the wind-only system. However, the wind–photovoltaic system demonstrated more reliable power production. The system purified 16,000 L/day, which is enough potable water for around 4,000 people at an estimated equipment cost of around $5,000. In Afghanistan, water is purified by ozone produced from electricity from wind–photovoltaic hybrid systems. These small systems use around 160 W to produce 2 g of ozone per hour. Treatment is on a batch basis of 500 L to produce 2,000 to 4,000 L/day. The system is powered by a 1 kW wind turbine, 280 W of photovoltaic energy, a small battery bank, and an inverter.

LINKS

A. Allderdice and J. Rodgers. 2002. Renewable energy for microenterprise. www.nrel.gov/docs/fy01osti/26188.pdf
All Small Wind Turbines. http://turbines.allsmallwindturbines.com
American Wind Energy Association. www.awea.org
Danish Wind Turbine Manufacturers Association. www.windpower.org
DOE Wind Energy Program. www1.eere.energy.gov/windandhydro
V. Gevorgina et al.. 1999. Wind–diesel hybrid systems for Russia's northern territories. www.nrel.gov/docs/fy99osti/27114.pdf
Global Wind Day. www.globalwindday.org/
Global Wind Energy Council. www.gwec.net org.
P. Lundsager and H. Bindner. 2002. Gaia 11 kW wind turbine operating in a diesel grid and as a stand-alone turbine. www.windpoweringamerica.gov/winddiesel/pdfs/2002_wind_diesel/gaia_lundsager.pdf
National Wind Coordinating Committee. www.nationalwind.org
National Wind Technology Center. www.nrel.gov/wind/
Powercorp. www.pcorp.com.au/index.php?option = com_content&task = view&id = 99&Itemid = 157
Small wind manufacturers and turbines. http://energy.sourceguides.com/businesses/byP/wRP/swindturbine/swindturbine.shtml
Wind energy database. http://www.windenergydatabase.pl/index.php?option=com_content&view=article&id=7&Itemid=9
Wind Technologies Market Report. 2011. www1.eere.energy.gov/wind/pdfs/2011_wind_technologies_market_report.pdf
Xiao Qing Dao Village Power Wind–Diesel Hybrid Pilot Project. www.nrel.gov/docs/fy06osti/39442.pdf

REFERENCES

1. The Wind Power. Wind turbines and wind farm database. www.thewindpower.net
2. Australia Wind Power. Wind power and wind farms in Australia. http://ramblingsdc.net/Australia/WindPower.html
3. European Wind Energy Association. www.ewea.org/statistics/offshore/ and RenewablesUK. Wind: state of the industry 2012. www.renewableuk.com
4. Dong Energy. Horns Rev offshore wind farm. www.hornsrev.dk/Engelsk/default_ie.htm
5. Kansas Energy Information Network. Wind energy. http://kansasenergy.org/wind.htm
6. Global Wind Energy Council. Global wind statistics 2011. www.gwec.net/uploads/media/Global_Wind_2011_Report_final.pdf
7. P. Shi. 2012. Opportunities and challenges facing Chinese wind power industry. Presentation to Chinese Wind Energy Association.
8. World Wind Energy Association. 2012. Small Wind Report. www.wwindea.org
9. China Wind Energy Association data.

10. U.S. Small Wind Turbine Market Report. 2011. www.awea.org/learnabout/smallwind/upload/AWEA_SmallWindReport-YE-2011.pdf
11. Small and Medium Wind. 2012. U.K. Market Report www.renewableuk.com/en/publications/reports.cfm/SMMR2012
12. National Wind Technology Center. www.nrel.gov/wind/smallwind/
13. American Wind Energy Association. 2002. www.awea.org/learnabout/smallwind/index.cfm
14. Small Wind Certification Council. www.smallwindcertification.org
15. NREL Regional Test Centers. www.nrel.gov/wind/smallwind/regional_test_centers.html
16. T. Forsyth and I. Baring-Gould. 2007. *Distributed wind market applications*. Technical Report NREL/TP-500-39851.
17. Utility Wind Integration Group Distributed Wind Impacts Project. www.uwig.org/ug.htm
18. NREL. Distributed wind case study. www.nrel.gov/docs/fy12osti/53626.pdf
19. L. Flowers. 2012. Community wind: one size doesn't fit all. http://awea.org/learnabout/smallwind/upload/CW-Case-Study-PPT.pdf
20. Windustry. Community wind. http://windustry.org/community-wind
21. *Community wind development handbook*. http://www.auri.org/2008/01/community-wind-study
22. P. Mazza. 2008. Community Wind 101: A Primer for Policymakers. http://iowa.sierraclub.org/Energy/CommWind.pdf
23. Juhl Wind. www.juhlwind.com/index.html
24. Windustry. Minwind III-IX. www.windustry.org/resources/minwind-iii-ix-luverne-mn-community-wind-project
25. U.S. Department of Energy. Wind for Schools Project. www.windpoweringamerica.gov/schools_wfs_project.asp
26. AEI Wind Test Center. http://windtestcenter.org
27. University of North Texas. Sustainability Wind Energy Grant. http://sustainable.unt.edu/node/66 and PublicView of performance; http://northernpower.kiosk-view.com/unt
28. Northern Power Systems. Wind for schools program. www.northernpower.com/wind-power-products/wind-for-schools-program.php
29. U.S. Department of Energy. America's schools use wind energy to further their goals. www.nrel.gov/docs/fy04osti/35512.pdf.
30. T. Galluzzo and D. Osterberg. 2006. Wind power and Iowa schools. www.iowapolicyproject.org/2006docs/060307-WindySchools.pdf
31. Wind Powering America. Spirit Lake School District case study. www.windpoweringamerica.gov/filter_detail.asp?itemid=3623
32. Golden Spread Electric Cooperative. www.gsec.coop
33. Fox Islands Wind Project. www.foxislandswind.com/index.html
34. Basin Electric Power Cooperative. www.basinelectric.com/Electricity/Generation/Wind/index.html
35. Berkshire Wind Power Co-op. Using wind wisely. www.berkshirewindcoop.org
36. Hull Wind. www.hullwind.org
37. Atlantic County Utilities Authority Renewable Energy Projects. www.acua.com/acua//content.aspx?id=486
38. Middelgrundens Vindmøllelaug. www.middelgrunden.dk/middelgrunden/?q=en
39. J.K. Jensen and J. Bech-Madsen. 2007. Utilisation of hydrogen and fuel cell technology for micro combined heat and power production. www.h2-lolland.dk/mediafiles/14/other/ehec_final.pdf
40. D. Biello. 2010. One hundred percent renewable? One Danish island experiments with clean power. *Sci. Am.*, January 10. www.scientificamerican.com/article.cfm?id=samso-attempts-100-percent-renewable-power&page=2
41. German Wind Energy Association. Community wind power: local energy for local people. www.wind-energie.de/en/infocenter/publications
42. Community-led wind power, how to plan, build and own a medium or large wind turbine in your community's backyard. www.communityplanning.net/index.php
43. Energy4All. Delivering community-owned green power. www.energy4all.co.uk/home.asp
44. Hepburn Wind. http://hepburnwind.com.au.
45. Denmark Community Windfarm, Western Australia. www.dcw.org.au/index.html
46. E. Eggleston and N. Clark. 1998. Wind/biodiesel performance of USDA hybrid system. www.biodiesel.org/resources/reportsdatabase/reports/gen/19980601_gen-268.pdf
47. E. Baring-Gould et al. 2004. Worldwide status of wind–diesel applications. Paper presented at Wind–Diesel Workshop. www.windpoweringamerica.gov/winddiesel/pdfs/2004_wind_diesel/101_status.pdf

48. R. Hunter and G. Elliot. 1994. *Wind Diesel Systems: A Guide to the Technology and Its Implementation.* Cambridge, UK: Cambridge University Press.
49. American Wind Energy Association. 1991. *Wind–Diesel Systems Architecture Guidebook.* AWEA Standard 1901. www.awea.org
50. E.I. Baring-Gould. 2008. Wind–diesel power systems: basics and examples. 2008 International Wind/Diesel Workshop. www.akenergyauthority.org/wind/02_Wind-diesel_power_systems_basics_01-01-2008.pdf
51. Powercorp, Australia. Wind–diesel. www.pcorp.com.au/index.php?option = com_content&task = view&id = 99&Itemid = 157
52. International Wind–Diesel Workshop. 2011. www.uaf.edu/acep/alaska-wind-diesel-applic/international-wind-diesel/ (has links for presentations from earlier workshops); Wind–Diesel Workshop. 2009. www.pembina.org/arctic/wind; Wind Powering America. www.windpoweringamerica.gov/winddiesel/past_workshops.asp (has proceedings from earlier workshops).
53. S. West and G. Siefert. 2002. Wind–diesel hybrid design experience: six years of Ascension Island wind farm operations. www.eere.energy.gov/windandhydro/windpoweringamerica/pdfs/workshops/2002_wind_diesel/ascension.pdf
54. G. Seifert and K. Myers. 2006. Wind–diesel hybrid power, trials and tribulations at Ascension Island. Paper presented at Proceedings of European Wind Energy Conference. www.ewec2006proceedings.info
55. P. Lundsager, H. Bindner, and J. Hansen. 2005. Hybrid systems for isolated communities. Risoe National Laboratory. www.e4d.net/UM-IFU%20seminar%2016June05/pelu%20hybrid.pdf
56. R. Stromberg. 2013. Alaska wind energy barriers, Alaska Wind Energy Workshop; Unalakleet wind–diesel data analysis; Alaska Energy Authority overview.
57. S. Drouilhet. 2004. Overview of high penetration wind–diesel system in Wales, Alaska. www.pcorp.com.au/index.php?option = com_content&task = view&id = 99&Itemid = 157
58. B. Reeve. 2004. Wales operational experiences. www.pcorp.com.au/index.php?option = com_content&task = view&id = 99&Itemid = 157
59. Alaska Wind–Diesel Applications Center. www.uaf.edu/acep/alaska-wind-diesel-applic/
60. Hydro Tasmania's King Island Renewable Energy Integration Project. www.kingislandrenewableenergy.com.au.
61. Powercorp. Wind–diesel www.pcorp.com.au
62. Australian Antarctic Division, Mawson station electrical energy. www.antarctica.gov.au/living-and-working/stations/mawson/living/electrical-energy
63. P. Larsonneur. 2010. Realities and promises of wind–diesel systems, European Wind Energy Conference. www.vergnet.com/pdf/vergnet_ewec.pdf
64. Danvest. Hybrid power: wind–diesel www.danvest.com/home.pp
65. C. Brothers. 2011. Wind–diesel and H_2 activities in Canada. www.uaf.edu/acep/alaska-wind-diesel-applic/international-wind-diesel/presentations/CarlBrothersAlaska_AdvancedWindDieselSystems_110310.pdf
66. Tennessee Valley Infrastructure Group. 2002. Economics of megawatt-scale wind–diesel hybrids.
67. Nalcor Energy. 2010. Ramea Report. www.nalcorenergy.com/uploads/file/nalcorenergyrameareport_january2010.pdf
68. International Wind–Diesel Workshop. 2009. Using hydrogen energy storage in remote communities. www.pembina.org/docs/arctic/Wind-Diesel-1-Robert-McGillivray.pdf
69. Ursira wind power and hydrogen plant. www.iphe.net/docs/Renew_H2_Ustira.pdf
70. WEIcan. Wind–Hydrogen village. www.weican.ca/projects/current_projects.php
71. NREL Wind to Hydrogen Project. www.nrel.gov/hydrogen/proj_wind_hydrogen.html
72. V.C. Nelson et al. 2001. *Wind hybrid systems technology characterization r*eport. West Texas A&M University and New Mexico State University. http://solar.nmsu.edu/publications/wind_hybrid_nrel.pdf
73. RETScreen International. Software and data. www.retscreen.net
74. National Wind Technology Center. Energy analysis: HOMER Hybrid 2. ViPOR. www.nrel.gov/analysis/analysis_tools_tech_wind.html
75. Global Village Energy Partnership. www.gvepinternational.org
76. Bergey Windpower. http://bergey.com/global-projects
77. C. Dou, Ed. 2008. *Capacity building for rapid commercialization of renewable energy in China.* CPR/97/G31, UNDP/GEF, Beijing.
78. V. Nelson, R.N. Clark, and R. Foster. 2004. *Wind Water Pumping.* Alternative Energy Institute, West Texas A&M University. CD.
79. J. A.C. Kentfield. 1996. *The Fundamentals of Wind-Driven Water Pumpers.* Amsterdam: Gordon & Breach.

80. J. van Meel and P. Smulders. 1989. *Wind Pumping: A Handbook*. Technical Paper 101. Washington: World Bank.
81. Agricultural Research Service and Alternative Energy Institute. www.windenergy.org
82. Turbex. www.turbex.co.za/index.htm
83. R.N. Clark and F.C. Vosper. 1984. Electrical wind-assist water pumping. In *Proceedngs of Third ASME Wind Energy Symposium*, p. 135.
84. R.N. Clark. 1985. Wind–diesel hybrid system for pumping water. In *Proceedings of Windpower Conference*, p. 221.
85. M. L.S. Bergey. 1990. Sustainable community water supply: a case study from Morocco. In *Proceedings of Windpower Conference*, p.194.
86. R. Thomas and D. Baldwin. 1981. The NASA–Lewis large wind turbine program. SERI/CP-635-1340. In *Proceedings of Fifth Biennial Wind Energy Conference and Workshop*, p. 39.
87. V. Nelson. 1984. SWECS industry in the United States. Report 84-2. Alternative Energy Institute; a history of the SWECS industry in the U.S., *Alternative Sources of Energy*, March/April, p. 20.
88. A.S. Laxson, S.M. Hock, W.D. Musial et al. 1992. An overview of DOE's wind turbine development program. In *Proceedings of Windpower Conference*, p. 426.
89. R. Lynette. 1992. Development of the WC-86 advanced wind turbine. In *Proceedings of Windpower Conference*, p. 450.
90. C. Coleman. 1993. Northern Power Systems advanced wind turbine development program. In *Proceedings of Windpower Conference*, p. 152.
91. Utility Wind Integration Group. www.uwig.org
92. U.S. Department of Energy. 2012. 2011 Wind Technologies Market Report. DOE/GO-102012-3472. www.eere.energy.gov
93. National Wind Technology Center. www.nrel.gov/wind/projects.html
94. Sandia National Laboratory. Wind energy. http://energy.sandia.gov/?page_id=344
95. V. Nelson. 2011. *Introduction to Renewable Energy*. Boca Raton, FL: CRC Press.
96. D. Castelvecchi. 2012. Gather the wind, *Sci. Am.* March, p. 48.
97. Gyuk. 2009. Energy storage status and progress. www.narucmeetings.org/Presentations/Gyuk.pdf
98. Sandia National Laboratory. Energy storage system projects. www.sandia.gov/ess/
99. International Renewable Energy Agency. 2012. Electricity storage and renewables for island power: a guide for decision makers. www.irnea.org
100. R. Foster, M. Ghassemi, and A. Cota. 2010. *Solar Energy: Renewable Energy and the Environment*. Boca Raton, FL: CRC Press.
101. J. Himelic and F. Novachek. 2011. Sodium sulfur battery energy storage and its potential to enable further integration of wind, final report. www.xcelenergy.com/staticfiles/xe/Corporate/Renewable%20Energy%20Grants/Milestone%206%20Final%20Report%20PUBLIC.pdf
102. PowerPedia.Vanadiumredoxbatteries.http://peswiki.com/index.php/PowerPedia:Vanadium_redox_batteries.
103. A. Vieira da Rosa. 2009. *Fundamentals of Renewable Energy Processes*, 2nd Ed. New York: Academic Press.
104. B.D. Vick et al. 2003. Remote solar, wind, and hybrid solar/wind energy systems for purifying water. *Journal of Solar Energy Engineering*, 125, 107.

PROBLEMS

1. What is a wind-assist power system?
2. Go to Global Wind Energy Council (www.gwec.net). Which country has the largest installed capacity of wind power? How much?
3. Estimate the wind installed capacity five years in the future for the world and for your country.
4. Estimate the world capacity of offshore wind farms today.
5. Are there any small wind turbines in your region? If yes, how many are rated power, stand-alone, or grid-connected?
6. Are there any wind distributed or community systems in your region? If yes, provide number of turbines and rated power for one project. If no, find information on the Internet for one system.
7. Should distributed wind systems receive incentives? If yes in your opinion, what incentives should they receive?

Applications and Wind Industry

8. What is the main difference between low and high penetration for wind–diesel systems?
9. Go to the National Wind Technology Center NREL website. What are the Center's current R&D programs?
10. Go to the National Wind Technology Center NREL website. What are the Center's non-R&D programs?
11. Find an example of a village project that is not cited in the text. What type and size systems were installed?
12. In your opinion, what are the major advantages and disadvantages of renewable village power?
13. Why are wind–electric water pumping systems installed instead of proven farm windmills?
14. Compare the annual efficiency of a farm windmill and a wind–electric system.
15. What factors contributed to the initial wind farm boom in California?
16. Which manufacturer is the largest supplier of megawatt wind turbines?
17. Go to two or three manufacturers' websites. State diameter, rated power, and tallest tower for their largest commercial wind turbines.
18. Does the utility from which you buy electricity have wind farms on its transmission lines? If yes, determine the number of turbines and rated power for one wind farm.
19. What are some applications of small wind turbines? Bergey Windpower has examples of applications or use an example from the text.
20. In your opinion, why does the U.S. not have more manufacturers of large wind turbines?
21. Find two examples of wind–diesel projects on the Internet. State specifications of the systems.
22. Compare two criteria for flywheel and battery storage systems.
23. What is the general efficiency of large storage systems for wind farms?
24. Find an example of a large battery system for wind farms. What are the general specifications and cost?
25. List two advantages and two disadvantages of hydrogen storage for wind farms.
26. Find the closest community wind project to your area. List size, general specifications, and performance data.
27. Go to the Small Wind Certification Council website. How many companies produce certified wind turbines?
28. What are the two primary objectives of the U.S. Wind for Schools program?
29. Go to any website for community and/or distributed wind installations that publish performance online. How much energy has the project produced over what time period?
30. For the three 100-kW units at North Texas (Reference 27), how many kWh have they produced since installation? What is the current power output? Note date and time determined.

11 Institutional Issues

The interconnection of wind turbines to utility grids, regulations covering installation and operation, and environmental concerns are the main institutional issues. The U.S. National Energy Act of 1978 was a response to the energy crisis caused by the oil embargo. The main purpose was to encourage conservation of energy and the efficient use of energy resources. The Public Utility Regulatory Policies Act (PURPA) covers small power producers and qualifying facilities (independent power producers) rated up to 80 MW [1,2]. Sections 201 and 210 of PURPA encourage the use of renewable energy. The main aspects of PURPA are:

- Utilities must offer to buy energy and capacity from small power producers at the marginal rate (avoided cost) the utility would pay to produce the same energy.
- Utilities must sell power to these small power producers at nondiscriminatory rates. Qualifying facilities are entitled to simultaneously purchase and sell. They have the right to sell all their energy to the utility and purchase all the energy needed.
- Qualifying facilities are exempt from most federal and state regulations that apply to utilities.

Public utility commissions, utilities, independent power producers, and the courts determined the implementation of PURPA. Determination of avoided costs was the main point of contention among small power producers, independent power producers, and utilities.

The National Energy Strategy Bill of 1992 covered wheeling power over utility transmission lines. The Federal Energy Regulatory Commission (FERC) can order the owners of transmission lines to wheel power at costs determined by FERC. The utilities are allowed to recover all legitimate, verifiable economic costs incurred in connection with the transmission and necessary associated services including an appropriate share of costs incurred for enlargement of transmission facilities.

From a wind power view, this legislation is very important because the Great Plains region is a major source of wind energy and transmission to major load centers is needed if that power is to be utilized. In 1997, FERC opened transmission access.

The state deregulation of the electric utility industry changed the competition for renewable energy. Deregulation essentially means that integrated electric utility companies are split into the areas of power generation, transmission, and distribution and consumers are free to buy from different power producers. The other aspects for increased use of renewable energy are green power and reduction of pollution and emissions from fossil fuel plants that generate electricity.

Cavallo [3] argued that wind energy could become a high-capacity system by wheeling power from the Great Plains to California, or from the Texas Panhandle to Dallas–Fort Worth. He conducted a paper study of a 2 GW wind farm in Kansas that could have a capacity factor of 60%. The first large wind plant (initially 40 MW, expanded to 80 MW) in Texas was in the western part of the state, and power was wheeled to the Lower Colorado River Authority area in central Texas. In 2010, the Electric Reliability Council of Texas (ERCOT) formed competitive renewable energy zones (CREZ) and as a result existing and new transmission lines will increase the transmission of power.

11.1 AVOIDED COSTS

Avoided costs were established by the public utility regulatory bodies in all states. The Federal Energy Regulatory Commission (FERC) defines avoided cost as the incremental or marginal costs to an electric utility of energy or capacity that the utility would have to generate or purchase from another source if it did not buy power from a qualifying facility. Avoided cost reflects the cost of new power plants, not the average cost of plants already installed. The avoided cost includes both present and future costs.

However, many utilities claimed they did not need new generation. In those cases, avoided costs were only fuel adjustment costs. A utility may set a standard purchase rate for qualifying facilities under 100-kW capacity. Contact your public regulatory body for more information about small or independent power production.

In the 1980s, the California Public Utilities Commission (PUC; now CEC) set the avoided costs and types of contracts for qualifying facilities [4]. Standard Offer 4 set the avoided costs for ten years, while Standard Offer 1 was variable, depending on the cost of fuel. One of the reasons wind farms started in California was the high avoided cost set by the PUC.

The fuel adjustment cost (avoided cost) for Southwestern Public Service in the Texas Panhandle in January 1994 was $0.02/kWh. The company was consolidated with a company in Colorado and Minnesota (now Xcel Energy). Until 2008, the avoided cost was still the fuel adjustment cost. In 2008, the fuel adjustment cost increased to around $0.05/kWh due to the increased price of natural gas. Since then, the fuel adjustment cost has declined due to the boom in natural gas produced from shale formations.

The reduction in the price of natural gas also makes the economic justification for new wind farms more difficult because the value of fixed contracts for wind power from independent producers declined. By the end of 2011, Xcel Energy had over 4,000 MW of wind on its system that represents around 10% of the energy provided. Due to renewable portfolio standards, primarily by states, utilities are actively using or building wind farms and some utilities consider wind a hedge against future volatility of natural gas costs. Because natural gas emits less carbon dioxide than coal, it is touted as the fuel for electricity and transportation and as a bridge toward renewable energy.

11.2 UTILITY CONCERNS

A few wind turbines on a large utility grid would present no problems with the amount of power. The power would be considered a negative load—a conservation device equivalent to turning off a load. For large penetration, 20% and greater, other factors such as the variability of the wind and dispatching become important. Utilities are concerned with safety and the quality of power from wind turbines on their grids.

11.2.1 SAFETY

Safety is a primary consideration particularly from energizing dead utility lines, grounding equipment, and lightning. The safety issue has been resolved and large numbers of wind turbines have been connected safely to utility lines. Induction generators have to be energized by the utility line and they do not operate when a fault occurs on the line. Inverters have sensors that disconnect them from utility lines in case of a loss of load.

Of course, safety during installations and operations is of concern, as it is for any industrial enterprise. High voltages, rotating blades and machinery, large weights, and work at heights of 50 to 100 m make for a hazardous workplace. Safety is the first consideration for working around wind turbines. One rule is to never climb a meteorological or turbine tower if you are the only person at a site. Although the large wind turbines have taller towers, climbing inside a tubular tower is easier than climbing on a truss or met tower, especially in inclement weather.

The Caithness Wind Farm Information Forum in the U.K. developed a summary and full list of wind turbine accidents up to December 12, 2012, that shows accident type, turbine, date, and location [5]. The summary notes 1,328 accidents that include ninety-nine fatalities.

The most common causes are falls from turbines. Eighty-three of the fatalities were industry workers, and fifty were public fatalities, most of which resulted from accidents during transportation of wind turbine components. Surprisingly, four people were killed when an airplane crashed into a wind turbine during a fog. The largest number of accidents arose from blade failure; the second most common cause was fire. Because of turbine height, a fire cannot be extinguished. It can only be watched until it burns out.

Safety data, especially early data, are not comprehensive. For example, the Alternative Energy Institute (AEI) at West Texas A&M University and the U.S. Department of Agriculture (USDA) tested more than eighty prototype or first production wind turbines and noted a number of failures, from lost blades to complete destruction, that are not included in the database. The longest distance for a blade failure was 56 m from a small (4 kW) wind turbine, which is quite a bit shorter than the documented 400-m distance noted in the summary. Another fatality that was found on the database occurred when the top of a forklift hit a high-voltage line while moving a rebar cage for a wind turbine foundation and Patrick Acker, the worker holding the cage, was electrocuted.

Wind-Works has a section on accidents and safety [6] plus a database tracking fatal accidents (80 through 2012) in the wind industry. It shows mortality of 0.034 deaths per TWh, which is a significant decline from the 0.4 deaths/TWh for the mid-1990s and 0.15 deaths/TWh by the end of 2000. Of course, some of the deaths related to the transport of wind turbines. The mortality statistics for the wind industry need to be compared to those of other energy industries such as coal.

11.2.2 Power Quality

Power quality involves harmonics, power factor, and voltage and frequency control. A number of wind turbines on the end of a feeder line may require extra equipment to maintain quality of power. Utility companies have to supply reactive power for induction generators, and in general, capacitors on wind turbines or at wind farm substations are required to maintain power factors.

11.2.3 Connection to Utility

A utility should be informed at the earliest possible stage of the intention to connect a wind turbine to its system. Information for the utility should include: (1) wind turbine specifications, (2) block diagram of electrical system, and descriptions of controls for handling loss of load by utility. Even if the arrangement is net energy billing, the utility may require a meter to measures energy flow in both directions.

Liability for damage is another concern of a utility. Most utilities want to be insured against all damage arising from wind turbine operation. Of course, a small power producer wants to be insured against wind turbine damage resulting from utility operations but that is impossible to obtain. Insurance should be available as part of a homeowner or business policy. Some electric cooperatives require proof of a $500,000 liability policy before allowing connection of a wind turbine to their systems.

A utility interconnection study will cost $30,000 to $100,000 and determines the effects of wind farm operation on transmission lines and existing generators. The American Wind Energy Association has a lot of information on utilities and wind power [7].

An example of onerous regulation of small wind turbines comes from the State of Washington, where the Department of Labor and Industries will not approve small wind systems without Underwriters Laboratory (UL) listing for all components. The state now requires a few specially registered electrical engineering firms to certify all existing and planned wind systems that are not UL listed at a cost of around $2,000. This action brought small wind equipment sales to a halt in Washington.

States that provide incentives and/or rebates for small wind installations require or will require certification. See Section 10.2 for more information about small wind turbine certification in the U.S.

11.2.4 ANCILLARY COSTS

Wind farms, especially as they provide more generation capacity on the grid, create other costs for utilities. Wind variability can increase operating costs for committing unneeded generation, scheduling unneeded generation, allocating extra load-following capability, violating system performance criteria, and increasing cycling operation on other generators. Estimates of these costs are $0.001 to 0.005/kWh [8] or even up to 0.0185/kWh. The wind integration impact becomes more significant at higher wind penetration into the grid [9].

In 2008, the Montana Public Service Commission set a rate up to $0.00565/MWh for integrated wind power into the Northwestern Energy utility from the Two Dot wind farm. The integration rate is subtracted from the amount Northwestern Energy pays the wind farm for power and thus reduces the utility payment for wind-generated electricity as low as $44.25/MWh.

A major storm in Spain with winds above the cut-out wind speed caused a major drop in output from wind farms of 7,000 MW compared to the predicted input to the utility grid operation for that day. In another case, wind farms produced 53% of the total demand in Spain for five hours (November 2009) when there was ample wind and low demand on the grid during early part of the day. In November 2012, the ERCOT system in Texas set a record wind power output of 8,521 MW—26% of the load.

11.3 REGULATIONS

Regulations for renewable projects vary by country, region, and state from simple reviews by a single agency to multiple complex reviews by various agencies and even multiple levels of government. Sometimes agencies pass conflicting regulations. National laws and policies may restrict connection of a renewable energy system to a utility grid. Most large projects require consideration of environmental impacts although enforcement practices may vary widely.

In the U.S., federal permitting requirements range from environmental restrictions to Federal Aviation Administration regulations on lights for tall towers and wind turbines taller than 200 ft. Industry maintains that regulations now represent major shares of their costs of doing business. In most cases, industry claims it cannot meet proposed regulations because they are uneconomical.

Permits are required for construction in residential areas and even in rural areas in some states. The major zoning issues are tower height, setbacks, noise, aesthetics, environmental impact, and safety. The probability of failure, for example, because of a thrown blade, is the most common objection, even though risks from cars and utility lines are accepted. Signs, trees, and even utility poles have failed in high winds or under conditions of icing.

Tower access and access to high voltage equipment must be controlled. One certain fact is that any incident that interferes with television reception will be unacceptable to the public. When the metal blades on the MOD-0A interfered with reception on Block Island, Rhode Island, the DOE had to install cable for the residents. Most locations have no specific zoning regulations for wind turbines. Before installing a small wind turbine, be prepared to educate public boards and residents [10–14].

11.4 ENVIRONMENT

Environmental issues surround all large projects and vary by location. Some may apply to small wind turbines. An expensive environmental impact analysis and/or study, particularly related to impacts on fish and wildlife will be required for any project (even small wind turbines) receiving federal funding in the U.S. At the end of a project lifetime, decommissioning, recycling, and disposal, especially of toxic components must be considered. The goal of the U.S. Fish and Wildlife Service is to protect wildlife resources, streamline site selection, and attempt to avoid environmental problems after construction. The service issued guidelines on minimizing the impacts of land-based wind farms on wildlife and habitats [15].

Land areas that are excluded from use because of environmental considerations include national and state parks, wetlands, and some wildlife refuges. In the U.S., environmental impact statements are required and the Environmental Protection Agency has jurisdiction over many aspects of wind farm location. Some states and even counties have environmental requirements that must be met before a wind turbine or farm can be constructed. The first step is to check local requirements.

A developer should consider the environment, permits, licenses, and regulatory approvals; threatened or endangered species; wildlife habitats; local avian and bat species; wetlands and other protected areas; and locations of known archaeological and historical resources. Geographical Information Systems are excellent tools for depicting environmental and land use constraints. Regulations on archeological sites differ by state and may not apply to private lands. A search for archeological site information should be performed for a medium or large wind farm site even if it is not mandatory.

After the first analysis of environmental issues, a more detailed analysis should address possible impacts and their mitigation. After a project is operational, mitigation of the impacts must be monitored regularly. Biological concerns are habitat loss, alteration or fragmentation of habitats, bird and/or bat collisions with wind turbines, electrocution of raptors, and effects on vegetation. Water, especially wetlands, soil erosion and water quality must be considered. For wind farms, the clearing of scrub brush for roads, sites, and laying underground wires is welcomed by ranchers. However, the cleared areas such as road shoulders must be seeded and monitored for growth, erosion, and noxious weeds. Another possibility in complex terrain is that maintained roads may be welcomed by land owners/operators and they can serve as firebreaks.

The main environmental issues are visual impacts, noise, birds, and bats. The visual impact can be detrimental, especially in locations near scenic areas, and people are in favor of renewable energy as long as it is "not in my backyard." Turbines should painted with non-reflective paint in drab colors. All the rotors should rotate in the same direction.

Some people are adamantly opposed to wind farms, most are neutral, and the rest are in favor. For those opposed, the visual impact is generally the greatest concern [16]. The photo gallery of Stopillwind has before-and-after photos of wind farms. Wind farms produce economic returns both directly and indirectly. A wind farm developer should educate the local community as soon as a project is planned. Economic development in rural areas is very powerful from a political standpoint.

Noise measurements have shown that wind turbines fall below the ambient noise levels. However, the repetitive sound from the blades is obvious and no one would want to live in the middle of a wind farm. The whine from gearboxes on some units is also noticeable. However, much noise has been reduced by larger wind turbines at higher hub heights and new airfoils. Farmers who live near the wind turbines at the White Deer wind farm in Texas (80 1-MW turbines) report that noise is not a problem. Some residents near the Waubra wind farm in Australia report health problems caused by living among wind turbines.

The rotor area for a 90-m diameter wind turbine is over 6,000 m^2, 10 rpm speed, and blade velocity differs from the root to the tip. Large wind turbines revolve more slowly than small wind turbines, but their tip speed ratios are similar. Both blades have sufficient velocity to kill a bird.

Parameters related to birds include fatalities caused by wind turbines, species, season, threats, and possible mitigation measures [17]. Collision rates per turbine per year vary from 0.01 to twenty-three. The twenty-three collision figure was for a coastal site in Belgium that had a large population of gulls, terns, and ducks. Annual average collision rates for other coastal sites in northwest Europe ranged from 0.01 to 1.2 birds per turbine. None of these examples has been associated with significant population declines. Flocks of geese and ducks entering an offshore wind farm decrease by a factor of 4.5. At night, more migrating flocks entered the wind farm but increased their distances from the wind turbines. Overall, fewer than 1% of the ducks and geese flew near enough to the turbines to be at any risk of collision [18].

In general, migratory birds fly well above the heights of wind turbines [19] although overcast and ground clouds may lower flight paths. Two large wind farms near the Texas coast south of Corpus

Christi use radar to monitor migrations of birds. The turbines are shut down if they pose threats to the birds.

Avian mortality became an issue at Altamont Pass (California) after wind turbines killed some raptors. Transmission line poles were capped to prevent the birds from using them as perches, extending their wings between the lines, and being electrocuted. Xcel Energy agreed to evaluate 90,000 miles of transmission lines in twelve states to fix any equipment likely to kill birds.

The primary areas of concern are (1) possible litigation resulting from killing even one bird protected by the Migratory Bird Treaty Act or the Endangered Species Act, and (2) the effect of avian mortality on populations. A number of projects [20] have been funded since 1994 to determine the effects of rotating blades on raptors and find methods to make them repel birds. Truss towers make natural perches since wind farms lack trees. One wind farm utilized tubular towers to prevent perching and most large turbines now have tubular towers. NREL has a section on avian studies [21].

Southern Spain is another area that experiences bird problems [22]. Tarifa is a temporary roosting area for birds migrating to and from Africa. Biologists believe the problem of avian mortality at the site is partly due to aerodynamics. The soaring birds travel the air currents that propel them up the ridges where the wind turbines are. The large birds do not have the maneuverability of the smaller birds.

Based on the situation at Tarifa, it is obvious that some locations should be off limits to wind farms. Wind farms should not be located next to refuges for endangered bird species such as whooping cranes. Although thousands of birds are killed by communication towers, buildings [23], hunters, and even cars, the Sierra Club and other environmental groups will become adversaries if birds are killed. Of the hundreds of millions of birds killed annually in the U.S., how many are killed by wind turbines? After bats became a problem in West Virginia, guidelines covering bird and bat impacts were developed [24,25]. Wildlife–wind interaction publications are available from the National Wind Coordinating Collaborative. One provides guidance on preconstruction utilization counts for making predictions and conducting post-construction fatality studies [26].

As expected, fatality rates for birds vary by the characteristics of a wind farm and the surrounding area. In Altamont Pass, raptors suffered the highest fatality rates. Outside California, studies at twelve wind projects estimated fatality rates from 0.63 to >10 per turbine per year at a fragmented mountain forest site in Tennessee. Bat fatality rates are estimated from a low of 1.5 per turbine per year for most of the U.S. to a high of forty-six for the eastern United States [27].

National Wind Coordinating Collaborative (NWCC) members are from utilities, state legislatures, state utility commissions, consumer advocacy organizations, wind equipment suppliers and developers, green power marketers, environmental organizations, and state and federal agencies. Permitting publications are available from NWCC [26].

The issues of regulatory framework, environment, and impact analysis and mitigation are covered in AWEA's *Siting Handbook* [28]. The information applies to projects of 5 or more megawatts but is still useful for smaller projects. Early in the siting process, a developer should conduct a critical analysis of the environmental issues such as permits, licenses, and regulatory approvals; threatened or endangered species and habitats; avian and bat species; identification of wetlands and other protected areas; and locations of known archaeological and historical resources. A constraints map is a useful tool for depicting environmental and land use constraints.

Regulations from federal to local levels play parts in every project. Federal permitting requirements for wind energy projects range from environmental to aviation matters. There is coordination at the federal level for recommendations and regulations [29]. A check on archeological sites is generally imperative during planning for wind farms. Regulations differ by state and some states exclude private land. A developer would be wise to learn about archaeological requirements. NREL in collaboration with the National Association of Counties created a helpful guide for county commissioners [30].

Institutional Issues 273

The visual impacts for large wind farms are very different from the impacts of from those of small wind turbines because of the numbers and heights of towers. Wind farms may be visible from 20 km and they are visible from all directions in the U.S. plains (Figures 11.1 and 11.2). Only the curvature of the earth limits viewing distance. At close range, wind turbines dominate the landscape (Figure 11.3).

In mountainous areas, wind turbines are arranged in lines on the ridges, but in general they are not visible from all angles because most roads are in the valleys and the views are blocked. The moving rotors make wind turbines more visible. The requirements to install lights on towers over 60 m make the turbines conspicuous at night, especially when the flashing red lights are

FIGURE 11.1 Visual impacts of various sizes of wind turbines at different distances. Photo taken in late afternoon looking south. Foreground: 3.2 km distance to 1 turbine, diameter 90 m, 3 MW, tower height 80 m. Middle at left: 6.4 km distance to 8 turbines, diameter 64 m, 1.25 MW, tower height 72 m. Background: 9.5 km distance to first row, 14.5 km to back row, 38 turbines, diameter 88 m. 2.1 MW, tower height 80 m.

FIGURE 11.2 Visual impact of wind farm. Near edge is 11 km, far edge is 17 km. Photo taken in late afternoon looking east. Wind turbine diameter 56 m, 1 MW, tower height 60 m.

FIGURE 11.3 Turkey Track wind farm south of Sweetwater, Texas, on Highway 153. Wind turbine diameter 77 m, 1.5 MW, tower height 80 m. Near wind turbine is around 1 km away. Wind turbines in background on horizon are on another wind farm.

synchronized to outline the wind farm. Shadow flicker happens and the high impact is generally located within approximately 300 m of a turbine.

The Denmark requirements include a limit on wind turbine shadow flicker on neighboring houses not to exceed ten hours per year. If the shadow limit is exceeded, a turbine owner may be required to shut down the wind turbine in critical periods. In a pasture with no trees in the summer, a rancher noticed that yearling calves at the New Mexico Wind Energy Center lined up in the shadows of the towers and moved to remain in the shade as the shadows moved. For information about the visual impacts of small wind turbines see Section 9.1.2.

The noise from large wind turbines is much less than in the past, and in general the permitted levels at property boundaries are established by most states and localities. The most prominent noises are caused by the movements of the blades and also from components, primarily gearboxes.

One issue that is often overlooked is the traffic from the large trucks hauling the wind turbines and the cranes to the project and then the numerous pickups, both during and after construction. Routes from source and delivery ports become important, and invariably local roads have to be improved and the question is who pays for the improvements. For mesas and complex terrain, most ranchers like the new roads because they are maintained by wind farms.

The amount of activity and number of people involved in the construction phase and the amount of space and equipment required are surprising to rural communities. During construction, cattle guards may have to be installed because opening and closing gates take too much time. Livestock may be injured or killed and damages will have to be paid. Finally, solid and hazardous wastes generated during construction and operation must be managed. At the end of a project's life, environmental problems may arise during dismantling.

For protection against liability, a developer should perform a screening assessment or an environmental site assessment before acquisition of a property. The American Society for Testing and Materials has screening tools and standards for environmental site assessment [31].

11.5 POLITICS

Every endeavor produces political issues and activities. To change a behavior, especially in an entrenched industry, you need incentives, penalties, and education. Someone estimated that the amount of each type of energy used is in direct proportion to the subsidies for that type of energy. Subsidies take the forms of taxes, tax breaks, and regulations and all these forms generally require legislation. Every company wants incentives for itself and penalties for its competitors. Industries also want governments to fund their research and development and even assist with commercialization.

Incentives are usually in the form of tax breaks, subsidies, mandates, and regulations. Public utility commissions now demand that utilities use integrated resource planning and must consider renewables and conservation in their planning processes. Can utilities make money on kilowatt hours saved? Should the consumers or shareholders take the risks? Three Mile Island in the U.S., Chernobyl in Russia, Fukushima in Japan, and the nuclear utility industry in general are good examples of political impacts from the local to the national level. The federal Price Anderson Act limited the amount of liability from nuclear accidents. Without that legislation, the nuclear industry could not have sold plants to utilities.

Penalties are generally in the form of taxes and regulations. U.S. environmental groups have already shown that utility planners will be held accountable for the risk of a carbon tax if they plan new coal plants. In other words, the environmental group opinion is that the shareholders (not consumers) should shoulder the risks.

Education creates public awareness of the possibilities and options and should provide a realistic cost and benefit comparison over the lifetime of an energy system. Entrenched industries try to paint adverse science as questionable or theoretical, for example, cancer caused by tobacco use and global warming caused by greenhouse gases. No one can fool Mother Nature and humans will pay one way or another.

Politics will continue to influence which and how many different energy sources are subsidized. Present energy policies include taxes or limitations on greenhouse gas emissions, rebates on equipment, and incentives for electrical energy produced from renewable energy, renewable portfolio standards, set prices for renewable energy (feed-in tariffs), and tax credits.

11.6 INCENTIVES

Energy subsidies produce serious effects and generally favor conventional fossil fuels and established energy producers. Subsidies for renewable energy between 1974 and 2000 amounted to over $20 billion worldwide. This compares with the $300 billion per year paid to conventional energy sources without accounting for the costs of infrastructure, safeguards, and military actions [32]. The privatization of the electric industry along with the restructuring into generation, transmission, and distribution opened some doors for renewable energy. For 2011, the estimated incentives were $88 billion for renewable energy and $523 billion for fossil fuels. It would be interesting to know what incentives benefitted nuclear power.

11.6.1 UNITED STATES

The major impetus to the wind industry came from federal tax credits, the National Energy Act of 1978, and the avoided costs set by the California Public Utilities Commission. The federal tax credits for wind turbines were available from 1980 to 1985. For small systems for personal use, the tax credits were 40% of the cost, up to a maximum of $4,000. For a business, the tax credits were 25% off the bottom line. During this period, tax shelters for California wind farms were the primary methods of financing.

The National Energy Strategy Act of 1992 provided a $0.015/kWh incentive for production of electricity by wind energy. An investor can claim the production tax credit (PTC) under Section 45 of the Internal Revenue Service Code [33] if the investor: (1) owns a wind facility that was placed in service between December 31, 1993 and July 1, 1999; (2) produces electricity at the wind facility; and (3) sells the electricity to an unrelated party.

The credit applies to production through the first ten years of operation. It is intended to serve as both a price incentive and a price support. The credit is phased out as the average national price exceeds $0.08/kWh, based on the average price paid during the previous year for contracts entered after 1989. Both values will be adjusted for inflation. The credit can be carried back three years and carried forward fifteen years to offset taxes on income from other years. The PTC was extended several times, now through 2013 and was a major factor in the large increase in wind power in the U.S. starting in the 1990s.

The provision of direct payments (Renewable Energy Production Incentives) to public utilities, co-operatives, and Native American tribes is equivalent to the PTC arrangement. Congress has to fund the incentive program every year. The funds provided may be less than requested and wind projects must compete with other renewable projects.

The federal government continues to support wind energy through the Department of Energy budget for Energy Efficiency and Renewable Energy [34]. As always, the budget for renewable energy is less than the budget for nuclear energy. In 1973, the amount designated for wind was $300,000, and that increased steadily to $67 million in 1980. During Reagan's term, the amount was reduced every year, and in 1988 the amount budgeted was $8 million. Increases have been granted since then and the budget for wind for the 2012 fiscal year was $93 million (not including projects funded by the American Recovery and Reinvestment Act). A major part of the funding went toward development of large HAWTs.

The tone or direction is set by the administration and changes under every president. The early direction was R&D plus demonstration projects that were supposed to lead to commercialization. During the Reagan years, *commercialization* was a bad word and private industry was supposed to commercialize wind turbines. Federal funding was for generic R&D topics such as aerodynamics and wind characteristics. Funding increased slightly during the George Bush term, and the advanced technology program was initiated in an attempt to recapture some of the market acquired by foreign wind turbine manufacturers.

Under Clinton, interest in renewable energy increased and the direction was commercialization. The Climate Change Action Plan moved DOE's focus on technology development to an active role in renewable energy commercialization. This initiative was backed with $72 million for fiscal year 1995 ($18 million for wind) and a total of $432 million through 2000. DOE started looking to wind to achieve emissions reductions from renewables since this was and is the most economical resource.

Under George W. Bush, the national energy plan first focused on increased production of oil and gas. With pressure from Congress, conservation, energy efficiency, and renewables became part of the package. However, a mandated increase in fuel efficiency for the automotive industry (the CAFE standards) did not pass until the last year of the G.W. Bush presidency.

When money is available, every university and federal laboratory wants some of it. New institutes and consulting groups proliferate. The wind money in the early years was divided among the following programs:

Large HAWTs (>100 kW)	NASA Lewis
Small wind turbines (<100 kW)	Rocky Flats, Rockwell International
Vertical axis wind turbines	Sandia Labs
Wind characteristics	Battelle Pacific Northwest Laboratory
Innovation wind turbines	Solar Energy Research Institute
Agricultural applications	U.S. Department of Agriculture

The Wind Energy Research Center at Rocky Flats was in charge of the small wind systems program. The location was chosen because of politics (too much publicity about environmental problems at the plutonium facility). Early in the program, the center purchased units for testing and started a field evaluation program [35] to install two units in every state and the territories—definitely a political plus. After forty units were installed, the program was abandoned due to costs and also because the wind turbines from small wind manufacturers were not ready.

The small wind machine program was transferred to the Solar Energy Research Institute (SERI). NASA–Lewis retired from the large HAWT program, transferring what was left to SERI. The president designated SERI as the National Renewable Energy Laboratory (NREL), on an equal footing with the other national laboratories originally created to study the development of nuclear weapons and high-energy physics. The expected outcome was that NREL would absorb all the other programs associated with renewable energy, although Sandia continued its VAWT program. As always, political infighting ensued. Today, NREL's National Wind Technology Center performs R&D and administers most wind energy programs.

A 1999 initiative was Wind Powering America [36], whose goals were to meet 5% of U.S. energy needs with wind (80,000 MW installed) by 2020; double the number of states with 20 MW of wind capacity to 16 by 2005; triple it to 24 by 2010; and increase wind's contribution to electricity production to 5% by 2010. Subsequently, the secretary accelerated the DOE 5% commitment to 2005. Achieving the 80,000 MW goal would entail a $100 billion investment along with $1.5 billion invested in rural economic development where wind resources are the greatest. Note for comparison that the 2012 installed capacity is 60 GW, thirty-five states exceeding 20 MW (fifteen exceeding 1,000 MW), and about 2% of U.S. electricity produced by wind.

11.6.1.1 State Incentives
States also compete for renewable energy as a way to offset importation of energy and also create jobs. The Database of State Incentives for Renewable Energy (DSIRE) is a comprehensive source of

information about state, local, utility, and selected federal incentives that promote renewable energy [37]. Overview maps and tables are classified by incentive types and policies.

Minnesota passed legislation requiring Northern States Power (now part of Xcel Energy) to acquire 425 MW of wind power by the year 2002 in exchange for permission from the legislature to store waste from its Prairie Island nuclear facility in dry casks outside the plant. In 2012, Xcel had over 4,000 MW of wind power on its system. After the success of the Renewable Portfolio Standard in Texas, twenty-nine states passed similar measures by 2012 and another eight passed renewable portfolio goals.

Texas legislation allowed the Lower Colorado River Authority to acquire renewable energy from plants located on state lands outside the authority's service territory. This paved the way for a 35-MW wind farm (1995) in the Delaware Mountains of the Trans-Pecos region, with an extension for another 200 MW; 40 MW has been installed.

Some states mandated deregulation of the electric utility industry. Besides giving the consumers choices of producers, most states impose a system benefits charge (SBC) that lets utilities recover stranded costs of power plants, primarily nuclear plants. In some states, part of the SBC is set aside for renewable energy. For example, in California, funds from SBC are available to offset part of the costs for small wind systems.

The wind farm boom in Texas was fueled by the production tax credit and a renewable portfolio standard (RPS), enacted in 1999 as part of electric restructuring. The mandate was for 2,000 MW of new renewables by 2009 in the following amounts and two-year steps:

Year	MW
2003	400
2005	450
2007	550
2009	600
Total	2000

Because so much wind power was installed, the RPS was increased in 2005 to 5,880 MW by 2015 (again by two-year steps), with a goal of 10,000 MW by 2025. The RPS also mandated 500 MW from other renewables. For the 2015 mandate, the amount to be produced by wind was surpassed in 2008, as wind farm capacity was 7,611 MW and the goal of 10,000 MW by 2025 was surpassed in 2011.

Another aspect of electric restructuring in Texas is that electric retailers must acquire renewable energy credits (RECs; 1 REC = 1 MWh) from renewable energy produced in Texas or face penalties of up to $50/MWh. Anyone may participate in the REC market: traders, environmental organizations, and individuals. The RECs are good for the year created and bankable for two years. The market opened in January 2002, and early prices were around $5/REC. In 2008, RECs sold for $5 to $8 but by 2012 they declined to around $2, primarily due to the large amount of wind installed. The price for RECs at the national level had fallen below $2.

As always, industries seek tax breaks at every level. States and local entities give tax breaks for economic development. Wind farm developers want property tax breaks or abatements on installed costs because taxes and installations are both major costs. Conventional power producers can deduct the cost of fuel, whereas renewable energy producers have no deductions because their fuel is free. Tax abatements have become common.

11.6.1.2 Green Power

Green power is the result of a voluntary consumer decision to purchase electricity supplied from renewable energy sources or contribute funds for utilities to invest in renewable energy development. Green power is an option in some states' policies and also has been driven by responses of utilities to customer surveys and town meetings. Green power is available to retail or wholesale customers in forty-two states [38] and eleven states mandated options for utility green power.

In the early 1990s, a small number of U.S. utilities began offering green power options to their customers. A consumer had to pay a premium of about $3 per month for a 100 kWh block or $0.03/kWh. This represented a powerful market support mechanism for renewable (mainly wind) energy development. By 2007, more than half of all U.S. electricity customers had options to purchase green power from more than 750 utilities (about 25%) nationally [39]. Of course, those numbers were larger in 2012.

It is interesting that some utilities lowered the rate premium on green power as traditional fossil fuel costs increased. As of December 2010, the low prices for green power at national level ranged from $0.0014 to 0.008l/KWh. As green power becomes cheaper than regular power, will those consumers who purchased green power pay less than the regular rate? NREL ranks the utility green power programs annually [40].

11.6.1.3 Net Metering

If the renewable energy system produces more energy needed on site, the utility meter runs backward; if the load is greater, the meter runs forward. The bill is determined at the end of the time period—generally one month. If the renewable energy system produced more energy over the billing period than was used on site, the utility company pays the avoided cost. Most states have net metering that ranges from 10 to 1,000 kW, with most in the 10- to 100-kW range [37].

In general, net metering did not increase sales of wind turbines because the small 10- to 50-W turbines are not cost competitive with retail electricity. Larger-sized wind turbines can be cost competitive for users with large loads where all the electricity is used on site. Because the value set for avoided cost is generally only equal to the fuel adjustment cost, a company wants to use that energy on site because it displaces energy at the retail rate. Also, if the time period could be set longer than one month, net metering would be more useful to producers. This is especially true for irrigation that creates large demand in the summer when most of the U.S. experiences low winds.

Of course, utility companies do not like net metering because it increases the billing problem. The utilities claim that one group of customers is subsidizing another group. With electric restructuring, utilities are worried that large customers will find cheaper electricity and rates will rise for residential customers. Does that mean that many residential customers are subsidized today?

11.6.2 OTHER COUNTRIES

Several European countries started wind energy programs in the 1980s, with most emphasizing megawatt wind turbines. The programs had little success. The manufacturers in Denmark produced small to larger units in steps and acquired around 50% of the early U.S. market and 66% of Europe's installed capacity in 1991. Today European manufacturers have captured a major share of the world wind farm market although China manufacturers are catching up due to their large domestic market.

Different policy options for renewable energy apply in the European Union (EU) [41–43]. The effectiveness and efficiency of current and future support for renewable energy for producing electricity were analyzed [44]. Free trade in renewables in the EU market is complicated by the fact that renewables are supported by mandates or fixed prices at different levels by country and even state. This support may be regarded as a substitute for a pollution tax on fossil fuels.

Promotion of wind energy in Europe was based on two models: (1) price support for kilowatt hour production (feed-in tariff) and (2) quota or capacity-based (Table 11.1). The quota concept is similar to a renewable portfolio standard. In general, the minimum base price led to the most installations [45,46].

Paul Gipe of WindWorks wrote several articles about electricity feed laws and feed-in tariffs that discuss renewable tariffs by county, reviews of books about feed-in tariffs, and relevant links [47]. Electricity feed laws (EFLs) permit the interconnection of renewable sources of electricity with utilities and specify how much will be paid for the electricity and for how long. EFLs are widely used in Europe where they resulted in large amounts of installed wind power, most notable

TABLE 11.1
Models of Utility Compensation in Europe

Country	2003 €/MWh	2011 €/MWh
Feed-In Tariff		
Netherlands	9.2	8.0
Germany	6.6 to 8.8	9.0
France	8.4	3.0
Portugal	8.1	7.0
Austria	7.8	10.0
Spain	6.4	6.6 to 7.9
Greece	6.4	8.7 to 10.0
Italy		13.5
United Kingdom		4.7 to 9.9
Quota		
Italy	13	
United Kingdom	9.8	

in Germany and Spain. Advanced renewable tariff values are classified by technology, application, project size, and/or resource intensity.

Denmark's windmill law requires electric utilities to purchase energy from private wind turbine owners at 85% of the consumer price plus ecotax relief of about $0.09/kWh. Electric utilities receive about $0.015/kWh as a subsidy for wind power. The development of wind power was tied to the Energy 21 goal of reducing CO_2 emissions 20% by 2005.

Germany accounted for half the European market after 1995. It adopted the Electricity Feed Law (EFL) in 1990 as a measure to protect climate, save fossil fuels, and promote renewable energy. The law requires utilities to buy renewable energy from independent power producers at a minimum price defined by the government. The minimum price is based on the average revenue of all electricity sales in Germany. The initial value in 1991 was €0.16/kWh. The EFL was modified in 1998 to set a regional cap of 5% for renewable electricity. Since the Renewable Energy Sources Act was enacted, electricity generation from renewable sources in Germany increased from 6% in 2000 to 25% in 2012, and much of that is generated by wind.

Earlier programs for promoting wind (100 MW and expansion to 250 MW) mandated kilowatt hour support. Because so much wind was installed, the figure was changed in 2004 to €0.085/kWh for five years and €0.055/kWh for the following fifteen years. A decrease of 2.5% per year meant in 2010 that the price would be €0.079/kWh for five years and then €0.05/kWh for fifteen more years. Some states in Germany also implemented 50% investment grants in the late 1980s and early 1990s. Special low-interest loans for environmental conservation measures were also available for financing wind projects. These factors contributed to the massive growth of wind in the 1990s in Germany—ranked first in the world in installed capacity.

China now ranks first due to favorable government policies and a law that requires energy companies to purchase all the electricity produced by the renewable sector. The feed-in tariff is now around €50/MWh. The wind industry relied on financing through the Clean Development Mechanism (Kyoto Protocol) that presented some problems for new projects.

As in other countries, transmission must be upgraded from windy areas to the load centers. Some regions experienced curtailments up to 20%. Presently China mandates that 70% of the components of wind turbines erected in the country be produced by factories in the country. To meet the goal of

3% of electricity from non-hydro renewables by 2020, 100 GW of wind must be installed. That goal will be reached before 2020 since China had 76 GW installed by the end of 2012.

India ranks fourth in installed capacity based on a favorable fiscal environment and policy measures. In the past ten years, wind power development in India has been promoted through R&D, demonstration projects, programs supported by government subsidies, fiscal incentives, and liberalized foreign investment procedures. The central government provides income tax holidays, accelerated depreciation, duty-free imports, energy capital, and interest subsidies. State government measures include buybacks, power wheeling and banking, sales tax concessions, electricity tax exemptions, demand cut concessions for industrial consumers that establish renewable power generation units, and capital subsidies. Tamil Nadu and several other state boards purchase wind energy at about $0.064/kWh.

11.7 EXTERNALITIES

Externalities are defined as social or external costs and benefits attributable to an activity, the costs of which are not borne by the parties involved in the activity. Externalities are not paid by the producers or consumers and are not included in market prices, although someone will pay or be affected by them at some point.

Social benefits (called subsidies) are paid by a third party and accrue to a group. An example is the Rural Electrification Act that brought electricity to rural U.S. An example of a positive externality (social benefit) is the cleaner air resulting from installation of wind farms. A good example of a negative externality is the use of coal in China. Every city of with a population of 100,000 or more has terrible smog because of all the coal used for heating, cooking, industrial production, and generation of electricity. China will face huge public health costs in twenty years when today's children are adults.

External costs can be classified as (1) hidden costs borne by governments including subsidies and R&D programs; (2) costs associated with pollution arising from health problems, environment damage from acid rain, destruction of the ozone, and unclean air, and lost productivity. Although global warming due to CO_2 emissions is disputed by many in industry and some scientists, it will have far-reaching effects. Government and other mechanisms for including externalities into the market fall into several categories.

Regulation — This historical approach created inefficient and monopolistic industries and made them inflexible and highly resistant to change. The current philosophies are deregulation and privatization of energy industries. However, if external costs are not included, short-term interests prevail. Regulations can require a mix or minimum use of energy sources with lowest life cycle costs that include externalities.

Pollution taxes —Governments can impose taxes on pollution generated. European countries utilize such taxes. Another possibility is allowing renewable energy credits for producing clean power. Pollution taxes and avoidance of pollution have the merit of simplicity. They exert only marginal effects on energy costs, but they are not true integrations of external costs into market prices. The taxpayer pays instead of the consumer. The pollution tax could be added to consumer bills and be paid based on energy use.

Integrated resource planning (IRP) — This model combines the elements of a competitive market with long-term environmental responsibility. An IRP mandate from the government would require the selection of new generating capacity to include all factors, not just short-term economic ones.

Subsidies — These measures promote research and development and production advances.

Many studies on externalities have been conducted. The European Union's six-volume *ExternE: Externalities of Energy* is probably one of the most systematic and detailed studies to evaluate the external costs for a range of fuel cycles [48]. In the estimates, external costs for production of electricity by coal can be as high as $0.10/kWh, and for nuclear power, $0.04/kWh.

TABLE 11.2
2009 Values of U.S. Generation of Electricity and Air Emissions

		Carbon Dioxide	Sulfur Dioxide		Nitrogen Oxides		
	MWh 10⁶	kg/MWh	Tons 10⁶	kg/MWh	Tons 10⁶	kg/MWh	Tons 10⁶
Coal	1764	950	1676	3.0	5.3	1.5	2.6
Oil	931	710	661	2.7	2.5	0.7	0.6
Natural gas	284	480	136	0.003	0.001	0.6	0.2
Total	2979		2473		7.8		3.4

Source: Megawatt-per-hour and metric ton data U.S. Department of Energy's Energy Information Administration. Emissions factors estimated from various sources.

Since 1995, U.S. companies have been trading sulfur dioxide (SO_x) and nitrogen oxide (NO_x) emissions. Both compounds are precursors of acid rain and contributors to ground-level ozone and smog. Essentially, industries trade allowance units that may be bought, sold, or banked for future use. Carbon dioxide trading [49] does not exists in the U.S., but some states are now passing laws to reduce CO_2 production and it is highly probable that CO_2 emissions will be covered by federal legislation. The largest emitters of carbon dioxide are China and the U.S. [50].

U.S. emissions from generation of electricity are primarily due to the burning of coal (Table 11.2). The emission factors in the table were adjusted somewhat to show total values of metric tons calculated by the Energy Information Administration [51]. For example, the average value for sulfur dioxide from coal is 3.0 kg/MWh and the worst coal plant in the U.S. produces 18 kg/MWh. New coal plants are equipped with scrubbers, but almost 40% of the coal plants do not follow the same pollution control standards because they were online before enactment of the Clean Air Act of 1970.

The average carbon dioxide emission is around 720 kg/MWh for all fuel types and of course coal emissions are higher—around 1,000 kg/MWh. Wind turbines reduce carbon dioxide emissions by one metric ton per megawatt hour when displacing coal generation, Furthermore, wind turbines do not require water to generate electricity. Natural gas production of electricity has increased in market share. Carbon dioxide emissions per megawatt hour are smaller, so the average has decreased. A Minnesota group (connected with the utility industry) estimated the external costs for carbon dioxide from coal as only $0.34 to $3.52/ton. In Europe, carbon dioxide emission reductions at one time were worth €30/ton in some countries. In 2012, the values on the trading market ranged from €10 to 20/ton [52].

11.8 TRANSMISSION

A major problem for wind farm development is that many load centers are far away from the wind resource, and wind farm projects can be brought online much faster than new transmission lines can be constructed (estimated at ten to twenty years). A number of large transmission projects have been proposed in the United States [53, 54]. New transmission lines will have to be constructed for the major wind farm developments in the Great Plains.

NREL's transmission grid integration section includes transmission and two of the main reports are Eastern and Western integration studies [55]. A National Interest Electric Transmission Corridor [56] is a geographic region designated by the Department of Energy (DOE) to help acquire rights of eminent domain for transmission lines. In 2007, the Mid-Atlantic and Southwest corridors were defined. A large transmission investment of $13 billon would increase a retail bill of $100 by about $1.

For states with electric restructuring, transmission is now a separate company and the question is jurisdiction (who pays for new lines). If curtailment is needed, who is curtailed and what are the curtailment priorities? Curtailment happens when wind farms produce more power than the transmission lines can carry; therefore, some or all the wind turbines on a farm have to be shut down.

In the McCamey area of West Texas and also in an area of large numbers of wind farms from Abilene to Snyder, curtailment of output from wind farms was a problem. Some of the electric restructuring in Texas in 2005 was intended to establish Competitive Renewable Energy Zones (CREZs) managed by the Public Utility Commission [57]. Project information maps show lines and substations and quarterly reports publish construction timelines and costs. Two of the zones in the Panhandle (major wind region) are not within the jurisdiction of Electric Reliability Council of Texas (ERCOT), but are in the Southwest Power Pool (SPP). The transmission lines between can only be connected by DC interties. The SPP [58] is also considering and constructing new transmission lines (Figure 11.4), again primarily for wind power.

The Panhandle of Texas will have wind farms that may be connected to either of the two major transmission systems. One of the primary objectives of CREZ was to bring wind power from West Texas and the Panhandle to the load centers in the ERCOT system. The new lines for ERCOT will have a capacity of 18.5 GW. This represents about 11 GW of wind capacity added to the ERCOT system (9 GW when CREZ was adopted). Construction of the high voltage transmission lines (345 kV) started in 2009 and will essentially be completed at the end of 2013. The progression from planning to the end of construction was about five years—much less than the national average because ERCOT is not under FERC regulation. Even with new transmission lines, wind farm development will still be limited by transmission capacity.

The Transwest Express Transmission Project will deliver Wyoming wind energy via a 600-kV high voltage, direct current (HVDC) line to the desert southwest region of the U.S. Construction of the 1,170-km (725-mile) line is expected to start in 2014 and take around three years. The estimated cost is $3 billion. The Western Grid Group [59] proposed more than twenty major transmission

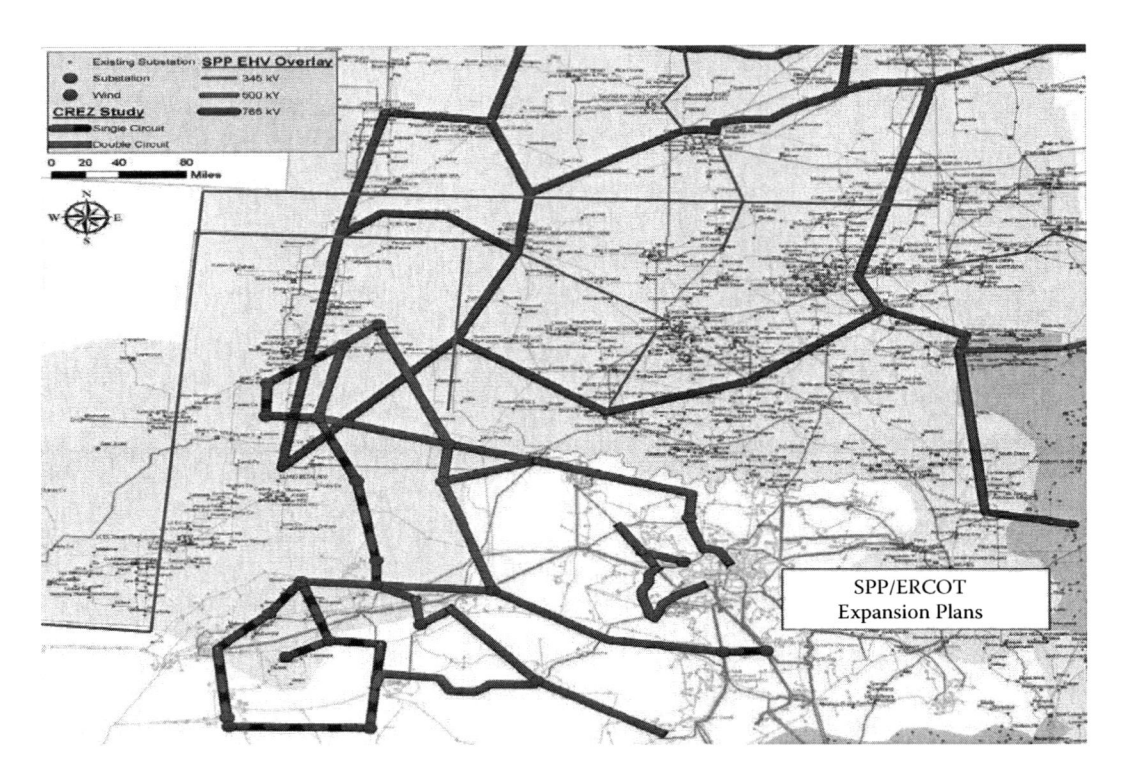

FIGURE 11.4 ERCOT transmission lines at route planning stage. Completion of construction scheduled for 2013. Transmission lines (northern set) for SPP at planning stage for some routes. Others at study stage.

Institutional Issues

projects around the western U.S. Each project has a website detailing size, purpose, and in some cases proposed routes.

Transmission is also a major issue in the European Union [60]. The electricity markets are not competitive for four reasons: (1) lack of cross-border transmission links; (2) dominant, integrated power companies; (3) biased grid operators; and (4) low liquidity in wholesale electricity markets. The conclusion is that wind power issues are determined more by economics and regulations than by technical or practical constraints.

China has a few wind farms constructed in locations before adequate transmission capacity was available. China is constructing an 800-kV HVDC transmission line to connect the Hami prefecture in eastern Xinjiang with the central city of Zhengshoum. The 2,210-km line will cost $3.7 billion. A second 740-kV NVDC line links Xinjiang with the main network of Northwest China (2,180 km, $1.5 billion). These projects will help transmit electricity from the west—one of the main wind power regions that also has major coal reserves.

LINKS

Energy Information Administration. www.eia.doe.gov/cneaf/solar.renewables/page/wind/wind.html

D. Koplow and A. Martin. 2005. Fueling global warming: federal subsidies to oil in the U.S. www.greenpeace.org/usa/press-center/reports4/fueling-global-warming

Reality of U.S. energy subsidies. http://www.awea.org/learnabout/publications/upload/Subsidies-Factsheet-May-2011.pdf.

A Silverstein. 2011. Transmission 101. www.naruc.org/grants/Documents/Silverstein%20NCEP%20T-101%2004202011.pdf

REFERENCES

1. D. Bain. 1980. An introduction to PURPA, sections 201 and 210. In V. Nelson, Ed., *Proceedings of Windpower Conference*, p. 33.
2. U.S. Federal Energy Regulatory Commission. Order 69. Final rule regarding the implementation of Section 210 of the Public Utility Regulatory Policies Act of 1978. Docket RM79-55.
3. A.J. Cavallo. 1994. High capacity factor wind turbine transmission systems. *Wind Energy* 15, 87.
4. J.E. Bryson. 1980. New directions for utilities. In V. Nelson, Ed., *Proceedings of Windpower Conference*, p. 68.
5. Caithness Windfarm Information Forum. Summary of wind turbine accident data to 31st December 2012. www.caithnesswindfarms.co.uk
6. P. Gipe. 2012. A summary of fatal accidents in wind energy. http://www.wind-works.org/cms/index.php?id=108
7. American Wind Energy Association. *Electrical guide to utility scale wind turbines*. http://awea.org/_cs_upload/documents/issues/5976_1.pdf
8. B. Parsons et al. 2003. Grid impact of wind power: a summary of recent studies in the United States. www.nrel.gov/docs/fy03osti/34318.pdf
9. E. Ela, B. Kirby, N. David et al. 2011. Effective ancillary services market design on high wind power penetration systems, NREL/CP-5500-53514. http://www.nrel.gov/docs/fy12osti/53514.pdf
10. D.M. Dodge and C. Lawless-Butterfield. 1982. *Small wind systems zoning issues and approaches*. RFP-3386, UC-60. Wind Energy Research Center, SERI.
11. Berkshire Regional Planning Commission. 2011. Model Large-Scale Wind Energy Facility Zoning By-Law. www.berkshireplanning.org/community/documents/FINALMODELWINDBYLAW.pdf
12. Minnesota Department of Energy and Economic Development. 1983. *Zoning for Wind Machines: A Guide for Minnesota Communities*.
13. M. Moorehead. 1984. *Reference Guide to Wind Energy Land Use: Issues and Actions*. Salem: Oregon Department of Energy.
14. American Wind Energy Association. *Small Wind Permitting Handbook*. www.awea.org
15. U.S. Fish and Wildlife Service. 2012. Final land-based wind energy guidelines. www.fws.gov/windenergy/
16. Stopillwind. www.stopillwind.org/index.php

17. A. Drewitt and R.H.W. Langston. 2006. Assessing the impacts of wind farms on birds. *Ibis* 148, 29. www.blackwell-synergy.com/doi/pdf/10.1111/j.1474-919X.2006.00516.x?cookieSet=1
18. M. Δεσηολμ ανδ ϑ. Καηλερτ. 2005. Αϖιαν χολλισιον ρισκ ατ αν οφφσηορε ωινδ φαρμ. *Biology Letters* 1, 296. www.pubmedcentral.nih.gov/articlerender.fcgi?artid=1617151
19. M. Τιδωελλ. Ραδαρ στυδιεσ σηοω προποσεδ ωινδ φαρμσ υνλικελψ το ιμπαχτ μιγρατορψ βιρδ ποπυλατιονσ. ωωω.χηεσαπεακεχλιματε.οργ/παγεσ/παγε.χφμ?παγε_ιδ = 97
20. K. C. Sinclair. 2001. Status of avian research at the National Renewable Energy Laboratory. In *Proceedings of Windpower Conference*. CD.
21. NWTC Library. Wind–wildlife impacts literature database (W(LD). www.nrel.gov/wind/wild
22. A. Lue, A. W. Hosmer, and L. Harrison. 1994. Bird deaths prompt rethink on wind farming in Spain. *Windpower Monthly* 10, 14.
23. Geographica. 2003. Photo in *National Geographic*, September.
24. AWEA/Audubon Workshop. 2006. Understanding and resolving bird and bat impacts. http://c2.mcbusiness.org/file_depot/0-10000000/10000-20000/16786/folder/88844/01_AWEA_Audubon_Proceedings_2_24.pdf
25. California Energy Commission. 2004. California guidelines for reducing impact to birds and bats from wind energy development. www.energy.ca.gov/windguidelines/index.html
26. National Wind Coordinating Collaborative. www.nationalwind.org/publications/default.htm
27. National Wind Coordinating Collaborative. 2010. Wind turbine interactions with birds, bats, and their habitats. www.nationalwind.org/publications/bbfactsheet.aspx
28. American Wind Energy Association. *Wind Energy Siting Handbook*. www.awea.org/sitinghandbook
29. Wind Turbine Guidelines Advisory Committee. www.fws.gov/habitatconservation/windpower/wind_turbine_advisory_committee.html
30. M. Costanti and P. Beltrone. 2006. Wind energy guide for county commissioners. www.nrel.gov/docs/fy07osti/40403.pdf
31. American Society for Testing and Materials, www.astm.org/Standards/E1528.htm
32. H. Scheer. 1998. Energy subsidies: a basic perspective. Paper presented at Second Conference on Financing Renewable Energies, Bonn.
33. E.T.C. Ling. 1993. Making sense of the federal tax code: Incentives for wind farm development. In *Proceedings of Windpower Conference*, p. 40.
34. U.S. Department of Energy. Energy efficiency and renewable energy. www1.eere.energy.gov/wind/index.html
35. V. Nelson. 1984. SWECS industry in the United States. Report 84-2, Alternative Energy Institute; A history of the SWECS Industry in the U.S., *Alternative Sources of Energy*, March/April, p. 20.
36. L. Flowers and P. J. Dougherty. 2001. Wind powering America. In *Proceedings of Windpower Conference*.
37. Database of State Incentives for Renewable Energy (DSIRE). www.dsireusa.org
38. Green Power Network. http://apps3.eere.energy.gov/greenpower/
39. L. Bird, C. Kreycik, and B. Friedman. 2008. *Green Power Marketing in the United States: A Status Report*, 11th ed. Technical Report NREL/TP-62A-44094. www.nrel.gov/docs/fy09osti/44094.pdf
40. National Renewable Energy Laboratory. NREL highlights utility green power leaders. www.nrel.gov/news/press/2011/1367.html
41. S. Krohn. 2000. Renewables in the EU single market: an economic and policy analysis. In *Proceedings of Windpower Conference*. CD.
42. V. Pollard. 2001. Developments in the European policy framework(s) for wind energy. In *Proceedings of Windpower* Conference. CD
43. European Wind Energy Association. 2009. *Wind Energy: A Guide to the Technology, Economics, and Future of Wind Power*. London: Earthscan.
44. M. Ragwits et al. 2007. *Assessment and optimization of renewable energy support schemes in the European electricity market*. Final report, Intelligent Energy Europe. www.optres.fhg.de/OPTRES_FINAL_REPORT.pdf
45. German Wind Energy Association. 2005. Minimum price system compared with the quota model: which system is more efficient? www.oregon.gov/energy/RENEW/Wind/docs/Feedlaw_vs_quota_GWEA.pdf
46. Renewable Energy Sources Act. 2007. Progress Report. www.erneuerbare-energien.de/fileadmin/ee-import/files/english/pdf/application/pdf/erfahrungsbericht_eeg_2007_zf_en.pdf
47. P. Gipe. 2012, Feed laws. www.wind-works.org/cms/index.php?id=86
48. *ExternE: Externalities of Energy*. www.externe.info
49. Carbon Dioxide Information Analysis Center. http://cdiac.esd.ornl.gov/home.html
50. Carbon Monitoring for Action. www.carma.org

Institutional Issues

51. Energy Information Administration. www.eia.doe.gov/cneaf/electricity/epa/epat5p1.html
52. Analysis of EU CO$_2$ market. www.co2prices.eu
53. U.S. Department of Energy. 2002. National transmission grid study. http://energy.gov/sites/prod/files/oeprod/DocumentsandMedia/TransmissionGrid.pdf
54. American Wind Energy and Solar Energy Industries Association. 2009. Green power superhighways: building a path to America's clean energy future. www.awea.org/documents/issues/upload/GreenPowerSuperhighways.pdf
55. National Renewable Energy Laboratory. Transmission grid integration. www.nrel.gov/electricity/transmission/
56. National Electric Transmission Corridor. http://nietc.anl.gov/nationalcorridor/index.cfm
57. D. Woodfin. 2008. CREZ transmission optimization study summary. www.ercot.com/meetings/board/keydocs/2008/B0415/Item_6_-_CREZ_Transmission_Report_to_PUC_-_Woodfin_Bojorquez.pdf
58. Southwest Power Pool. 2008. Oklahoma Electric Power Transmission Task Force Study. www.spp.org/publications/OEPTTF%20Report_FINAL_4_22_08_updated.pdf
59. Western Grid Group. Proposed western transmission projects. www.westerngrid.net/proposed-western-transmission-projects/
60. European Wind Energy Association. 2010. Powering Europe: wind energy and the electricity grid. www.ewea.org/index.php?id = 196

PROBLEMS

1. What type of incentives should be devised for renewable energy, particularly wind energy? Give a brief explanation for your choices.
2. For energy production from a wind facility, what is the avoided cost that the utility will pay?
3. What are ancillary costs?
4. List two environmental issues for installing a large wind farm in your area.
5. How much support should the U.S. government provide for wind energy? Why?
6. What types of projects should the federal government support? (Some examples are R&D, prototype, demonstration, turbine verification, and commercialization.)
7. Should state and local governments provide incentives for wind energy? If yes, list your choices and explain why.
8. What type of education would be most effective for promoting renewable energy? At what level and to whom?
9. What are the major environmental concerns if a renewable energy system is planned for your area?
10. List three externalities for electricity from coal power plants.
11. How many U.S. states have net energy metering of 100 kW or greater?
12. What is the longest period for net energy billing?
13. What incentives does your state grant for residential size wind systems?
14. Does your electric utility offer green power. If yes, what is it? If no, briefly describe the green power program of Austin Energy in Texas.
15. Go to www.dsireusa.org. How many states have renewable portfolio standards? How many states have rebate programs for purchasing wind systems?
16. What are present market values of renewable energy credits in Texas?
17. Should a pollution tax be imposed on electricity produced by fossil fuels? If yes, how much per metric ton?
18. Calculate the carbon dioxide that wind displaced for the world. Use kilograms per kilowatt hour from coal power electric plants for comparison. Use a 35% capacity factor for wind plants in estimating annual energy production.
19. Calculate the carbon dioxide that wind displaced for the European Union and the United States. Use kilograms per kilowatt hour from coal power electric plants for comparison. Use a 35% capacity factor for wind plants in estimating annual energy production.
20. What is the cost per kilometer to build a major transmission line, 300 kV or larger?

21. Compare the fatality rates for birds and bats from wind turbines.
22. What is the leading cause of death in the wind industry? What are the leading causes of wind turbine accidents?
23. What is the status of new high voltage transmission lines for the Great Plains of the United States?
24. What are the values of feed-in tariffs for renewable energy in Canada?

12 Economics

The most critical factors in determining whether installing wind turbines is financially worthwhile are the initial cost of the installation and the annual energy production. In determining economic feasibility, wind energy must compete with the energy available from competing technologies. If a system produces electrical energy for a grid, the price for which the electrical energy can be sold is also critical. Wind farms today are essentially competitive with all new power plants, except for combined-cycle natural gas turbines because the price for natural gas has declined from peak values due to the exploitation of shale formations. Oil prices were around $100/bbl in 2012. As natural gas replaces coal for producing electricity and begins to replace oil for transportation, natural gas prices will rise. To increase market penetration of wind systems, the return from the energy generated must exceed all costs in a reasonable time.

All values for electricity produced by wind turbines depend on wind resources and cover a range. Installed weighted average costs for wind farms declined to $1,000/kW by 2003—a value of electricity produced of $0.04 to 0.06/kWh. Operation and maintenance (O&M) costs for wind farms were around $0.01/kWh. In the U.S., contracts for selling electricity from wind farms in 1995 were signed for $0.04/kWh and dropped to less than $0.03/kWh in 2002. Since then, the prices of steel, cement, and copper increased, and the installed cost increased to $1,180 to 3,500/kW (weighted average = $2,155/kW) in 2010, and dropped slightly in 2011 [1]. At an installed cost of $2,000/kW, that translates to a value of electricity of $0.07 to 0.09/kWh.

The earlier U.S. Department of Energy (DOE) goals for wind turbines for wind farms to produce electricity at $0.03/kWh for class 6 lands (6.7 m/sec annual average at 10 m height) by 2004 and $0.03/kWh for class 4 lands (5.8 m/sec annual average at 10 m height) by 2010 were not met. These values include O&M at $0.005/kWh. However, the cost of electricity from new power plants using fossil fuels will also increase, so electricity from wind farms is still competitive.

Systems of 1 kW are not cost effective when connected in parallel to a utility grid, even for single residences but people purchase them for other reasons. Residences connected to a utility grid need 5- to 10-kW systems. Farms, ranches, and businesses need a minimum size of 25 kW (about 10-m diameter) or larger. In general, installed costs for small wind turbines are around $2,500 to $5,000/kW, which translates to a value of electricity produced of $0.15 to 0.30/kWh.

The sizes of wind turbines for residences, farms, ranches, and rural applications depend on the amount and price of electricity from a grid if net metering is available and also the local infrastructure. The kilowatt hours consumed can be determined from a monthly electric bill or by calling a local utility. To maximize the return on a wind system, most of the energy should be used on site because it is worth the retail rate. However, net energy billing allows larger systems. A system can be sized to produce all the energy needed within a billing period.

As stated in Chapter 11, economics is intertwined with incentives and penalties, so actual life cycle costs [2] are hard to determine, especially when external factors like pollution control and government support for R&D for competing energy sources are not included.

12.1 FACTORS AFFECTING ECONOMICS

The following list includes most of the factors to consider when purchasing a small wind energy system for home, business, farm, or ranch use.

1. Load (power) and energy (calculated by month or day for small systems)
2. Cost of energy from competing energy sources to meet need

3. Initial installed costs (equipment purchase, shipping, installation of foundation, utility connection, labor, and land)
4. Production of energy (wind turbine types and sizes, warranties, vendor reputation and history, reliability, and availability); wind resources including annual and more frequent variations
5. Selling price of energy produced or unit worth of energy; anticipated energy cost changes (escalations) of competing sources
6. Operation and maintenance costs encompassing general operation, ease of service, emergency services and repairs, insurance, infrastructure, availability of trained service personnel
7. Cost of money (fixed and variable interest rates)
8. Inflation (estimated for future years)
9. Legal fees (contract negotiations, land conveyances or leases, easements, and permits)
10. Depreciation (only for business)
11. National and state incentives

12.2 GENERAL COMMENTS

The general uncertainties surrounding future energy costs, dependence on imported oil, reduction of pollution and emissions, and availability provided the driving force to develop renewable sources. The prediction of energy cost escalation is a hazardous endeavor because energy cost is driven primarily by oil cost. Oil cost $15 to 25/bbl in the 1990s. Predictions in the late 1990s were for gradual increases to $30/bbl by the 2020. However, oil reached the $30 level in 2003, soared to $130 in 2008, and is about $100/bbl in 2012. Price increases have not been and will not be uniform based on time or geography.

At the point when demand exceeds production, the price of oil will increase further. Some experts predict that the peak of world oil production will occur in this decade; others predict the peak will be from 2020 into 2040. The most important factors are the estimated total reserves and amounts recoverable. As prices increase, it becomes economic to recover more from existing reservoirs and extract oil from more difficult sources (polar ice, deep seas, tar sands, and even oil shale).

Every effort should be made to benefit from all national, state, and other incentives for installing wind turbines. The cost of land is a real even for operators that use their own land. This cost is often obscured because it is, in essence, lost income. Wind turbines occupy space and decrease the amount of land available for farming or ranching. The land taken out of production for a wind farm can range from 0.5 to 1.5 hectares (ha) per turbine.

Wind turbine availability is important in determining the quantity of energy produced. For optimum return, the equipment must operate as much of the time as possible, consistent with safety considerations. Performance and failure data should be obtained and used to estimate downtime. Availability for earlier machines was low, but recent availability reached 98%. The distribution of energy throughout the year can affect its value. Energy produced during a time of increased demand on a utility or when energy is needed at the site is clearly more valuable.

Wind turbines can produce electricity for consumption at or near the site, sell it to a utility, or use it and sell it. The higher the selling price, the more economically feasible a project becomes. Generally, an owner of one or a few wind turbines will use some of the energy and sell the excess to a utility. The electricity used on site displaces electricity at the retail rate. For states that use net energy billing (usually limited to small wind turbines), even the energy fed back to the utility is worth the retail rate. If more energy was produced than used during a billing period, it is sold for avoided cost. For locations where retail rate is higher than the avoided cost paid for excess energy fed back to a utility, economic feasibility improves with increasing on-site consumption. The price paid by a utility is negotiated, set by law, or decided by a public regulatory agency.

Economics

Example 12.1

A wind turbine produces 2,000 kWh in a month and has two meters. One measures energy purchased from a utility company (3,000 kWh), and the second measures energy fed back to the grid (1,200 kWh). The energy displaced by the wind turbine is 800 kWh (2,000 − 1,200). Retail rate (from grid) is $0.08/kWh. The value of the excess energy sold to the grid (avoided cost set by the state) is $0.04/kWh.

Clearly net billing is preferable, because all the energy produced by the wind turbine is worth the retail rate, up to the point where the meter reads no difference from the previous month.

The costs of routine operation and maintenance for individuals represent time and parts costs. Until system reliability and durability are better known for long periods, repair costs will be difficult to estimate. It is important for an owner to clearly understand the manufacturer's warranty and the manufacturer should have a good reputation. Estimates should be made of the costs of repairing the most probable failures. Insurance costs may be complicated by companies that are uncertain about the risks posed by a comparatively new technology. However, the risks are fewer than those associated with operating a car.

Inflation will have its principal impact on expenses incurred over the lifetime of a system. O&M costs, especially for unanticipated repairs, fall into this category. On the other hand, cheaper dollars may be used to repay fixed-rate loans.

12.3 ECONOMIC ANALYSIS

Both simple and complicated economic analyses provide guidelines. Simple calculations should be made first. Commonly calculated quantities are (1) simple payback, (2) cost of energy (COE), and (3) cash flow. A wind turbine is economically feasible only if its overall earnings exceed its overall costs within a period up to the lifetime of the system. The time at which earnings equal cost is the payback time. The relatively large initial cost means that the payback time could be several years and in some cases earnings may never exceed the costs. Of course, a short payback is preferred, and five to seven years is acceptable. Longer paybacks should be viewed with caution.

How do you calculate the overall earnings or value of energy? If you had no source of energy for lights, television, and other uses, a cost of $0.50 to $1.00/kWh might be acceptable for the benefits received. Many people are willing to pay more for green power because it produces less pollution. Past green power premiums were around $0.03/kWh for a 100-kWh block but are lower today (see Section 11.6.1.2). Few people want to be completely independent of a utility grid, no matter what the cost.

12.3.1 SIMPLE PAYBACK

A simple payback calculation can provide a preliminary judgment of economic feasibility. The difference between borrowing money for a system and lost interest if an owner or operator has enough money to pay for the system is usually about 3 to 7%. In 2003 and again after 2008, lost interest rates were very low. The easiest payback calculation is cost of the system divided by cost displaced per year, assuming that O&M needs and maintenance minimal and will be done by the owner.

$$SP = IC/(AEP \times \$/kWh) \tag{12.1}$$

where SP = simple payback in years; IC = initial cost of installation in dollars; AEP = annual energy production in kilowatt hours per year; and $/kWh = price of energy displaced.

Example 12.2

You purchase a 300-kW wind turbine for battery charging. Installed cost = $850. The unit produces 220 kWh per year at $0.50/kWh (estimated cost for remote electricity).

$$SP = \$900/(220 \text{ kWh/year} \times 0.50 \text{ \$/kWh})$$

$$SP = 900/110 = 8 \text{ years}$$

The next calculation includes the value of money, borrowed or lost interest, and annual operation and maintenance costs:

$$SP = \frac{IC}{(AEP * \frac{\$}{kWh} - IC * FCR - AOM)} \quad (12.2)$$

where FCR = fixed charge rate, per year and AOM = annual operation and maintenance cost in dollars per year.

Example 12.3

You purchase a 3.5-kW wind turbine with inverter to connect to the grid. Installed cost = $14,000. Unit produces 6,000 kWh per year. You are losing interest at 4% on the installed cost. Retail rate of electricity is $0.11/kWh. Assume AOM of $50 per year.

$$SP = 14,000/(6,000 \times 0.11 - 14,000 \times 0.04) = 14,000/(660 - 560) = 140 \text{ years}$$

You would think twice before purchasing this system on an economic basis and the payback does not consider O&M.

The FCR could be the interest paid on a loan or the value of interest that would have been received from money displaced from savings. An average value for a number of years (5) will have to be assumed for cost per kilowatt hour for displaced electricity because the future costs of electricity from a utility may be difficult to estimate. In general, electric rates do not fluctuate much or increase rapidly. The one change with deregulation is that fuel adjustment cost can change quickly.

Example 12.4

You purchase a 50-kW wind turbine. Installed cost = $120,000. Unit produces 120,000 kWh per year. AOM = 0.01 × IC = $1,200 per year. FCR = 0.07. Retail rate of electricity is $0.11/kWh.

$$SP = 120,000/(120,000 \times 0.11 - 120,000 \times 0.07 - 1,200) = 120,000/(13,200 - 8400 - 1200)$$

$$SP = 33 \text{ years}$$

Equation (12.2) involves several assumptions: the same kilowatt-hours are produced each year, the value of the electricity is constant, and no inflation occurs. More sophisticated analyses would include details such as escalating fuel costs of conventional electricity and depreciation. In general, these factors reduce the payback.

12.3.2 COST OF ENERGY

The cost of energy (value of the energy produced by the wind turbine) gives a levelized value over the life of the system (assumed to be twenty to twenty-five years). The cost of energy (COE) is primarily driven by the installed cost and the annual energy production.

$$COE = (IC \times FCR + AOM)/AEP \quad (12.3)$$

Economics

The COE is one measure of economic feasibility and is compared to the price of electricity from other sources (generally a utility company) or the price for which wind-generated energy can be sold. If you are purchasing a wind turbine to displace electricity on site, the COE should be compared with a projected average cost of electricity from the utility company over the next ten years. The cost of energy for small systems is higher than for wind farms, with some economies of scale for larger sizes of wind turbines (Table 12.1). In general, the AOM is around $0.005/kWh. In Equation (12.3), major replacement costs are included in the annual O&M costs.

Example 12.5

You purchase a 50-kW wind turbine. Installed cost = $120,000. Unit produces 120,000 kWh per year. AOM = 0.03 × IC = $1,200 per year. FCR = 0.08. Retail rate of electricity is $0.11/kWh.

$$COE = (120{,}000 \times 0.08 + 3600)/120{,}000 = \$0.11/\text{kWh}$$

A sensitivity analysis (Figure 12.1) shows how the different factors in Equation (12.3) affect the cost of energy. The most important factors are installed cost and annual energy production. The cost of energy formula from the Electric Power Research Institute (EPRI) [3] is similar to Equation 12.3. There are additions for levelized replacement costs (major repairs) and fuel costs for conventional power plants. Since the cost of fuel for wind energy is zero, that term will be left out:

$$COE = \frac{(IC * FCR) + LRC + AOM}{AEP} \qquad (12.4)$$

TABLE 12.1
Range of Energy Costs for Small Systems (Wind Class 4 to 2)

System (kW)	$/kWh
1	0.12 to 0.20
10	0.11 to 0.18
50	0.10 to 0.15
100	0.10 to 0.15

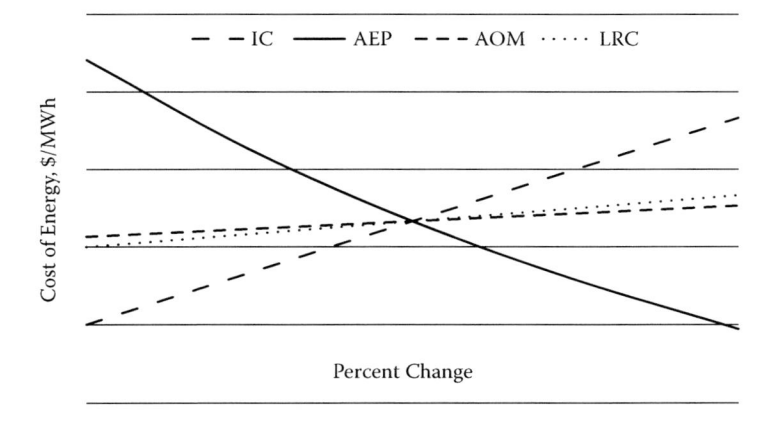

FIGURE 12.1 Sensitivity analysis for cost of energy for wind turbine.

where LRC = levelized replacement cost (dollars per year) and AEP = net annual energy production (megawatt hours or kilowatt hours per year). The COE can be calculated for cost per kilowatt hour or cost per megawatt hour and the last term could be separate as AOM/AEP, cost per kilowatt hour, or cost per megawatt hour. With histogram data and power curves to calculate annual energy production, the cost of energy can be calculated. A first estimate for levelized replacement costs could be 4 to 5% of installed cost.

Example 12.6

Unit is a 1-MW wind turbine. Installed cost = $1,600,000. FCR = 0.07. AEP = 3,000 MWh per year. LRC = $80,000 per year. AOM = $8/MWh = $0.008/kWh.

$$COE = \frac{(1,600,000 * 0.08) + 80,000}{3000} + 8 = \$77/MWh = \$0.077/kWh$$

That cost of energy must be compared to all expected net income from the wind farm including incentives, depreciation, and expected rate of return.

Levelized replacement cost distributes the costs for major overhauls and replacements over the life of a system. For example, storage batteries in a village power system must be replaced every five to seven years. The levelized replacement cost can be calculated with the equation below. The result will be an estimate for future replacement costs based on present costs of components.

1. Year in which replacement is required (n)
2. Replacement cost including parts, supplies, and labor (RC)
3. Present value of each year's replacement cost (PV)

The present value for replacement costs is calculated as

$$PV(n) = PV(n) \times RC(n) \tag{12.5}$$

where $PVF(n)$ = present value factor for year; $n = (1 + I)^{-n}$; I = discount rate of 0.07; and $RC(n)$ = replacement cost in year n. The levelized replacement cost is the sum of present values multiplied by the capital recovery factor (CRF):

$$LRC = CRF * \sum_{n=1}^{20} PV(n) \tag{12.6}$$

where CRF = 0.093.

Example 12.7

This spreadsheet can be used to calculate levelized replacement cost (LRC) for 50-kW turbine.

Component	Years	RC	I	PVF	PV
Bearings	10	6,500	0.07	0.508	3,304
Blades	10	5,000	0.07	0.508	2,542
Subtotal					5,846
LRC ($/yr)		544			

COE = (120,000 × 0.08 + 544)/120,000 = 0.01 = $0.095/kWh. This value can be compared with the value in Example 12.5. The problem is the determination of major repairs, year, and replacement cost.

12.3.3 Value of Energy

Another formula [4] for estimating the value of energy is

$$\frac{f_0}{c} \geq \frac{(1+r)^L \alpha r L}{[(1+\alpha)^L][(1+r)^L - 1]} \tag{12.7}$$

where f_o = value of energy saved per year (dollars); c = initial installed cost (dollars); L = years to payback; α = fuel inflation rate; and r = interest rate.

Because there is no factor for O&M, the interest rate should be increased by 1 to 2%. Equation (12.7) can be solved by iteration using different values of L to calculate the right side and comparing that to the left side of the equation. As interest rates increase, payback times increase, and as fuel inflation factors and electricity costs increase, payback times decrease.

12.4 LIFE CYCLE COSTS

A life cycle cost (LCC) analysis reveals the total cost of a system including all expenses incurred over the life of the system and salvage value, if any [2,5,6]. Two reasons to conduct an LCC analysis are to (1) compare different power options and (2) determine the most cost-effective design. The competing options to small renewable energy systems are batteries or small diesel generators. For these applications, the main concerns are the initial cost of the system, the infrastructure to operate and maintain the system, and the price people pay for the energy. However, even if small renewable systems are the only options, LCC analysis can be helpful for comparing costs of various designs and determining whether a hybrid system would be a cost-effective option. An LCC analysis allows a designer to study the effects of different components with different reliabilities and lifetimes. For instance, a less expensive battery may be expected to last four years, while a more expensive battery may last seven years. Which is the best buy? This type of question can be answered with an LCC analysis.

$$LCC = IC + M_{PV} + E_{PV} + R_{PV} - S_{PV} \tag{12.8}$$

where LCC = life cycle cost, IC = initial cost of installation, M_{PV} = sum of all yearly O&M costs, E_{PV} = energy cost (total of all yearly fuel costs), R_{PV} = sum of all yearly replacement costs, and S_{PV} = salvage value (net worth at end of final year), usually 20% for mechanical equipment.

Future costs must be discounted because of the time value of money, so the present worth is calculated for costs for each year. Life spans for wind turbines are assumed to be twenty to twenty-five years but replacement costs for components must be calculated. Present worth factors are given in tables or can be calculated. Life cycle costs are the best bases for purchasing decisions and show that many renewable energy systems are economical.

The financial evaluation can be done yearly to determine cash flow, break-even point, and payback time. A cash flow analysis will be different in every situation. Cash flow for a business will be different from a residential application because of depreciation and tax implications. The payback time is easily seen by graphing data. Bergey Windpower has a spreadsheet that can be downloaded on its website for calculating cash flow analysis.

Example 12.8

The example is a residential application with rebate. IC = $25,000, down payment = $7,000, loan = $18,000 at 10% (payment = $4,000/year), O&M = 2.5% × IC = $500 per year, energy production = 50,000 kWh per year (75% consumed directly, displacing 8 cents/kWh electricity and 25%

sold to the utility at 4 cents/kWh with utility escalation at 3%/year). Cash flow analysis should be done on a spreadsheet.

Year	0–1	2	3	4	5	6	7	8	9	10
Down payment	7000									
Principle left	18,000	15,800	13,380	10,718	7,790	4,569	1,026	0		
Principal paid	2200	2420	2662	2928	3221	3543	3897	1128		
Interest	1800	1580	1338	1071.8	778.98	457	103	0		
O&M	500	500	500	500	500	500	500	500	500	500
Insurance	50	50	50	50	50	60	60	60	60	60
Property tax	70	70	70	70	70	70	70	70	70	70
Costs	7620	4620	4620	4620	4620	4630	4630	1758	630	630
Value energy used	3000	3090	3183	3278	3377	3478	3582	3690	3800	3914
Value energy sold	500	515	530	546	563	580	597	615	633	652
Rebate	4000									
Income	7500	3605	3713	3825	3939	4057	4179	4305	4434	4567
Cash flow	−120	−1015	−907	−795	−681	−573	−451	2546	3804	
Cumulative		−1135	−1922	−1702	−1476	−1253	−1023	2096	6350	

In this analysis, the payback time is year 8. There are a number of assumptions about the future in the analysis. A more detailed analysis would include inflation and increases in O&M costs as equipment ages.

A cash flow analysis for a business with $0.02/kWh tax credit on electric production and depreciation of the installed costs would give a different answer. Also, all operating expenses are business expenses. The economic utilization factor is calculated from the ratio of the costs of electricity used at a site to the costs of the electricity sold to the utility.

The RETScreen tool [7] consists of standardized and integrated renewable energy project analysis software that can be used to evaluate energy production, life cycle costs, and greenhouse gas emission reductions for several renewable energy technologies: wind, small hydro, PV, passive solar heating, solar air heating, solar water heating, biomass heating, and ground-source heat pumps. The Hybrid2 software package [8] includes economic analysis. The cost of energy for wind, PV, and solar thermal decreased dramatically since 1980 (Figure 12.2). The range of values around $3 to

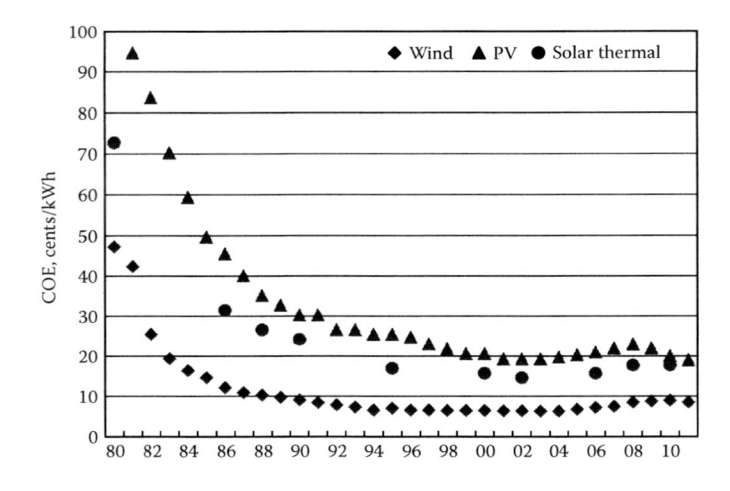

FIGURE 12.2 Cost of energy for generation of wind, photovoltaic, and solar thermal energy (2011 dollars).

$10/MWh is due primarily to the value of the renewable resource and the installed costs. The average cost of energy for wind in Europe was around $95/MWh and it was lower in China.

An economic cash flow model, cost of renewable energy spreadsheet tool (CREST), and a simplified levelized cost of energy calculator are available from NREL. CREST is a suite of four analytic tools, for solar (photovoltaic and solar thermal), wind, geothermal, and anaerobic digesters.

12.5 PRESENT WORTH AND LEVELIZED COSTS

Money increases or decreases with time, depending on interest rates for borrowing or saving and inflation. Many people assume energy costs in the future will increase faster than inflation. The same mechanism of determining future value of a given amount of money can be used to move money backward in time. If each cost and benefit over the lifetime of the system were brought back to the present and then summed, the present worth can be determined:

$$PW = \frac{(\text{cost total for year } S) - (\text{financial benefit total for year } S)}{(1+d)^M} \quad (12.9)$$

where cost total = negative cash flow, S = specific year in the wind system lifetime, M = years from the present to year S, and d = discount rate. The discount rate determines how the money increases or decreases over time. Therefore, the proper discount rate for any life cycle cost calculation must be chosen with care. Sometimes the cost of capital (interest paid to a bank or lost opportunity cost) is appropriate. Possibly the rate of return on a given investment perceived as desirable may be used as the discount rate. Adoption of unrealistically high discount rates can lead to unrealistic life cycle costs. The cost of capital can be calculated from

$$CC = \frac{1 + \text{loan interest rate}}{1 + \text{inflation rate}} - 1$$

If the total dollars are spread uniformly over the lifetime of the system, this operation is called levelizing.

$$\text{annualized cost} = \frac{PWd(1+d)^P}{(1+d)^P - 1} \quad (12.10)$$

where P = number of years in the lifetime. One further step has been utilized in assessing renewable energy systems versus other sources of energy such as electricity. This step is the calculation of the annualized cost of energy from each alternative. The annualized cost calculated from Equation (12.8) is divided by the net annual energy production of that alternative source.

$$COE = \text{annualized cost}/AEP$$

It is important that annualized costs of energy calculated for renewable energy systems are compared to annualized costs of energy from other sources. Direct comparison of annualized cost of energy to current cost of energy is not rational. Costs of energy calculated in the above manner provide a better basis for the selection of sources of energy.

RETFinance is an Internet-based cost of electricity model [9] that simulates a twenty-year nominal dollar cash flow for a variety of renewable energy power projects. It is difficult to compare cost and COE for different years without considering the effects of inflation. A number of sites on the web show inflation calculations from past years to the present [10]. As an example, installed costs for wind farms in 2003 were $1,000/kW, which is equivalent to $1,237/kW in 2012. Of course, the amount of inflation in the future is a guess, but most people assume the number will be positive.

12.6 EXTERNALITIES

Externalities now play a role in integrated resource planning (IRP) as future costs for pollution, carbon dioxide, and other factors are added to life cycle costs. Values for externalities range from zero (past and present value assigned by many utilities) to as high as $0.10/kWh for steam plants fired with dirty coal. Again, values are assigned by legislation and regulation (public utility commissions).

As always, both sides will litigate. The Lignite Energy Council petitioned the Minnesota Public Utilities Commission to reconsider its interim externality values. The council represents major producers of lignite, investor-owned utilities, rural electric cooperatives, and other entities. It focused its protest on values assigned to CO_2 emissions based on an acknowledged lack of reliable science proving that CO_2 emissions are harmful to society. In Europe, different values assigned to CO_2 emissions make wind energy more cost competitive.

Wind turbines have three main beneficial externalities: local sources of energy, no requirement for water to aid generation, and no emissions of greenhouse gases. In Texas, fossil fuel power plants use 1,670 L/MWh [11] of water, so the 12 GW of wind power in Texas (2012) will save about 60×10^6 m^3 of water per year. An average of 700 kg of CO_2/MWh is emitted from coal and natural gas power plants. The 12 GW of wind in Texas will reduce CO_2 emissions by around 25×10^6 metric tons per year. The present value for CO_2 trading in Europe is $20 to $30/metric ton, which is equivalent to $15 to $20/MWh, and about $30/MWh if replacing coal plants.

12.7 WIND PROJECT DEVELOPMENT

The three most important considerations for development of wind farms are (1) land with good to excellent wind resources, (2) contract to sell produced electricity, and (3) access to transmission lines (proximity and carrying capacity). The American Wind Energy Association [12,13] and Wind Powering America [14] provide information on project development. Some wind manufacturers publish information about project development on their websites. *Wind Energy Engineering* has a chapter on planning and execution of wind projects [15]. The project development list covers many areas but it is based on economics because the final decision to proceed is based on economics. Much of the information came from Disgen [16].

PROJECT DEVELOPMENT

1. Site selection
    1.1. Evidence of significant wind resource
    1.2. Preference for privately owned remote land
    1.3. Proximity to transmission lines (based on 69 kV; up to twenty-five miles for good site and 135 kV) and potential for future transmission lines.
    1.4. Reasonable road access
    1.5. Few environmental concerns
    1.6. Receptive community
2. Land
    2.1. Term: Expected life of turbine (early, twenty to thirty years, ten-year option; later, thirty to fifty years, multiple ten-year options)
    2.2. Rights for wind, ingress, and egress; transmission right of way for wind farm
    2.3. Owner compensation: percentage of revenue, per turbine, or combination
    2.4. Assignable financing requirement
    2.5. Indemnification
    2.6. Reclamation provision

    2.7. Bond to remove wind turbines at end of project
    2.8. Wind energy easements and other legal issues
3. Wind resource assessment
    3.1. Lease (dollars per acre, 1 to 1–5 years) or flat fee
    3.2. Corollary data (state, national and other wind maps, NWS and other data)
    3.3. Meteorological tower installation at hub height or at least 50 m
    3.4. Collection of ten-min or hour wind speed and direction data for one to two years (one year minimum)
    3.5. Meteorology quality report
    3.6. Output projections for several turbine types
    3.7. Delivery of meteorology report and output projections if developer does not exercise option for installation
4. Environmental
    4.1. Cursory review for endangered species
    4.2. Biological resources
        4.2.1. Wildlife habitat
        4.2.2. Loss of vegetation
    4.3. Avian studies
        4.3.1. Raptors
        4.3.2. Migratory birds
        4.3.3. Review with interested parties: Audubon Society, federal, state, local agencies
        4.3.4. Required studies and reports
    4.4. Bats
    4.5. Archeological sites
    4.6. Noise
    4.7. Visual impact
    4.8. Soil erosion and water quality
    4.9. Solid and hazardous wastes
    4.10. Active compliance monitoring
5. Economic modeling
    5.1. Output projections (see Section 3.6)
    5.2. Turbine costs
    5.3. Turbine installation
    5.4. Roads, substations, transmission requirements
    5.5. Communication and control
    5.6. Sales, income, and property taxes; depreciation schedule; tax abatement possibilities
    5.7. O&M estimates
    5.8. Finance assumptions such as production tax credit, accelerated depreciation, equity rate of return, state and local incentives, debt rate and term (coverage ratios), debt/equity ratio
    5.9. Insurance and legal requirements
6. Interconnection studies
    6.1. Interconnection request and electric reliability council assistance
    6.2. Capacity limitation
    6.3. Load flow analysis
    6.4. Voltage controls
    6.5. System protection

7. Permits
    7.1. Local, state, federal
    7.2. Public involvement at early stage
    7.3. Public land, private land
        7.3.1. Land use permit
        7.3.2. Building permit
8. Sale of energy and/or power
    8.1. Energy or power purchase agreement
        8.1.1. Long-term contract with utility
        8.1.2. Green power market
        8.1.3. Market, avoided cost
        8.1.4. Renewable energy credits
        8.1.5. Future income, emission trading
    8.2. Kilowatt hours (real or nominal levelized)
    8.3. Capacity in kilowatts
    8.4. Term
    8.5. Credit-worthy buyer
    8.6. Facility sales agreement
    8.7. Turnkey price
9. Financing
    9.1. Source of equity; rate of return, 15 to 18%
    9.2. Source of debt
    9.3. Market rates
    9.4. Term of debt
    9.5. Assignable documents
    9.6. Third-party due diligence
10. Turbine purchase
    10.1. Power curve (output projection)
    10.2. Turbine cost
    10.3. Turnkey construction cost
    10.4. Warranties, equipment, and maintenance
    10.5. Construction financing
    10.6. Past history of manufacturer
    10.7. Date of turbine availability
11. Construction (turnkey)
    11.1. Roads
    11.2. Water and gravel
    11.3. Turbine foundations (excavation, concrete)
    11.4. Interconnection to utility (substations, transformers, wiring)
    11.5. Turbine assembly and erection (cranes)
    11.6. Commissioning
    11.7. Environmental restoration
        11.7.1. Road widths
        11.7.2. Grass
        11.7.3. Control of noxious weeds
        11.7.4. Assembly area
12. Maintenance
    12.1. Fixed cost per turbine per year
    12.2. Availability warranties

Economics

12.3. Penalties for non-performance
12.4. Types of costs
12.5. Labor
12.6. Management
12.7. Insurance and taxes
12.8. Maintenance equipment: cranes, vehicles, other
12.9. Parts on hand
12.10. Non-recurring costs and major repairs
12.11. Roads including maintenance and landowner access

The following example shows the main points of a contract signed by the Permanent University Fund, State of Texas, for a Woodward Mountain wind farm (32 MW) near McCamey (year 2000).

Area: 602 ha (1,487 acres)
Term: twenty years, with option to terminate early
Installation bonus: $2,000/MW plus security deposit
Royalty: 4%, years 1 through 10; 6%, years 11 through 20; minimum annual royalty projected income stream
Wind turbines: 48 Vestas V47 660-kW
RECs: Royalty paid if any value realized
Removal bond: Mutual agreement
Hunting: Company indemnified
University audits: Independent outside auditor
Meter calibration: Every three years
Curtailment: Shared by all landowners

A wind farm landowner may receive one or more offers, and the lease terms (Table 12.2) will differ by region, wind resources, and access to transmission. Some landowners are forming associations for dealing with wind farm developers. In general, a landowner in the U.S. has three options for payment of electricity generated by a wind farm: (1) percent royalty on net production; (2) per turbine payment of $4,000 to 6,000 (minimum value); (3) larger value of (1) or (2) for each year.

If a landowner signs a lease and a wind farm is constructed, there is no guarantee that a wind turbine will be placed on that land, so the only payment received would be for roads and/or transmission lines. Some developers guarantee at least one wind turbine for each landowner and some farmer

TABLE 12.2
Representative Lease Terms for Wind Farm

Resource	1 to 3 yr
Flat fee or	$10,000
$/acre/yr	$1 to $4
Contract	30 yr
Option	2 (10 yr)
Construction, road, etc.	$3 to $4/m
or flat fee	$4,000/MW
Income/yr	
Royalty and/or	4 to 6%
per turbine (minimum)	$4,000 to $6000/MW
Escalation	0.5% every 5 yr

associations have provisions for sharing payments across the association. In the U.S., wind turbines can be installed on land currently under the Conservation Reserve Program (CRP), but penalties or reimbursements may be decided by the CRP district.

Other considerations for the landowner are who certifies the energy meter and how often. The landowner should share future revenue from pollution credits. In countries where national or state governments control the land, the questions concern present occupants. In addition to determining who receives payments (once or annually), the amount to be paid for land removed from previous use must be settled.

The development of a project will take 1.5 to three years; the construction phase will take six to twelve months, but total development time from land selection to commissioning may take up to six years (Table 12.3). Problems of economics, financing, and access to adequate transmission can extend the time beyond 6 years. Wind farms can be installed much faster than transmission lines can be built. In addition to production tax credits, a limiting factor that began in 2007 was the demand for wind turbines that exceeded production. Lead times for delivery after ordering were two to three years. Since the 2008 recession, delivery has not been a problem.

12.7.1 Costs

The installed costs in 2010 for onshore wind farms ranged from $1,300 to $1,450/W in China and India, $1,850 to 2,100/kW in Europe, and $2,000 to 2,200/kW in North America. Offshore cost in Europe was $4,000 to $4,500/kW [17]. O&M costs for onshore wind farms range from $0.01 to 0.025/kWh and for offshore installations range from $0.027 to 0.048 because of the difficulties of the environment. The levelized cost of energy at good wind sites was $0.06 to 0.14/kWh.

The installed costs [18] for wind farms in the United States (Figure 12.3) increased from around $1.3 million/MW in 2003 to $1.7 to 3.5 million in 2010 (in 2011 dollars). The fairly wide range of installed costs depends somewhat on project size and region. Prices started to increase after 2004 because of the increases in the prices of steel, copper, and cement. The slight decrease in 2011 and preliminary data for 2012 indicate the trend is continuing. Also during that period, the price increased because the world demand for wind turbines was higher than production. In general, 30 MW is required to achieve economies of scale, primarily for installation. For example, the cost of a crane to install a single 3-MW wind turbine is not economical.

The average installed cost in the U.K. was $2.7 million per megawatt for projects installed from 2010 to 2012. The average installed cost for offshore wind farms was estimated at $4.7 million per megawatt. The world largest offshore wind farm, the London Array (630 MW, commissioned in spring 2013) cost an estimated $4.4 million per MW. The installed cost for offshore wind farms in Europe is around twice the cost of wind farms on land. However, because the offshore winds are better and land costs are high, offshore wind farms are economical in Europe.

TABLE 12.3
Representative Wind Farm Project Timeline

Site Evaluation	Permitting and Negotiation	Construction, Commission
Identify site, conduct preliminary evaluation, secure land options 5–8 months	Permits, land use, transmission Negotiate power purchase agreement, interconnect 12–36 months	Construction 6–12 months
Install anemometers, collect and analyze data 12–24+ months	Turbine purchase agreement 12–24 months	Commission 1–2 months
←	Developer (36–72 months)	→
	← Turbine supplier (12–24 months) →	

Economics

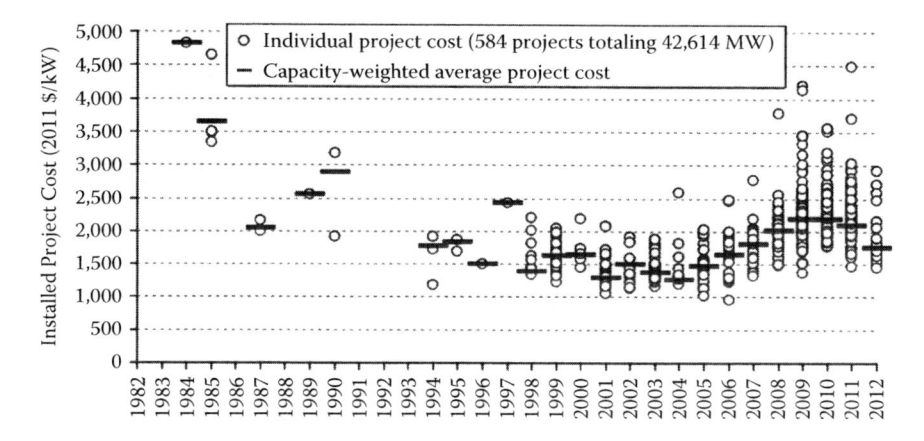

FIGURE 12.3 Installed costs for wind projects in the United States. The 2012 data represent preliminary cost estimates for a sample of twenty projects totaling 2.6 GW (already built or to be built in 2012) for which substantive cost estimates were available. (*Source:* Lawrence Berkeley National Laboratory.)

TABLE 12.4
Wind Farm Installation Component Costs (%)

Component	%
Turbine	74 to 82
Foundation	2 to 5
Electric	2 to 7
Connection to the grid	3 to 7
Finance	1 to 5
Land	1 to 3
Roads	1 to 5
Consultants	1 to 3

A comparison of the estimated components of the cost of energy formula shows, as expected, that capital cost is the major component and the primary capital cost of a project is for wind turbines (Table 12.4). The figure indicates that the average turbine price for 2008 to 2011 in the U.S. was $1.5 million per MW for orders under 100 MW and $1.4 million per MW for 100 MW and greater sizes. These numbers for prices of wind turbines are reflected in the project costs shown in Figure 12.4.

12.7.2 Benefits

Wind farms represent rural economic development. The primary benefit to landowners is long-term stable income (no fluctuations typical of commodity prices). Representative numbers are for a wind farm (30 MW or greater) using capacity factors of 30% in wind class 3 and 35% in wind class 4. A 50-MW wind farm would require 1,200 ha (1 ha = 2.5 acres) and the site could include ten to thirty landowners. Around 1 to 3% of the land will be removed from production, primarily for roads. The return from a wind farm on land removed from previous use is around $10,000 to 16,000/ha/year—a far greater return per hectare than farming or ranching. In contrast to oil and gas leases, the return on a wind farm is lower, but it presents the big advantage of a non-depletable resource.

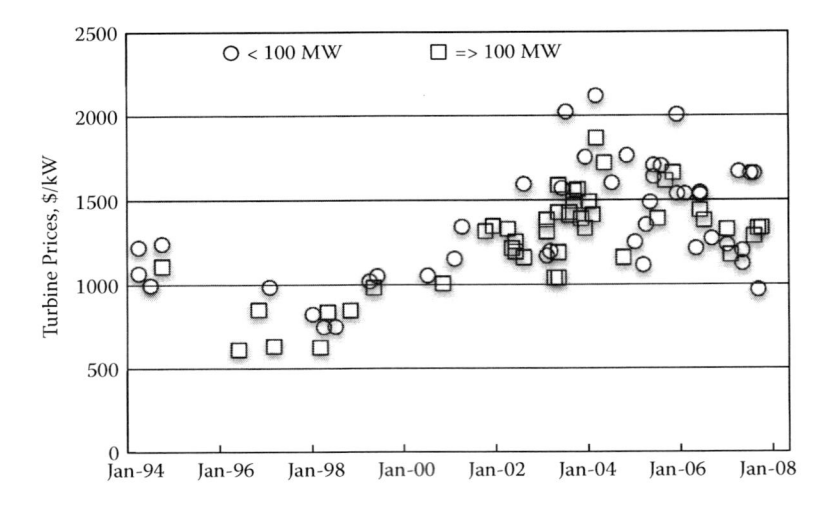

FIGURE 12.4 Wind turbine transaction prices in the United States. (*Source:* Lawrence Berkeley National Laboratory.)

Rural economic development involves construction and then operation. Construction will create 100 to 200 jobs for four to eight months (about one man year per megawatt). The administration, operation, and maintenance of a wind farms require six to ten full-time jobs per 100 MW. The long-term economic impacts at county level across the Great Plains of the U.S. [19] indicate a personal income of approximately $11,000 per megawatt from 2000 through 2008. This shows why state legislatures and local entities promote wind power and also promote area manufacturing of turbines and components.

The Colorado Green Wind Power Project near Lamar is an example. Construction started in summer 2003. The 162-MW project consists of 108 GE wind turbines (1.5 MW) on a lease of 4,450 ha from fourteen landowners. The footprint from the wind farm is about 2% of the land. Construction created 200 to 300 jobs and operation after completion involved about fifteen local jobs. The wind farm pays around $2 million per year in property taxes. After construction, the project was purchased for $212 million by Shell Wind and PPM Energy from GE Wind.

12.7.3 Sales of Electricity

The crunch number for a wind farm project is the sale price of electricity generated. For some older contracts in Texas, the sale price was below $25/MWh for fifteen years. The only way this could happen was utilizing the production tax credit, accelerated depreciation, tax abatements, and renewable energy credits (RECs). For wind farms installed today in the U.S., the production tax credit is still the main driver. Sale contracts are higher, and some wind farms sell electricity in the wholesale and merchant markets. One selling price is the mandatory avoided cost. The minimum value that should be paid to the wind farm is the fuel adjustment cost of the utility. For wind farms in Texas in 2012, new contracts were in the range of $35 to 45/MWh in the ERCOT system and $28 to 35/MWh in the Southwest Power Pool.

The levelized COE is estimated for a 50-MW wind farm in the Panhandle of Texas with class 4 winds. The wind turbines are rated at 1 MW and sited on 70-m towers. The installed cost (2007 dollars) is around $1,600/kW, and from Example 12.6, the COE is $77/MWh. A production tax credit (PTC) of $20/MWh plus other factors such as accelerated depreciation assist in the return. Therefore the wind farm developer would need to obtain around $35/MWh. The value to the landowner can be estimated as

$$AEP = 50 \times 3 \times 10^6 \text{ kWh/year} = 1.5 \times 10^8 \text{ kWh/year}$$

The $35/MWh (landowners will not receive PTC) generates $5.2 million per year and at 4% royalty, the landowners receive $210,000/year. At $4,000/MW, the minimum would be $200,000/year. At 0.5 ha per turbine taken out of production, 20 ha are lost. The value at 4% royalty is $16,000/ha/year. This is much more than a farmer or rancher would earn from crops and livestock.

The wind farm will also pay property taxes. However, in many cases, they try to obtain tax abatements for some period based on economic development. Instead, the wind farm will pay in lieu of taxes, primarily for schools.

Quarterly data on megawatt hours generated, income, and rates paid to wind farms can be obtained from the Federal Energy Regulatory Commission (http://eqrdds.ferc.gov/eqr2/frame-summary-report.asp). However the reports cover all types of electricity generation of electricity so a user must know the reporting name of a wind farm to access specific data. The capacity factor can be calculated from the megawatt hours generated and the installed capacity of the wind farm. Also, the type of sale can be determined from the rate (power purchase agreement at fixed rate, power purchase agreement with peak and off-peak values). For market sales, the database shows high and low values plus averages. As an example, for 2011, the Wildorado Wind Ranch received $18.1 million for 644 GWh from an average power purchase agreement of $28.12/MWh. Since the farm has an installed capacity of 161 MW, the capacity factor for the quarter was 45.7%.

12.8 HYBRID SYSTEMS

When wind is added to an existing diesel generation plant, the cost of the turbine and controls is compared to the dollars saved on diesel fuel. In 2004 for villages (under 1,000 people) in Alaska, the average price was $0.38/kWh for a village electric cooperative powered by diesel gensets. The expense breakdown is as follows:

	2004 Percent	2008 Percent
Fuel	46	77
Operation and maintenance	21	9
Renewal and replacement	19	8
General and administration	14	6
Total	100	100

Since then, the cost of diesel fuel increased significantly and the percent cost of fuel and electricity ($0.55/kWh) increased accordingly. This is the reason for the renewed interest in wind turbines. For villages in Nunavik, Canada, served by Hydro Quebec, diesel fuel represented 54% of the operation cost and that will increase. In some remote communities in Canada, the cost of electricity is over $1.00/kWh. In the Maldives Islands, diesel cost ranged from $1.10 to 1.22/L, so at a conversion rate of 3.65 kWh/L, the fuel cost for producing electricity was $0.33 kWh. On small islands that have little infrastructure, diesel fuel is delivered in barrels.

On Ascension Island, the simple payback was estimated at seven years for the addition of two 900-kW wind turbines in a high-penetration system. This saves an additional 2.4 million L of fuel per year. For diesel fuel at $1.50/L, the savings would be $3,600,000/year and the simple payback would be around three years. Most wind–diesel system results are not so dramatic. High-penetration systems should also save on diesel maintenance, since the diesel gensets will not operate as many hours but more on and off switching could increase O&M costs.

Three 100-kW wind turbines produce around 675,000 kWh/year as part of a wind–diesel plant at Toksook Bay, Alaska. The turbines displace 196,000 L of diesel fuel per year. If the bulk price of diesel is $1.50/L or even more, the annual savings are $300,000 per year. If the installed cost for the wind turbines was around $1.5 million, the simple payback would be five years. In May 2008, bid price for bulk diesel in remote Alaska was as high as $1.90/L.

At St. Paul Island, Alaska, the installed cost was $905,000 (1999 dollars) for a wind–diesel system that provided power to an industrial complex (no grid). The high-penetration, no-storage system consisted of one 225-kW wind turbine and two 150-kW diesel generators. The cost of energy from the system was $0.15/kWh, compared to diesel grid costs of $0.43/kWh (2004 dollars). Since then, two more turbines have been added to support economic development and generate enough power for residential consumption.

Two wind turbines (225 kW) were installed on Thursday Island between Australia and Papua New Guinea. They produced around 1,700 MWh per year and saved around 434,000 L of diesel fuel annually so the payback time is estimated at 7 years. A cost breakdown for the project is given [19].

Costs for renewable village power systems vary widely as shown in the table below.

Company	Size (kW)	Wind (kW)	PV (kW)	Battery (kW)	Inverter (kW)	Energy (kW/yr)	Cost
Bergey	10.1	7.5	2.6	84	6	12,000	57,000
Bergey	1.2	1.0	0.18	10.6	1.5	1,200	7,800
Southwest	1.3	0.40	0.88			750	

Most systems consist of components from various suppliers and manufacturers and are located in remote areas. The best example is China's SDDX project (2002–2005) consisting of 721 PV, wind, and PV–wind renewable village power systems (15,540 kW), 292 small hydro stations (113,765 kW), and 15,458 small single-household units (1,103 kW) with an installed capacity of 130,408 kW. The total investment was 4.7×10^9 Yuan (about $570 million), or an average of $4,370/kW [20, chap. 6].

The cost was $178,000 for one village hybrid system (Figure 12.5) in a remote region of China (2003 dollars) for all systems including power generation and mini grid transmission lines. The configuration is two 10-kW wind turbines, 4-kW PV power, 30-kVA diesel generation, 1,000-Ah battery bank, and a 38-kVa DC–AC inverter. At 54 kW, the installed cost was $3,300/kW—very reasonable for a remote location. The renewable part of the system produces around 150 kWh per day. The unknowns in calculating the cost of energy are percent of energy supplied by the diesel

FIGURE 12.5 Hybrid (wind–PV–diesel) renewable village power system for Subashi, Xinjiang Province, China. (Photo courtesy of Charlie Dou.)

Economics

generator, cost of diesel fuel, levelized replacement costs, and O&M costs. A known major cost is battery bank replacement every five to seven years.

Small hybrid systems that can be set up as modular units are available. Most manufacturers do not supply prices on their websites. Shipping and installation to remote locations will increase the cost, sometimes to double the cost of the energy components. Energy cost can be estimated from the initial cost and projected energy production.

For village power, the source choices are wind, PV, or hybrid wind–PV. A life cycle cost analysis for a hybrid system can determine the ratio of wind to PV energy. The advantages of PV are the lack of moving mechanical parts and placement at ground level. For comparison, suppose the local resources for both wind and solar are good and a 20-kW system is needed for village power. The capacity factor for wind is 25%; the solar capacity factor is 4 hr/day at peak power, 80% sunshine. The estimated yearly production for wind is 43,000 kWh and for PV is 6,000 kWh. Installed costs are lower for wind power than for PV, so the reason for choosing wind power is obvious. That is also the reason that hybrid systems generate more wind than PV power—five times or more—although PV prices dropped dramatically by 2012.

12.8 SUMMARY

Wind farms are the cheapest renewable energy sources for generating electricity. The cost of energy (COE) from wind turbines decreased from over $0.50/kWh in the 1970s to 0.06/kWh (Figure 12.6) based on a 2000 NREL analysis. The COE projections for 2005 and later were too low, primarily because of the increased costs of materials and oil. Since 2003, the COE for wind farms rose to $0.07 to 0.12/kWh in 2011. The earlier numbers in Figure 12.6 represent cost of energy for a class 6 wind resource. Starting in 1995, the change was made to class 5 and class 4 winds. New power generation from other energy sources will show similar cost increases for the same reasons.

Wind is cheaper than other renewable sources of energy (except hydro) for producing electricity (Figure 12.7) and is competitive with new fossil fuel plants (except combined-cycle natural gas turbines at $3/mcf). Notice that PV is forecast to continue to be cheap ($150/MWh compared to $190/MWh in 2011; Figure 12.2). Also the numbers are averages for a range. The range for renewables depends primarily on wind resources and installed costs. The wind farm business is much like the oil and gas business, except it is much easier to prospect for wind and the resource is non-depletable. As externalities are added to fossil fuel costs, wind energy becomes the most economical method of generating electricity. Of course, wind cannot provide all the electricity needed because of its variability. If an effective and economical storage system becomes available, the markets for all types of new power plants will change.

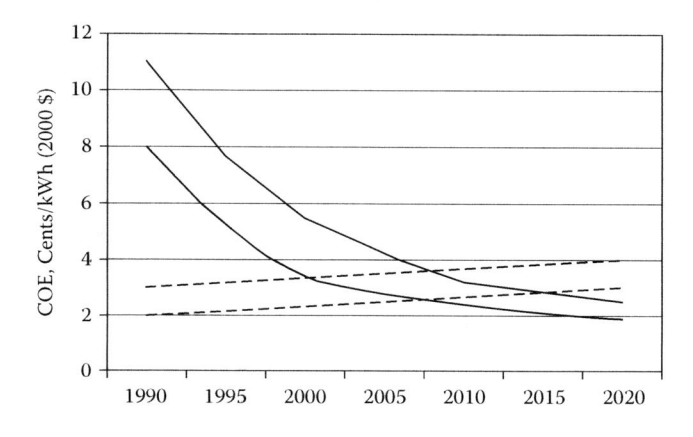

FIGURE 12.6 Cost of electricity from wind turbines to 2002 and projected future costs. Solid lines for high and medium wind regimes, dashed lines, bulk power generation. Values from NREL graph.

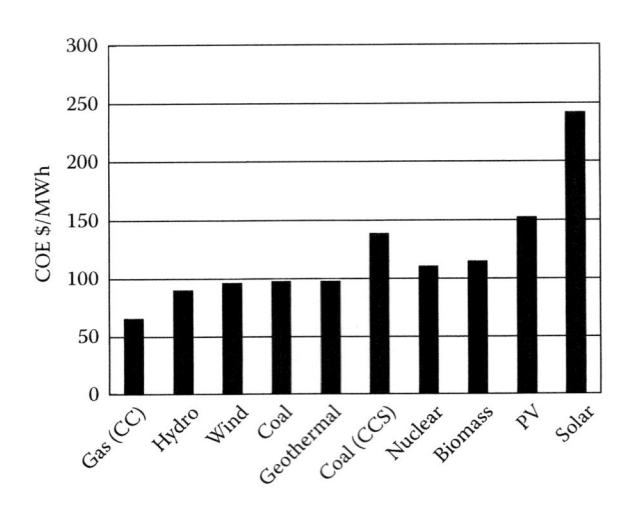

FIGURE 12.7 Estimated cost of energy for new power plants for generation of electricity that are entering service in 2017. Data from U.S. Energy Information Administration www.eia.gov/forecasts/aeo/electricity_generation.cfm (2/19/2013).

Economic considerations, legal requirements, and voluntary acceptance will lead to more use of renewable energy. Traditional energy sources have the advantage that fuel costs are not taxed but renewable energy incurs no fuel costs. The issue for renewable energy is high initial cost. Most consumers would rather pay for fuel as they use it. In 2012, small (10 kW) wind turbines were not generally cost-competitive with electricity from grids. However, if life cycle costs are used or rebates are available, wind turbines will become more competitive.

Green pricing is now available from many utilities. The premium was around $0.03/kWh for a block of 100 kWh per month; however, rate premiums continue to drop. In the U.S., 2012 utility green power sales exceeded 35.6×10^6 MWh. Approximately 1.8 million customers participate in utility green power programs.

Another major driving force for renewable energy is economic development and jobs creation at local or state levels. Renewable energy is produced locally; it does not have to be shipped from another state or country.

The capacity of existing transmission lines and the curtailment of wind farms are major problems. Another issue is the distance of wind resources from major loads. New transmission lines will have to be built. The questions arising from deregulation are decisions about who will finance construction and who will overcome right-of-way problems. The values of externalities range from zero (past and present value assigned by many utilities) to as high as $0.10/kWh for steam plants fired with dirty coal. Again, values are assigned by legislation and regulation and litigation may have to decide which parties will pay external costs.

12.9 FUTURE DEVELOPMENTS

Predictions about the future are always risky and generally wrong about specifics, but some trends are fairly clear. For example, a prediction of the price of oil at $200/bbl by 2020 is questionable; however, the price of oil will increase over the next decade. Here are other observations about the future of wind energy.

1. A distributed wind market very similar to the present farm implement business will develop. A farmer, rancher, or agribusiness owner will obtain a bank loan for a wind

turbine of 25 to 1,000 kW. He will expect a payback of five to seven years and the system will make money for him for the next fifteen years. One great advantage of earnings from wind-generated energy, the value of energy displaced (retail rates), and the avoided cost for electricity is that unlike other agriculture commodities, the benefits will not fluctuate.
2. Major transmission lines will be built from the windy U.S. plains areas to load centers. Lines will also be built in other countries that install large wind projects to produce electricity. Within five to ten years, wind power will compete with fuel adjustment cost without production tax credits, primarily due to value received for reducing carbon dioxide emissions.
3. The U.S. will implement carbon dioxide trading similar to the system of trading in NO_x and SO_x and wind energy will becomes the cheapest source of electricity. Shell Oil is now buying wind farms; Probably for the same reason, European countries are buying South American forests: to reduce carbon dioxide emissions. The La Venta II wind farm (83 MW) in Oaxaca, Mexico, displaces 205,380 tons of carbon dioxide annually. The CO_2 credit for the first seven years goes to the Spanish Carbon Fund that helped finance the project. The value of wind energy will increase by $0.03 to 0.04/kWh if the avoided CO_2 is worth $30/ton.
4. Cooperative wind plants of one to ten units will become common. Because of economies of scale, groups of farmers will form cooperatives to buy larger wind turbines.
5. My low estimate for global wind capacity for large turbines is 600 GW by 2020 and 1,000 GW by 2030, based on an average linear increase of 40 GW per year. A higher estimate is 1,400 GW by 2030 based on goals for U.S., China, and Europe. Past global and national forecasts for wind capacity were always underestimated, but the past exponential growth rates of 28% per year (1995 through 2012) will decrease (19% for 2012) and at some point change to linear growth. At the end of 2012, world installation was 282 GW.

Figure 1.15 shows that exponential growth. The Global Wind Energy Council forecasts for a moderate scenario are 493 GW by 2016, 832 GW by 2020, and 1,776 GW by 2030. The International Energy Agency projects 415 GW by 2020 and 572 GW by 2030. The U.S. has a goal of 20% of electricity from wind by 2030 [21], which would require a capacity of 304 GW. China has goals of 200 GW by 2020 and 400 GW by 2030. Europe [21] has goals of 180 GW by 2020 and 300 GW by 2030.

As stated in Chapter 2, the world faces a tremendous energy problem and a number of organizations have sounded the warning and suggested solutions [24,25]. The first priorities are conservation and energy efficiency, followed by the increased use of renewable energy. Wind has now become part of national energy policies, which is reflected in the large growth rate in wind capacity across the world.

LINKS

American Wind Energy Association. Subsidies. www.awea.org/_cs_upload/blog/5400_1.pdf
European Wind Energy Association. Economics of wind energy. www.ewea.org/index.php?id=201
National Renewable Energy Laboratory. Energy analysis. www.nrel.gov/analysis
National Renewable Energy Laboratory. Image gallery. http://images.nrel.gov (photos of wind turbines and projects: small systems, grid connections, village power, and hybrid systems).
National Renewable Energy Laboratory. *Power Technologies Energy Data Book,* 4th ed. www.nrel.gov/analysis/power_databook
U.S. Department of Energy. Energy efficiency and renewable energy planning, budgeting, and analysis. www1.eere.energy.gov/office_eere/bo_budget_main.html; www1.eere.energy.gov/wind/budget.html; www.cfo.doe.gov/crorg/cf30.htm; www.cfo.doe.gov/budget/13budget/content/volume3.pdf

REFERENCES

1. R. Wiser, E. Lantz, M. Bolinger et al. 2012. Recent developments in the levelized cost of energy from U.S. wind power projects. http://eetd.lbl.gov/ea/ems/reports/wind-energy-costs-2-2012.pdf
2. R.J. Brown and R.R. Yanuck. 1980. *Life Cycle Costing: A Practical Guide for Energy Managers*. Atlanta: Fairmont Press.
3. J.M. Cohen et al. 1989. A methodology for computing wind turbine cost of electricity using utility economic assumptions. *Proceedings of Windpower Conference*. CD.
4. H.C. Wolfe, Ed. 1975. Efficient use of energy. In *Proceedings of American Institute of Physics Conference*.
5. W.R. Briggs. 1980. *SWECS cost of energy based on life cycle costing*. Technical Report RFP-33120/3533/80/13, UC-60. Wind Energy Research Center, NREL.
6. J.M. Sherman, M.S. Gresham, and D.L. Ferguson. 1982. *Wind systems life cycle cost analysis*. RFP-3448, UC-60. Wind Energy Research Center, NREL.
7. RETScreen International. Clean energy project analysis tools. www.retscreen.net.
8. Hybrid2, Wind Energy Center, University of Massachusetts. www.umass.edu/windenergy/research.topics.tools.software.php
9. National Renewable Energy Laboratory. Market analysis. www.nrel.gov/analysis/analysis_tools_market.html; click on RETFinance.
10. U.S. Department of Labor. Inflation calculator. www.westegg.com/inflation/; Consumer price indices. http://data.bls.gov/cgi-bin/cpicalc.pl
11. P. Torcellini, N. Long, and R Judkoff. 2003. *Consumptive water use for U.S. power production*. NREL/TP-550-33905.
12. American Wind Energy Association. Ten steps to developing a wind farm. Wind energy fact sheet. www.awea.org/learnabout/publications/upload/Ten_Steps.pdf
13. American Wind Energy Association. 2008. *Wind Energy Siting Handbook*. www.awea.org/issues/siting/index.cfm
14. Wind Powering America. Wind energy finance calculator. www.windpoweringamerica.gov/pdfs/software_wef.pdf
15. P. Jain. 2011. *Wind Energy Engineering*. New York: McGraw Hill.
16. DISGEN. www.disgenonline.com
17. International Renewable Energy Agency. Wind power, 2012. Renewable Energy Technologies: Cost Analysis Series, 1, 5/5. Ihwww.irena.org/DocumentDownloads/Publications/RE_Technologies_Cost_Analysis-WIND_POWER.pdf
18. R. Wiser and M Bolinger. 2011 Wind technologies market report: energy efficiency and renewable energy. www.windpoweringamerica.gov/pdfs/2011_annual_wind_market_report.pdf
19. U.S. Department of Energy. The impact of wind development on county-level income and employment: review of methods and an empirical analysis. www.nrel.gov/docs/fy12osti/54226.pdf
20. C. Dou, Ed. 2008. Capacity building for rapid commercialization of renewable energy in China. CPR/97/G31, UNDP/GEF, Beijing, PR China.
21. U.S. Department of Energy. 2008. Twenty percent wind energy by 2030. www1.eere.energy.gov/wind/pdfs/41869.pdf
22. European Wind Energy Technology Platform. 2006. Wind energy: a vision for Europe in 2030. www.windplatform.eu/fileadmin/ewetp_docs/Structure/061003Vision_final.pdf
23. CADDET. Wind power for a remote island community. www.soe-townsville.org/strandwindproject/data/ER121.PDF
24. L.R. Brown. 2009. *Plan B 4.0: Mobilizing to Save Civilization*. New York: W.W. Norton.
25. R.E. Smalley. 2003. Nanotechnology, energy and people. www.americanenergyindependence.com/energychallenge.aspx

PROBLEMS

1. What are the two most important influences on the cost of energy?
2. Calculate the simple payback for a Bergey 1-kW wind turbine on a 20- or 60-m tower. Go to www.bergey.com to find the price. The turbine produces 2,000 kWh per year. Assume O&M and FCR = 0.

3. Calculate the cost of energy according to Equation (12.3) for a 400-W Air X wind turbine (Southwest Windpower). Installed cost of $2,000 includes 10-m tower and battery. Annual energy production is 400 kWh. Assume FCR and AOM = 0.
4. Calculate the cost of energy according to Equation (12.3) for a Bergey 10-kW wind turbine on a 30-m tower in a good wind regime. You can use a simple method to estimate annual kilowatt hours.
5. Calculate the cost of energy from Equation (12.4) for a 50-kW wind turbine that produces 120,000 kWh per year. The installed cost is $200,000; fixed charge rate is 6%; O&M represents 1% of installed cost; and levelized replacement cost is $4,000 per year.
6. Estimate the years to payback using Equation (12.7). IC = $150,000, r = 8%, AEP = 120,000 at $0.09/kWh. Assume a fuel escalation rate of 4%. This problem must be done numerically. Assume L, calculate, and then modify L in terms of your answer and calculate again.
7. Explain life cycle costs for a renewable energy system.
8. In 2011, the COE for wind was around $90/MWh. What is the estimated COE for electricity generation (large plants) for photovoltaic, solar thermal, biomass, and geothermal energy?
9. The estimated cost of energy from a wind farm is around $0.08/kWh. Make a comparison to proposed new nuclear power plants. What is the COE (retail rate) for the newest nuclear plants installed in the U.S.? (Do not calculate; use an estimate from any source.)
10. What are today's values for fuel inflation, discount rate, interest rate? What is your estimate for five years into the future?
11. A 100-MW wind farm (100 turbines, 1 MW) is installed in a class 4 wind regime. The production is around 3,000 MWh per turbine annually. The utility company pays about $40/MWh for the electricity produced. Estimate the yearly income from the wind farm. If the landowners receive 4% royalties, how much do they receive per year?
12. For the previous problem, assume installed costs are $1,600/kW, FCR = 6%, capacity factor = 35%, and AOM = 0.008/kWh. Calculate the COE using Equation (12.4). You will need to estimate the levelized replacement costs or calculate LRC using Equations (12.5) through (12.7). Compare your answer to the $0.05/kWh estimated price the wind farm receives. How can the wind farm make money?
13. A number of new wind farms are being installed in the U.S. The wind farm boom in Texas led to installation of more than 10,000 MW from 2005 to 2012. Why? Explain in terms of economics.
14. What is the price of oil (dollars per barrel) today? Estimate the prices of oil for 2015, 2020, and 2030. Compare the estimate to the U.S. Energy Information Administration projections for the same years. Place results in a table.
15. Estimate the price for oil (dollars per barrel) if the costs for the U.S. military to keep the oil flowing from the Middle East are added.
16. Why were wind turbines installed primarily in California from 1981 through 1985? Discuss in terms of economics.
17. How much should the U.S. government fund for conservation and efficiency, renewable energy, and wind energy? Compare your answer to the fiscal year 2012 budget for the same items. What did the budget allocate for fossil fuels and nuclear energy (nuclear fusion counts)? See the Links section at the end of the chapter.
18. At what dollar level should your national government fund renewable (wind) energy? What should it fund for fossil fuel and nuclear energy? Compare your results to the national budget for the current fiscal year or the latest year for which information is available.
19. Estimate the cost of energy for a Bergey Windpower 10.1-kW, hybrid (PV–wind) system. You will have to estimate FCR and O&M.
20. Estimate the cost of energy for a Southwest Windpower, 1.3 kW, hybrid (PV/wind) system. You will have to estimate FCR and O&M.

21. Estimate the cost of energy for the three 100-kW, wind turbines at Toksook Bay, Alaska. You will have to estimate FCR, O&M, and LRC.
22. A village power system in China consists of 10 kW wind plus battery bank and inverter. IC = $4,500/kW, energy production = 50 kWh per day, FCR = 0.03, and AOM = $0.01/kWh. Calculate the cost of energy.
23. A renewable village power system in China consists of 20 kW of wind and 10 kW of PV. Use an average cost of $4,300/kW, annual energy production of 65,000 kWh, FCR of 0.03, and AOM of $0.01/kWh. Calculate the cost of energy. How does the cost compare to the present rate you pay for electricity?

    For the following problems use data reported to the U.S. Federal Energy Regulatory Commission (http://eqrdds.ferc.gov/eqr2/frame-summary-report.asp). Pick any wind farm or pick Llano Estacado Wind (White Deer, installed capacity = 80 MW).
24. What is the rate of the power purchase agreement?
25. Installed cost was $1 million/MW or $80 million. For 2011, what was the income generated? Assume that 2011 was an average year. What is the time of simple payback?
26. For Problem 25, consider an additional $20/MWh return for the production tax credit. Now what is the payback time?
27. Calculate the capacity factor for the wind farm for 2007.
28. Find a wind farm in the U.S. sells electricity at the market rate. For the latest quarter, what are the high, low, and average rates (dollars per megawatt hour)?
29. What is the world installed wind capacity? What is your estimate for 2020?
30. What is the goal for your country for wind power for 2020?

Index

A

Accuracy and precision of instruments, 78–79
Aerodynamics, wind turbine, 90–92, 115–116
    performance prediction, 122–128
Ailerons, 130
Air, compressed, 256
Airfoils, 115–116, 126
Alta wind farm, 220
Alternative Energy Institute (AEI), 269
Alternators, permanent magnet, 156
American Wind Energy Association (AWEA), 61, 199, 224, 229–233, 248–249, 272, 296
American Wind Power Center and Museum, 227
Ancillary costs, wind farms, 270
Andreau, Edouard, 9
Anemometers, 73–74
    cup and propeller, 77
Applications
    battery, 253–259
    community wind, 14, 229–234
    distributed systems, 227–229
    small wind turbines, 222–227
    storage, 252–260
    utility scale, 219–222
    village power, 238–242
    water pumping, 242–245
    wind-diesel generation, 234–238
    wind industry and, 246–252
    wind turbine, 107–111
Armature, 153
Automobiles. *See* Fossil fuels
Availability, 167–168
Avoided costs, 267–268

B

Battelle Pacific Northwest Laboratory (PNL), 61
Batteries, 253–259
Bergey Excel wind turbine, 179–181
Berkshire Wind Power Project, 232
Bernoulli's theorem, 120
Bird fatalities, 271–272
Blades, wind turbine, 131–135
    flow visualization, 190–192
    performance, 186–192
    wake effects, 176–178
Boundary layer control, 189

C

Caithness Wind Farm Information Forum, 268
Calculated annual energy for wind turbines, 101–102
California Energy Commission (CEC), 170
California Public Utilities Commission, 238, 275
California wind farms, 170–171, 172, 247

Canadian Wind Energy Association, 199
Capacitance, 148
    phase angle and power factor, 150–152
Capacity factors, 170, 178
Carbon dioxide, atmospheric, 38–39, 281
Carter 25 wind turbine, 190–192
Chargers, wind, 5–6
Chernobyl accident, 274
China, renewable village power systems in, 240–242, 279–280
Climate Change Action Plan, 276
Coal, 33–34
Community wind, 14, 229–234
Competitive Renewable Energy Zones (CREZs), 282
Compressed air energy storage (CAES), 256
Computer codes, wind turbine, 135–136
Computer processing units (CPUs), 159
Connection to utility, 269
Connect time, 168
Conservation of energy, 23, 24
Conservation of momentum, 119–120
Consumption, energy
    exponential growth in, 26–28, 35–39
    global, 17–20
    lifetime of finite resource and, 36–38
Control, wind turbine, 93–99, 159–162
Controllers, electronic, 159–162
Conversion, voltage, 162–163
Cooperatives, electric, 232
Corporate average fuel economy (CAFE), 26
Costs. *See also* Economics
    ancillary, 270
    avoided, 267–268
    of energy (COE), 290–293, 302, 305
    life cycle (LCC), 293–295
    operating, 287
    present worth and levelized, 295
    wind project development, 300–301
Cup and propeller anemometers, 77
Current, 147
    phase angle and power factor, 150–152

D

Damping ratio, 78–79
Danish Wind Industry Association, 135
Danvest Energy, 238
Darrieus wind turbines, 105, 139, 199
Database of State Incentives for Renewable Energy (DSIRE), 276–277
Data loggers, 82–83
Design, water pumping system, 243–244
Design, wind turbine
    aerodynamic performance prediction and, 122–128
    aerodynamics, 115–116
    blades, 131–135

312 Index

construction, 131–138
drag device, 117–118
evolution, 138–139
lift devices, 118–122
mathematics, 116–117
measured power and power coefficient, 128–131
rotation, 121–122
small, 139–141
Diesel generators, 185–186
Digital Elevation Model (DEM), 205, 206
Digital Line Graph (DLG), 205
Digital maps and siting, 204
Direct-drive generators, 156
Direction, wind, 53–54, 77
Distance constant, 78
Distributed systems, 227–229
Distributions, wind speed, 59–61
Doppler anemometers, 73–74
Doubling time, 36
Doubly fed induction generators, 156
Drag devices, 87, 88, 117–118
Duration curve, 57–58
Dutch windmills, 1–2

E

Economics, 21, 305–306
analysis, 289–293
cost of energy (COE) and, 290–293, 302, 305
factors affecting, 287–288
future developments and, 306–307
general comments, 288–289
hybrid systems, 303–305
life cycle costs (LCC) analysis, 293–295
present worth and levelized costs, 295
simple payback, 289–290
value of energy and, 293
wind project development and, 296–303
Efficiency, energy, 24–26, 39–40
Electric cooperatives, 232
Electric field, 148
Electricity. *See also* Utilities
community wind and, 14, 229–234
Faraday's law of electromagnetic induction and, 150
feed laws (EFLs), 278–279
fundamentals, 147–152
generators, 150, 152–158
inverters, 163
lightning, 163–164
phase angle and power factor, 150–152
power quality and, 158–159, 269
project development and sales of, 302–303
transmission, 281–283
village power, 238–242
wind turbines and, 107–108
Electric Reliability Council of Texas (ERCOT), 61, 282–283
Electric-to-electric systems, 184–185
Electromagnetic induction, Faraday's law of, 150
Electronics
controllers, 159–162
power, 162–163
Enercon, 157
Energy

battery storage of, 253–259
conservation of, 23, 24
costs of (COE), 290–293, 302, 305 (*See also* Economics)
definitions of power and, 22–23
dilemma and laws of thermodynamics, 24–26
efficiency, 24–26, 39–40
flywheels and storage of, 256–257
fundamentals, 23–24
incentives and subsidies, 275–280
increase in global consumption of, 17–20
nuclear, 34–35, 274
production by wind turbines, 99–101, 109–110
renewable, 20–21, 238–242, 277–278
storage, 252–260
thermal, 23–24
types, 17
value of, 293
wind turbine calculated annual, 101–102
Energy Information Administration (EIA), 170
Energy Research and Development Agency (ERDA), 246
Enertech wind turbines, 178–179, 231
Environmental issues and utilities, 270–274
European Union, the
community wind in, 233–234
transmission in, 283
utilities, 219–222
utility incentives in, 278–279
wind farms, 174–176
wind maps, 67, 69, 70
European Wind Atlas, 67, 70, 73
Exponential growth, 26–28
mathematics of, 35–39
Externalities, utility, 280–281, 296
Extractable limits of wind power, 45, 47

F

Faraday's law of electromagnetic induction, 150
Farms, wind. *See* Wind farms
Farm windmills, 2–5, 182–184, 242–243
water pumping, 181–185
Faults, wind turbine, 98–99
Feather position, wind turbine, 94–95
Federal Aviation Administration, 54, 75, 270
Federal Energy Regulatory Commission (FERC), 169–170, 267, 303
Flash Earth, 214
Flettner rotors, 8
Flow batteries, 259
Flow visualization, 190–192
Flywheels, 256–257
Fossil fuels, 26, 28
coal, 33–34
natural gas, 32–33
petroleum, 29–32
FRP blades, 131–133
Fuel cells, hydrogen, 259–260
Fukushima accident, 274
Fundamentals, energy, 23–24

G

Gas, natural, 32–33
Generators

Index

comparisons, 156
diesel, 185–186
electric, 150, 152–158
examples, 156–158
induction, 153–158
size, wind turbine, 99–100
vortex, 190
Geographic information systems (GIS), 205, 207
Giromill, 88, 105
Global circulation of wind, 45, 46
Global Village Energy Partnership, 238
Global warming, 38–39
Global Wind Energy Council, 307
Golden Spread Electric Cooperative, 232
GoogleMaps, 204, 214
Greenhouse gases, 38–39
Green power, 277–278, 306
Gulf Coast, Texas, 71–73

H

Hamilton-Standard WTS-4, 139
Histograms, wind speed, 56–57
History of wind energy, 1–11
Horizontal-axis wind turbines (HAWT), 88, 90, 92
Horns Rev wind farm, 220
Hüttner, 9
Hybrid systems, wind, 111
economics, 303–305
performance, 185–186
Hydrogen fuel cells, 259–260

I

Incentives, energy, 275–280
Inductance, 148
phase angle and power factor, 150–152
Induction generators, 153–158
Industry, wind, 246–247
1980–1990, 247–248
1990–2000, 248–250
2010 onward, 251–252
2000–2010, 250–251
Instrumentation, 73–77
characteristics, 78–79
cup and propeller anemometers, 77
measurement, 79–80
vegetation indicators, 80–82
wind direction, 77
Integrated resource planning (IRP), 280, 296
Introduction to Renewable Energy, 252
Inverters, 163

K

Kotzebue Electric Association (KEA), 185–186

L

Laws of thermodynamics, 24–26
Lead-acid batteries, 257–258
Levelized costs and present worth, 295
Life cycle costs (LCC), 293–295
Lifetime
of finite resource, 36–38
measure of performance, 168
Lift devices, 87–88, 118–122
Light detection and ranging (LIDAR), 75
Lightning, 163–164
Lignite Energy Council, 296
Lithium ion batteries, 258
Loggers, data, 82–83
Long-term reference stations, wind farm, 203
Lower Colorado River Authority, 277

M

Magnetic field, 148–149
Magnetism, 147
Magnus effect, 6, 8
Maps, wind, 65–66
European Union, 67, 69, 70
siting and, 204
United States, 66–67, 68
Mathematics
of exponential growth, 35–39
wind turbine design, 116–117
Maximum theoretical power, 121
MBB Monopteros and Flair designs, 87
Measured power and power coefficient, wind
turbine, 128–131
Measurement, 79–80
performance, 167–169
Mechanical energy and wind turbines, 109–110
MesoMap, 66, 210
Metal-air batteries, 259
Metering, net, 278
Micrositing, 210–215
Microsoft VE, 214
Minnesota, 231
MOD-O design, 138
Momentum, conservation of, 119–120
Montana Public Service Commission, 270
Motor, electric, 149
Municipal and city operations, 232–233

N

National Climatic Center, 54
National Energy Act, 275
National Energy Strategy Act, 267, 275
National Renewable Energy Laboratory (NREL), 61, 65, 66, 67, 224, 238, 252
airfoils designed by, 126
digital maps, 204
Gulf Coast and, 71–73
hybrid systems, 111
incentives and, 276
innovation and, 103
National Wind Coordinating Collaborative (NWCC), 272
National Wind Technology Center (NWTC), 61, 66, 67, 135, 224
Natural gas, 32–33
Netherlands, the, 1–2
Net metering, 278
New Mexico Wind Energy Center, 274
Nickel-cadmium batteries, 259
Noah wind turbine, 106

Noise, 200–201
Northern States Power, 277
Nuclear energy, 34–35, 274
Numerical models for predicting winds, 210
Nysted wind farm, 177, 220

O

Ocean winds, 71–73, 215
Ohm's law, 147
Oil production, 29–32
Orientation of rotor axis in wind turbines, 88

P

Pacific Northwest Laboratory, 65, 66, 207–208
Payback, simple, 289–290
Performance
  Bergey Excel wind turbine, 179–181
  blade, 186–192
  boundary layer control and, 189
  electric-to-electric systems, 184–185
  Enertech 44, 178–179
  flow visualization and, 190–192
  measures, 167–169
  prediction, aerodynamic, 122–128
  vortex generators, 190
  wake effects and, 176–178
  water pumping, 181–185
  wind-diesel and hybrid systems, 185–186
  wind farm, 169–176
Permanent magnet alternators, 156
Petroleum, 29–32
Phase angle, 150–152
Pitch control system, wind turbine, 94–95
Planform, 124–126
Politics and utilities, 274–275
Pollution taxes, 280
Popular Science, 102
Potential, wind power, 54–55, 58–59
Power
  coefficient, 91–92, 128–131, 242
  curve of wind turbines, 95–96, 179–181
  definitions of energy and, 22–23
  electric, 147–148
  electronics, 162–163
  extractable limit of wind, 45, 47
  factor and phase angle, 150–152
  green, 277–278, 306
  maximum theoretical, 121
  potential, wind, 54–55, 58–59
  quality and electricity, 158–159, 269
  surface roughness of blades and, 187–188, 189
  village, 238–242
  of wind, 47–49
  wind turbine measured, 128–131
Predicting Offshore Wind Energy Resources project, 73, 74
Present worth and levelized costs, 295
Production tax credit (PTC), 275, 302–303
Programmable logic controllers (PLCs), 159
Project development, wind, 296–300
  benefits, 301–302
  costs, 300–301
  sales of electricity and, 302–303

PROPID, 126
Public Utility Regulatory Policies Act (PURPA), 267
Pumping, water, 181–185, 242–243
  design, 243–244
  large systems, 245

Q

Quality, power, 158–159, 269

R

Rayleigh distribution, 59–61, 101
Regional Wind Test Center, 134
Regulation, utility, 267, 270, 280
Reliability, 168
REmapping the World, 66
Renewable energy, 20
  advantages and disadvantages, 20–21
  credits, 302
  green power incentives and, 277–278
  net metering and, 278
  village power, 238–242
Renewable portfolio standard (RPS), 277
RenewableUK, 199
Resistance, 147
RETFinance, 295
RETScreen tool, 294–295
Rotation, 121–122
Rotational kinetic energy, 256–257
Rotor area and wind map, wind turbine, 100–101
Roughness parameter, 53
Rural Electrification Act, 280
Rural Energy for America Program (REAP), 228

S

Safety concerns for utilities, 268–269
Sail wing design, 105
Satellite images, 214
Savonius rotor, 87–88, 105, 123
Schachle-Bendix turbine, 139
Schools, colleges, and universities, 231–232
Screening, wind resource, 206–210
Shear, wind, 49–53
Simple payback, 289–290
Sindal Report, 169
Siting
  geographic information systems (GIS) and, 205, 207
  micro-, 210–215
  noise and, 200–201
  numerical models for predicting wind and, 210
  ocean winds and, 215
  small wind turbines, 197–200
  visual impact and, 201–202, 273–274
  wind farms, 203–204
  wind resource screening and, 206–210
Siting Handbook, 272
Small systems, 13–14, 139–141
  wind measurement for small wind turbines, 83–84
Small Wind Certification Council (SWCC), 224–225
Small wind turbines, 139–141. *See also* Wind turbines
  applications, 222–227
  noise, 200–201

Index

siting, 197–200
  visual impact, 201–202
Smith-Putnam wind turbine, 8
Sodium-sulfur batteries, 258–259
Solar Energy Research Institute (SERI), 103, 276
Solidity, 87
Sonic detection and ranging (SODAR), 75
Southwestern Public Service, 238
Specific output, 168–169, 171, 172
Speed, wind, 49–53, 54–55
  distributions, 59–61
  duration curve, 57–58
  histograms, 56–57
State incentives, 276–277
State Wind Working Group Handbook, 210
Statistics, wind, 169
Storage, energy, 252–260
Subsidies, energy, 275–280
Supervisory control and data acquisition (SCADA) units, 159, 162
Surface roughness and blade performance, 187–188, 189

T

Taxes
  pollution, 280
  production tax credit and, 275, 302–303
Texas, 61, 251, 282
  estimated wind power, 207–210
  Gulf Coast, 71–73
Thermal energy, 23–24
  wind turbines and, 111
Thermodynamics, laws of, 24–26
Thomas, Percy, 9
Three Mile Island, 274
Thunderstorms, 163–164
Time, connect, 168
Tip speed ratio, 87
Topozone, 204
Towers, 136–137
Transmission, electricity, 281–283
Transwest Express Transmission Project, 282–283
Turbines. *See* Wind turbines
Turbulence, 55–56

U

Ultraviolet water purifiers, 261
Underwriters Laboratory (UL), 269
United States
  community wind, 229–233
  energy incentives in, 275–278
  wind maps, 66–67, 68
Unit Juggler, 22
U.S. Department of Agriculture, 108, 133–134, 243
U.S. Department of Energy, 103, 105, 275, 281, 287
U.S. Fish and Wildlife Service, 270
U.S. Geological Service Terrain Elevation Data, 205
U.S. Geological Survey, 204
U.S. National Weather Service, 54
Utilities, 6–11, 219–222. *See also* Electricity
  avoided costs and, 267–268
  connection to, 269
  environmental issues and, 270–274
  externalities, 280–281, 296
  incentives, 275–280
  politics and, 274–275
  power quality, 158–159, 269
  regulation, 267, 270, 280
  safety concerns, 268–269
  transmission, 281–283

V

Value of energy, 293
Vegetation indicators, 80–82
Vertical-axis wind turbines (VAWT), 88, 92, 157, 247, 249–250, 252
Vestas turbine, 139, 156, 220
Village power, 238–242
Visual impact and siting, 201–202, 273–274
Visualization, flow, 190–192
Voltage, 147
  conversion, 162–163
  phase angle and power factor, 150–152
Vortex generators, 190

W

Wake effects, 176–178
WAsP, 210
Water pumping, 181–185, 242–243
  design, 243–244
  large systems, 245
Waubra wind farm project, 215
Wind
  community, 14, 229–234
  direction, 53–54, 77
  duration curve, 57–58
  extractable limits of power from, 45, 47
  global circulation of, 45, 46
  hybrid systems, 111
  industry, 246–252
  measurement for small wind turbines, 83–84
  ocean, 71–73, 215
  power, 47–49
  power potential, 54–55, 58–59
  project development, 296–303
  shear, 49–53
  speed, 49–53, 54–55
  speed distributions, 59–61
  speed histograms, 56–57
  statistics, 169
  turbulence, 55–56
Wind Atlases of the World, 66, 69
Wind chargers, 5–6
Wind-diesel generation
  applications, 234–238
  and hybrid systems performance, 185–186
Wind energy
  community wind, 14
  generation of electricity for utilities, 6–11
  history of, 1–11
  small systems, 13–14
  wind chargers, 5–6
  wind farms, 11–13
Wind Energy Engineering, 296
Wind Energy Institute of Canada, 238

Wind Energy Research Center, 276
Wind Energy Resource Atlas, 66
Wind farms, 11–13
    ancillary costs, 270
    California, 170–171, 172, 247
    micrositing, 210–215
    outside California, 172–174
    outside the United States, 174–176
    performance, 169–176
    project development, 296–303
    siting, 203–204
    visual impacts of, 201–202, 273–274
Wind for Schools Program, 231
Wind GIS Data Layers, 204
Wind-hydrogen systems, 238
Wind industry, 246–247
    1980–1990, 247–248
    1990–2000, 248–250
    2010 onward, 251–252
    2000–2010, 250–251
Wind maps
    around the world, 69, 71
    European Union, 67, 69, 70
    siting and, 204
    United States, 66–67, 68
Windmills
    Dutch, 1–2
    farm, 2–5, 182–184, 242–243
    water pumping performance, 181–185
Wind Powering America, 276, 296
Wind resource assessment
    data loggers, 82–83
    instrumentation, 73–82
    maps, 65–71
    ocean, 71–73
Wind Resource Assessment Handbook, 75
Wind resource screening, 206–210
Wind Site Assessment Dashboard, 204
WindStats Newsletter, 169
Wind turbines. *See also* Small wind turbines
    aerodynamic performance prediction, 122–128
    aerodynamics, 90–92, 115–116
    applications, 107–111
    blades, 131–135
    calculated annual energy, 101–102
    construction, 131–138
    control, 93–99, 159–162
    drag devices, 87, 88, 117–118
    economics, 287–289
    electrical energy produced by, 107–108
    energy production by, 99–101, 107–111
    evolution, 138–139
    faults, 98–99
    generator size, 99–100
    hybrid systems, 111
    induction generators, 154–155
    innovative wind power systems, 102–106
    lift devices, 87–88, 118–122
    lightning strikes to, 163–164
    manufacturer's curve, 101
    mathematical terms, 116–117
    measured power and power coefficient, 128–131
    measurement for small, 83–84
    mechanical energy produced by, 109–110
    normal operation, 95–97
    orientation of rotor axis, 88
    power curve, 95–96, 179–181
    rotation and angular momentum, 121–122
    rotor area and wind map, 100–101
    safety, 268–269
    specific output, 71, 168–169, 172
    system description, 89–90
    thermal energy produced by, 111
    utilities use of, 6–11
    visual impacts of, 201–202, 273–274
Wind-Works, 269
Winglets, 104
World Bank, 251

Z

Zinc-bromide batteries, 259